Springer Series in Materials Science

Volume 340

The Springer Series in Materials Science covers the complete spectrum of materials research and technology, including fundamental principles, physical properties, materials theory and design. Recognizing the increasing importance of materials science in future device technologies, the book titles in this series reflect the state-of-the-art in understanding and controlling the structure and properties of all important classes of materials.

Suprakas Sinha Ray · Lesego Tabea Temane · Jonathan Tersur Orasugh

Graphene-Bearing Polymer Composites

Applications to Electromagnetic Interference Shielding and Flame-Retardant Materials

Springer

Suprakas Sinha Ray
Centre for Nanostructures and Advanced Materials
Council for Scientific and Industrial Research
Pretoria, Gauteng, South Africa

Department of Chemical Sciences
University of Johannesburg
Johannesburg, Gauteng, South Africa

Lesego Tabea Temane
Centre for Nanostructures and Advanced Materials
Council for Scientific and Industrial Research
Pretoria, Gauteng, South Africa

Department of Chemical Sciences
University of Johannesburg
Johannesburg, Gauteng, South Africa

Jonathan Tersur Orasugh
Department of Chemical Sciences
University of Johannesburg
Johannesburg, Gauteng, South Africa

Centre for Nanostructures and Advanced Materials
Council for Scientific and Industrial Research
Pretoria, Gauteng, South Africa

ISSN 0933-033X ISSN 2196-2812 (electronic)
Springer Series in Materials Science
ISBN 978-3-031-51923-9 ISBN 978-3-031-51924-6 (eBook)
https://doi.org/10.1007/978-3-031-51924-6

This Springer imprint is published by the registered company Springer Nature Switzerland AG
The registered company address is: Gewerbestrasse 11, 6330 Cham, Switzerland

Dedicated to Our Parents

Preface

Modern society is encompassed with diverse challenges arising from the very state-of-the-art development in various sectors, especially materials science. Technology has negatively impacted humanity, including depression and other mental health issues, lack of sleep, increased obesity, rise in attention deficit hyperactivity disorder (ADHD), learning obstacles, cyberbullying, fire risk, diminished closeness and communication, lack of confidentiality, electromagnetic (EM) waves-based health and electronic device problems.

The monograph presents recent developments with respect to the existing literature and brings forth the knowledge gap and potential solutions for electrically conducting graphene-bearing polymer composites for electromagnetic interference (EMI) shielding and flame-retardant applications. Various materials have been employed in EMI shielding and flame-retardant applications. Graphene and graphene-bearing materials-containing polymer composites score high because of their superior performance.

In the past decade, significant progress has been made in developing graphene and graphene-bearing materials-containing polymer composites for various applications, from loading bearing to biomedical applications. A few reviews highlight different aspects of these materials; a monograph that provides the fundamentals of EMI shielding and flame retardancy with a special focus on graphene-bearing polymer composites and their processing techniques and application-specific developments is scarce. Throughout this book, graphene represents graphene nanoplatelets.

Furthermore, the tailoring of electrical properties by modifying blend morphology and selective localization of the graphene-based nanofiller in one of the phases/at the interphase has yet to be discussed in review articles or edited books. This book attempts to bridge this gap. It provides a comprehensive and critical insight into the state-of-the-art graphene-bearing polymer composite materials suitable for EMI shielding and flame-retardant applications.

By providing a holistic overview of the fundamental concepts related to graphene-bearing conducting composites and the intricacies of related research, this book is expected to provide the path for future advances in the development of advanced polymer composite materials. This monograph focuses on processing advanced functional composite materials for applications in EMI shielding and flame retardancy. This monograph particularly emphasizes molecular understanding and interpretation of obtained results.

The following are key features of this book:

- An exclusive information on graphene and its derivatives.
- The history of graphene-bearing polymer nanocomposites as EMI shields and flame-retardant materials.
- The importance of graphene-bearing polymer nanocomposites as EMI shields and flame-retardant materials.
- Other functional nanoparticles adopted in preparing polymer nanocomposite-based EMI shields and flame-retardant materials.
- Diverse processing techniques used for the fabrication of graphene-bearing polymer nanocomposite EMI shields and the preferred ones are discussed.
- A detailed discussion of how the processing techniques affect the final properties of the fabricated EMI shields and/or flame-retardant systems is discussed.
- Application of graphene-bearing polymer nanocomposite as EMI shields and flame-retardant materials is herewith discussed holistically with emphasis on the polymers as well as their blends.
- Up-to-date information on several characterization techniques adopted for the study of these materials is discussed.
- Diverse challenges faced in these materials' fabrication or application have been discussed.
- A holistic conclusion followed by future prospects on applying graphene-bearing polymer nanocomposite as EMI shields and flame-retardant materials are presented.

This book is ideal for polymer scientists and engineers, material scientists, researchers, engineers, including under- and postgraduate students interested in this exciting field of research. This book will also help industrial researchers and R&D managers to bring advanced graphene-based advanced polymer composite products into the market.

Pretoria, South Africa Suprakas Sinha Ray
Pretoria, South Africa Lesego Tabea Temane
Johannesburg, South Africa Jonathan Tersur Orasugh

Acknowledgments The authors would like to thank the Department of Science and Technology and the Council for Scientific and Industrial Research, South Africa, for financial support. We express our sincerest appreciation to all colleagues, postdoctoral fellows, and students for their valuable contributions as well as the reviewers for their critical evaluation of the proposal and chapters. We also thank the authors and publishers for their permission to reproduce their published works. Our special thanks go to Dr. Zachary Evenson at Springer Nature for his encouragement, cooperation, suggestions, and advice during various phases of preparation, organization, and production of this book.

Contents

About the Authors

Suprakas Sinha Ray is a Chief Research Scientist and Manager of the Centre for Nanostructures and Advanced Materials, DSI-CSIR Nanotechnology Innovation Centre, Council for Scientific and Industrial Research, Pretoria, South Africa. He received his Ph.D. degree in Physical Chemistry at the University of Calcutta, India, in 2001 and was a recipient of the "Sir P. C. Ray Research Award" for the best Ph.D. work. His current research focuses on the applications of advanced nanostructured and polymeric materials. He is one of the most active and highly cited authors in the field of polymer nanocomposite materials, and he has recently been rated by Thomson Reuters as being one of the top 1% most impactful and influential scientists and top 50 high-impact chemists.

He is the author of five books, co-author of five edited books, 32 book chapters on various aspects of polymer-based nanostructured materials and their applications, and author and co-author of 455 articles in high-impact international journals and 30 articles in national and international conference proceedings. He also has six patents and seven new demonstrated technologies shared with colleagues, collaborators, and industrial partners to his name. So far, his team has commercialized 19 different products. His honors and awards include South Africa's most *Prestigious 2016 National Science and Technology Award* (NSTF); Prestigious 2014 CSIR-wide Leadership Award; Prestigious 2014 CSIR Human Capital Development Award; *Prestigious 2013 Morand Lambla Awardee* (top award in the field of polymer processing worldwide), International Polymer Processing Society, USA. He is also appointed as Extraordinary Professor, University of Pretoria and Distinguished Professor of Chemistry, University of Johannesburg.

Lesego Tabea Temane is a distinguished Polymer Characterization Specialist renowned for her fervor in unraveling the intricate properties of polymeric materials. With a profound understanding of polymer science and a rich background in analytical techniques, she has emerged as a leading force in advancing the field of polymer characterization. Beyond her technical expertise, she is recognized for her collaborative spirit, seamlessly working with cross-functional teams to translate research findings into practical applications. Her inclusive approach has played a pivotal role in the successful localization of innovative polymer-based materials boasting enhanced performance characteristics. Her commitment to excellence and her ability to bridge the gap between theoretical knowledge and practical applications mark her as a key contributor in the dynamic realm of polymer science.

Jonathan Tersur Orasugh received his Ph.D. in Polymer Science and Engineering, from the University of Calcutta, India. Currently, he is working as a Postdoctoral Fellow at the Department of Chemical Sciences, University of Johannesburg and associated as a senior researcher at the Centre for Nanostructures and Advanced Materials, DSI-CSIR Nanotechnology Innovation Centre, Council for Scientific and Industrial Research, Pretoria, South Africa. His research work focuses on the processing and characterization of biopolymer-based immiscible polymer blends for biomedical applications, particularly drug delivery, and tissue engineering.

Abbreviations

2D	Two-dimensional
3D	Three-dimensional
Φ	Filler content
Φ_C	Percolation threshold
ε_∞	Relative dielectric constant
ε_S	Static dielectric constant
τ	Polarization relaxation time
ε''	Imaginary permittivity
ε'	Real permittivity
ε_r	Relative permittivity
μ_r	Relative permeability
σ	Electrical conductivity or shear stress
σ_t	Thermal stress
σ & σ_0	Represent the shear stress and yield stress
μ_p	Plastic viscosity
K_{oc}	Casson yield stress
λ	Radiation's wavelength
θ	Diffraction angle
μ	Magnetic permeability
Υ	Yield stress
ω	Angular velocity/frequency
δ	Skin depth
ρ	Density (D)
δ_μ	Magnetic loss angle
μ''	Imaginary permeability
μ'	Real permeability
η	Viscosity
d	Thickness
F_z	Normal force
h_0	Starting gap
h	The actual gap

Γ	Reflection coefficient
ξ	Debye length
ε_v	Chemical shrinkage
f(x,y)	Sample surface relative to the center plane
L_x & L_y	Surface dimensions
K_δ	Particle interaction
R_{max}	Maximum peak to valley height
Z_a	Average height values within the given area
Z_i	The height value
I_c	Crystalline segment
I_a	Amorphous region
A_{eff}	Effective absorbance
AEMA	Aminoethyl methacrylate
AEPZ	N-aminoethyl piperazine
AgNW	Silver nanowires
AIT	Autoignition test
AFM	Atomic force microscopy
AC	Alternating current
AMLR	Average mass loss rate
AMS	Aerospace Material Specification
ASTM	American Society for Testing and Materials
ANI	Aniline
AP	Aluminum hypophosphite
APP	Ammonium polyphosphate
ATH	Aluminum hydroxide
BE	Binding energy
B.C.	Before Christ
BPO	Benzoyl peroxide
BSI	BRITISH Standards Institution
c	Speed of light
C	Concentration
CA	Contact angle
CACC	Controlled atmosphere cone calorimeters
CB	Citizens band radio
CC	Cone calorimetry
CF	Carbon fiber
CFA	Charring-foaming agent
CHS	Copper hydroxystannate
CLC	Combined loading compression
CNF	Carbon nanofiber
CNT	Carbon nanotube
C_p	Specific heat
CP	Chlorinated paraffins
CPS	Chip package interaction
CRGO	Chemically reduced GO

CO_2	Carbon dioxide
CS	Chitosan
CTE	Coefficient of thermal expansion
Cu	Copper
CVD	Chemical vapor deposition
CSFR	Chitosan (CS) FR
DBP	Dibutyl phthalate
DETA	Diethylenetriamine
DL	OD per meter
DOP	Dioctyl phthalate
DAR	Dispersing agent requirement
dB	Decibels
D	Diffusion coefficient
DDM	4,4-diaminodiphenyl methane
DC	Direct current
DCRP	Dechlorane plus
DCM	Dichloromethane
DIC	Digital image correlation
DMF	Dimethylformamide
DSC	Differential scanning calorimetry
DTG	Differential thermogravimetric
DM	Moisture diffusivity
DMA	Dynamic mechanical analysis
EC	European Community
ECHA	European Chemicals Agency
EM	Electromagnetic
EMC	Electromagnetic compatibility
EMI	Electromagnetic interference
EMW	Electromagnetic waves
EMA	Ethylene methyl acrylate
EC	Electrical conductivity
EG	Expandable graphite
EIS	Electrochemical impedance spectroscopy
EDX	Energy dispersive X-ray analysis
EFM	Electrostatic force microscopy
EP	Epoxy
ESEM	Environmental SEM
EVA	Ethylene vinyl acetate
E_a	Absorbed waves
E_t	Transmitted electric field intensity
E_{in}	Incident electric field intensity
E_c	Moduli of reinforced composite
E_p	Moduli of the matrix polymer
h	The thickness of the sample
FCC	Federal Communications Commission

f-BN	Amino-functionalized boron nitride
FESEM	Field emission SEM
FGO	Functionalized GO
FLG	Few layer graphene
FM	Fluorescence microscopy
FMM	Force modulation microscopy
FPT	Flame propagation test
FR	Flame retardancy
FPVC	Flammable PVC
FTIR	Fourier-transform infrared spectroscopy
G	Permittivity of free space
GDY	Graphdiyne
GO	Graphene oxide
GNP	Graphene nanoplatelet
GP or GF	Graphite
GPO	Graphite oxide
G'	Storage modulus
G''	Loss modulus
G'_i	Entire storage modulus
HCCP	Hexachlorocyclotriphosphazene
HEA	Hydroxyethyl acrylate
HRR	Heat release rate
HRTEM	High-resolution TEM
HDPE	High-density polyethylene
H_t	Transmitted magnetic field intensity
H_{in}	Incident magnetic field intensity
HVLP	High volume low pressure
IFR	Intumescent flame-retardant
IC	Integrated circuit
IMFP	Inelastic mean free path
IR	Infrared
IrGO	In situ rGO
ISO	International Organization for Standardization
k	Thermal conductivity
KE	Kinetic energy
KPS	Potassium persulfate
KTN	Knowledge transfer network
LDPE	Low-density polyethylene
LIG	Laser-induced graphene
LOI	Limiting oxygen index
MMT	Montmorillonite
MFI	Melt flow index
MgST	Magnesium stearate
MLG	Multilayer graphene
MPP	Melamine polyphosphate

MWCNT	Multiwalled carbon nanotubes
M_t	Absorbed water mass at time t
M_s	Mass of saturated water
NBR	Nitrile butadiene rubber
N_A	Avogadro's number
NC	Nanoclay
Ni	Nickel
N_2	Nitrogen
NIST	National Institute of Standards and Technology
NMP	N-methyl-2-pyrrolidone
NMR	Nuclear magnetic resonance spectroscopy
NMT	Nanomaterial
NP	Nanoparticle
NT	Nanotube
OD	Optical density
OEMs	Original equipment manufacturers
OPFRs	Organophosphate flame-retardants
PP	Polypropylene
PA	Phytic acid
PAM	Polyamide
PA6	Polyamide 6
PBT	Poly(butylene terephthalate)
PDPFDE	Ferrocene-based polymer/nonphosphorus polymer based on ferrocene
PER	Pentaerythritol
PHDDT	Poly(hydroxybenzate-co-DOPO-benzenediol dihydrodiphenyl ether terephthalate
PHRR	Peak heat release rate
PI	Polyimide
PLA	Polylactic acid
PVA	Polyvinyl alcohol
PPY/PPy	Polypyrrole
PET	Polyethylene terephthalate
PVC	Poly(vinyl chloride)
PANI	Polyaniline
PCA	Polychlorinated alkanes
PC	Polycarbonate
PE	Polyethylene
PEI	Polyethylenimine
PDMS	Polydimethylsiloxane
PM	Photomultiplier
PMMA	Poly(methyl methacrylate)
PAA	Poly(amic acid)
PS	Polystyrene
PSA	Particle size analysis

PNC	Polymer nanocomposites
PEDOT	Poly(3,4 ethylenedioxythiophene)
PSS	Poly(styrenesulfonate)
PTGCAs	PDMS/rGO/SWCNT composites
PU	Polyurethane
PVDF	Polyvinylidene fluoride
P_{max}	Maximum load before failure
P_t	Transmitted power intensity
P_{in}	Incident power intensity
Q	Crystallinity
R	Gas constant
REACH	Registration, Evaluation, Authorisation and Restriction of Chemicals
RFI	Radiofrequency interference
RMS	Mean square roughness/root mean square
RP	Red phosphorus
RS	Raman spectroscopy
RC	Rate of combustion
RPT	Radiant panel test
R_u	Universal gas constant
Q_{act}	Activation energy
S_{11} or S_{22}	Port 1 to port 1 or port 2 to port 2 reflection parameters
S_{12} or S_{21}	Port 1 to port 2 or port 2 to port 1 transmission parameters
SAN	Poly(styrene-co-acrylonitrile)
SE_A	SE absorption
SE_M	Multiple reflections
SE_R/R_L/RL	Reflection loss
SE_T	Total SE
SSE	Specific SE
SC	Swelling coefficient
SCS	Salicylaldehyde-modified chitosan
SDC	Smoke density chamber
SLG	Single-layer graphene
SOJ	Small-outline J-leaded
SCTA	Sample controlled thermal analysis
SAXS	Small-angle X-ray scattering
SE	Secondary electrons
SEM	Scanning electron microscopy
SE	Shielding effectiveness
SWCNT	Single-walled carbon nanotube
TTI	Time to ignition
T	Temperature
t	Time
T_g	Glass transition temperature
TGA	Thermogravimetric analysis

THR	Total heat release
TOC	Total oxygen consumed
TOTM	Trioctyl trimellitate
TRGO/TrGO	Thermally reduced GO
TEM	Transmission electron microscopy
TV	Television
UBM	Under bump metallization
USAXS	Ultra-small-angle X-ray scattering
UTM	Universal tensile testing machines
UTS	Ultimate tensile strengthF_u
UV	Ultraviolet
V_f	Volume % of the reinforcing filler
VPR	Vapor phase reflow
WAXS	Wide-angle X-ray scattering
WEEE	Waste Electrical and Electronic Equipment
WXRD	Wide-angle XRD
WPU	Waterborne polyurethane
x	Cartesian location
XLPE	Cross-linked polyethylene
XRD	X-ray diffraction
XPS	X-ray photoelectron spectroscopy
Z	Shield impedance
Z_0	Free space impedance

Chapter 1
Polymer Nanocomposites

1.1 Introduction

The book provides an overview of nanocomposites in general, graphene, historical progress of polymeric nanocomposites as EMI shields but also flame-retardants, the principle of polymeric EMI shields as well as their flame retardancy, their characterization, measurement approaches for EMI shielding but also flame retardancy, their preparation techniques, and a range of current and emerging applications along with the current challenges and way forward in these. These present substantial opportunities for creating novel materials with distinct features that are more easily processed than traditional composite materials.

Here, it is crucial to start by clarifying what "nanotechnology" is as regards "composite materials," as well as momentarily describing how material characteristics at the nanoscale could differ from those seen at other micrometers to macroscales. We first discuss the variety of thermal, mechanical, electrical, but also optical properties that can be achieved using nanocomposite materials, and then we compare the characteristics of the GNPs reinforcement used to create polymeric (nano)composites, including other functional NPs, layered materials, and fibers (such as carbon CNTs and CNFs). XRD, thermal analysis, electron microscopy, and other commonly used analytical methods will also be described.

Since there is no one technology or field of science that falls under the umbrella of "nanotechnology," the term might be misleading. Instead, nanotechnology is a collection of methods, substances, uses, and ideas that are all characterized by scale [1]. Nanotechnology uses materials that are nanometer scale, or between 1 and 100 nm in size. As we will see in a moment, the physical characteristics of a given substance, such as carbon, GP, silicon, carbon black, or metals, are different at the nanoscale than they are at the macroscale. A nm is one millionth of a mm (0.000000001 m). For example, human hair ranges from 10,000 to 100,000 nm in diameter, while viruses range in size from 10 to 100 nm, red blood cells are about 5000 nm in diameter, and DNA molecules are roughly 2 nm in diameter.

S. S. Ray et al., *Graphene-Bearing Polymer Composites*, Springer Series in Materials Science 340, https://doi.org/10.1007/978-3-031-51924-6_1

A substance created artificially and made up of many phases is called a "composite material." These phases differ chemically from one another, and the material's chemical/physical phases clearly interact at their interfaces [2]. The other phase(s) distributed in this matrix is referred to as the "reinforcement or filler," as shown in Fig. 1.1, with the continuous phase being referred to as the "matrix." In order to create composite or nanocomposite materials, a variety of matrices (such as polymers, carbon, metals, as well as ceramics) along with other reinforcements (such as particles, fibers, and layered materials) have been used [3]; however, the focus of this book will only be on polymer matrix composites because of their particular technological interest in EMI shielding and flame retardancy applications.

Conducive polymeric nonwoven textile materials and their composites are receiving more attention due to increased concern over the avoidance of ecocide from high-frequency electromagnetic (EM) radiation. Since the spectrum of EM radiation includes various wave sources and uses, it is simple for someone to be exposed to EM radiation and become harmed.

Most electronic equipment, if not all of them, produce electromagnetic radiation waves that could be communicated from one electronic device to another through the atmosphere, space, or any other material. This phenomenon is called EMI [4, 5]. Electromagnetic contamination, or EMI, has significantly increased due to technological innovation and the widespread use of wireless communication. The passage

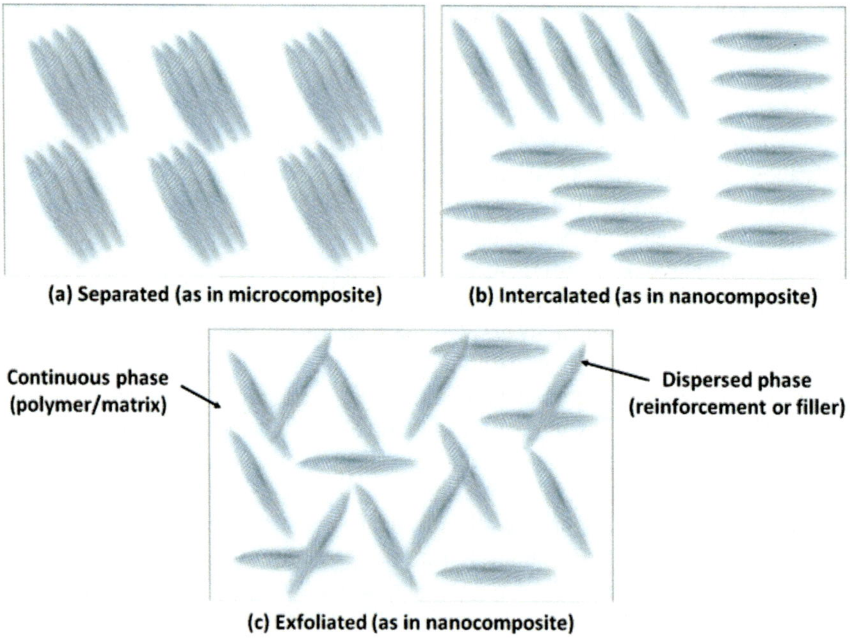

Fig. 1.1 Typical GNP-reinforced (nano)composite setup. The language is the same for all (nano)composite systems (especially nanoplatelets reinforced (nano)composites)

of electromagnetic waves and interference can impair the functionality of electronic devices, causing unfavorable reactions, operational failure, and a risk to safety and health in some sectors and applications [6, 7]. Automobile, consumer electronics, military and security, aerospace, 5G telecommunications, major medical equipment, wearable electronics, and applications for smart packaging are among the industries that are most susceptible to EMI [8]. Confidential data that is transferred through computer equipment is stored in security sectors and must be kept secure.

The incredible combination of properties, low weight, and processing ease that polymer materials provide is what promotes their widespread use in daily life. As a result, polymeric materials are also recognized for their fairly high flammability. Combustion can result in the generation of hazardous or corrosive gases as well as smoke. Because of this, the fire-retardant performance of polymeric matrices is a significant obstacle to their usage in the majority of applications. Although many flame-retardant additives, like halogenated systems, are actually being phased out due to their known or suspected antagonistic effects on the environment, the protection requirements in terms of polymer reaction to fire and their fire resistance performances have increased. The development of active and environmentally sensitive flame-retardant systems is thus the joint challenge for polymer materials. Scientific and technical literature describes numerous effective methods for increasing polymer fire resistance [9]. These methods mostly depend on the type and chemical makeup of the polymer in question, how it decomposes, how fire-safe the material must be, and how widely used the final products are. Flame retardants are employed in rubbers, plastics, and fabrics to prevent or delay the initial stages of a developing fire. As a result, these chemicals can be found in a wide range of products, including electronics like TVs and computers as well as construction materials, carpets, and upholstery. The application and fire safety requirements determine the type and capacity of flame-retardants utilized. Typically, flame-retardant additives can be found in polymeric materials in amounts between 5 and 30 wt.%. Although there are numerous cutting-edge uses for graphene, some of them include nanoelectronics, energy technology that has improved energy storage systems (e.g., highly effective batteries), medical applications (e.g., antibacterial agents), and the creation of composite materials and sensors.

The size of the reinforcement and its distribution within the matrix, at the macroscale, are the main determinants of the material properties of a polymer composite. For instance, "unidirectional" carbon fiber composites, in which all of the reinforcement is oriented in the same direction, have different mechanical properties than "dual-directional" composites, in which the GNP is oriented in successive laminates in separate directions [10]. The characteristics of two otherwise matching composites made using carbon fibers of various sizes will also vary. At the sub-micron scale, the reinforcement's size and the "interphase," or interface created between the reinforcement and the adopted matrix, are the primary determinants of the material properties [1]. However, as one or additional reinforcement's sizes slant toward the nanoscale, it is more and more the reinforcement's diameter and surface chemistry that dictate the material properties of the finally fabricated composite system, currently known as a "nanocomposite." A characteristic CF-reinforced polymer composite contains an interphase between the fiber and matrix that is larger than a complete NP, illustrating the huge length scale difference between traditional polymer composites and nanocomposites.

The book provides an overview of polymer nanocomposites' composition, characteristics, and variety of uses. The main approaches used to formulate and characterize these (nano)composite systems will be detailed after describing how material properties at the nanometer scale differ from those reported at longer scales. With examples of recent applications, the variety of mechanical, electrical, and thermal properties that can be obtained from nanocomposite materials will be discussed. Since they are currently of the greatest technological relevance, the creation of these GNP-based composites will be explored. The book will finish by analyzing potential barriers to the commercialization of polymer nanocomposites as well as the forecast for their further research and development.

This book stands out as the only book intercreatively discussing the current progress in GNP-bearing polymeric blends with an inclusive presentation of their preparation approaches, characterization, their application in EMI shielding but also fire retardancy along with deep consideration of the challenges and future outlook. It is of great importance to emerging researchers within the niches covered herewith.

1.2 Summary

In conclusion, the investigation into graphene-bearing polymer composites, with a focus on applications in EMI shielding and flame-retardant materials, underscores the immense promise of these advanced materials for addressing critical challenges in diverse industries.

The incorporation of graphene into polymer matrices has proven to be a transformative strategy, enhancing the electrical conductivity, thermal stability, and flame-retardant properties of the resulting composites. The comprehensive exploration of synthesis methods, dispersion techniques, and characterization methodologies has paved the way for the development of high-performance materials tailored to meet the stringent requirements of EMI shielding and flame retardancy.

The applications of graphene-bearing polymer composites in EMI shielding have shown remarkable effectiveness in attenuating electromagnetic radiation across a broad spectrum. The exceptional electrical conductivity of graphene, combined with its lightweight and mechanically robust nature, positions these composites as promising candidates for shielding electronic devices and systems from unwanted electromagnetic interference.

Simultaneously, the integration of graphene has demonstrated a notable impact on the flame-retardant properties of polymers. The improved thermal stability and reduced flammability of these composites showcase their potential for enhancing safety in various applications, including automotive, aerospace, and electronics.

As with any evolving field, challenges remain, such as scalability, cost-effectiveness, and optimization of composite properties. Addressing these challenges will be essential for the widespread adoption of graphene-bearing polymer composites in real-world applications.

Looking ahead, the prospect of tailored materials with dual functionalities for EMI shielding and flame retardancy opens exciting possibilities for multifunctional and high-performance materials in sectors where safety and reliability are paramount. Continued research, collaboration, and innovation will undoubtedly play a pivotal role in unlocking the full potential of graphene-bearing polymer composites for a range of practical applications.

References

1. E.T. Thostenson, C. Li, T.-W. Chou, Nanocomposites in context. Compos. Sci. Technol. **65**(3–4), 491–516 (2005)
2. W.D. Callister, *Fundamentals of Materials Science and Engineering*, vol. 471660817 (Wiley London, 2000)
3. S.S. Ray, A. Geberekrstos, T.S. Muzata, J.T. Orasugh, *Process-Induced Phase Separation in Polymer Blends: Materials, Characterization, Properties, and Applications* (Carl Hanser Verlag GmbH Co KG, 2023)
4. D. Wanasinghe, F. Aslani, G. Ma, D. Habibi, Review of polymer composites with diverse nanofillers for electromagnetic interference shielding. Nanomaterials **10**(3), 541 (2020)
5. J.T. Orasugh, S.S. Ray, Functional and structural facts of effective electromagnetic interference shielding materials: a review. ACS Omega **8**(9), 8134–8158 (2023)
6. S. Brzeziński, T. Rybicki, I. Karbownik, G. Malinowska, K. Śledzińska, Textile materials for electromagnetic field shielding made with the use of nano-and micro-technology. Open Phys. **10**(5), 1190–1196 (2012)
7. M. Tian, M. Du, L. Qu, S. Chen, S. Zhu, G. Han, Electromagnetic interference shielding cotton fabrics with high electrical conductivity and electrical heating behavior via layer-by-layer self-assembly route. RSC Adv. **7**(68), 42641–42652 (2017)
8. C.-H. Huang, P.-W. Hsu, Z.-W. Ke, J.-H. Lin, B.-C. Shiu, C.-W. Lou, J.-H. Lin, A study on highly effective electromagnetic wave shield textile shell fabrics made of point polyester/metallic core-spun yarns. Polymers **14**(13), 2536 (2022)
9. F. Laoutid, L. Bonnaud, M. Alexandre, J.M. Lopez-Cuesta, P. Dubois, New prospects in flame retardant polymer materials: from fundamentals to nanocomposites. Mater. Sci. Eng. R. Rep. **63**(3), 100–125 (2009)
10. F.L. Matthews, R.D. Rawlings, *Composite Materials: Engineering and Science* (Woodhead Publishing, 1999)

Chapter 2
Graphene Nanoplatelets in Brief

2.1 Introduction

GNP is a multilayer nanomaterial comprising individual graphene layers. At the University of Manchester, graphene was uncovered by Andre Geim and Konstantin Novoselov, who were awarded the 2010 Nobel Prize in Physics for their work [1]. Graphene is currently primarily a research material; nonetheless, it is already demonstrating significant technical awareness because of its extraordinarily high electrical and/or thermal conductivity, with material properties comparable to those of SWCNT in Table 2.1 [2, 3].

Since its discovery, a variety of preparation techniques as well as formulation techniques for polymer nanocomposites incorporating GNPs have been developed [2, 3]. By employing techniques comparable to those used to generate CNTs, such as CVD, arc discharge, and/or chemical reduction of carbon monoxide, graphene nanosheets are grown from carbon-rich gases in so-called bottom-up procedures. There have also been methods described for "unzipping" the NTs to produce graphene nanosheets from the CNT pioneering material. These bottom-up methods produce the best, most reliable material qualities, but it is challenging to scale them from the lab to mass production.

As a result, substitute "top-down" techniques better suited for scaling up have been developed. Here, obtaining graphene sheets involves separating layers of GP or GPO starting material. The simplest way to scale up is by exfoliating GPO and then reducing it chemically. GP has a significant benefit as a beginning material because it is inexpensive, costing only about $825 per ton. Moreover, researchers from Trinity College Dublin have proposed an intriguing bottom-up technique to produce graphene cheaply [4]. They exfoliated GO in NMP solvent to produce graphene nanosheets using a laboratory-scale sonic probe and have since claimed it is possible to achieve similar results using a kitchen blender and dishwashing liquid. According to Bergin [5], the technique was given a license to be improved for commercial manufacturing.

S. S. Ray et al., *Graphene-Bearing Polymer Composites*, Springer Series in Materials Science 340, https://doi.org/10.1007/978-3-031-51924-6_2

Table 2.1 Graphene characteristics performance compared with SWCNT

Property	Bulk density	Surface area	Elastic modulus	Tensile strength	Electrical conductivity	Thermal conductivity	Thermal stability	Gas permeability
Graphene	~0.004 g cm^{-3}	Up to 2630 m^2 g^{-1} (theoretical limit)	1 TPa	130 GPA	6000 Scm^{-1}	5000 Wm^{-1} K^{-1} at room temperature	< 200 °C (after functionalization)	Impermeable to all gases
SWCNT	1.33–1.40 g cm^{-3}	~2400 m^2 g^{-1}	1.2 TPa	2 GPa	Varies with chirality	6000 W m^{-1} K^{-1} at room temperature	2800 °C (vacuum); 750 °C (air)	Resistant to O_2, hydrocarbons, etc

GO readily exfoliates in water and protic solvents when in solution, making it feasible to mix GO with polymeric solution(s). In addition, GO could be exfoliated in an aprotic solvent if it has been treated with an amine and/or isocyanate, causing it to be compatible with polymeric matrices like PS, PU, PMMA, etc. After the GO has been exfoliated, the solvent can be removed; however, compared to nanoclays (exfoliated), GNPs are more likely to reassembly because hydrogen bonds are forming between the graphene sheets [2].

Through the intercalation of liquid monomer between the graphene layers and polymerizing it to create the completed polymer nanocomposite, as is done with other nanoplatelets like NC, in situ, polymerization processing of GNPs can be accomplished. Alternately, monomer and/or reactive functionalities may be grafted on the GNPs and then go through a cross-linking procedure to produce the polymer. Melt processing is also possible by utilizing customary methods. Even when compared to other nano-reinforcements, GNP has a relatively low bulk density and produces melt mixtures with a high viscosity, which makes it challenging to disperse and combine with the polymer during melt processing.

2.2 Pristine GNP

A carbon atoms monolayer organized in a honeycomb lattice serves as the foundation for graphene, a class of carbon-based nanomaterials. Graphene, a revolutionary sheet of sp^2 hybridized carbon atoms has received a lot of interest recently due to its exceptional and amazing qualities, including good electrical conductivity, a significant theoretical specific surface area, and high mechanical strength. Due to its excellent electronic low-noise and other exceptional electrical, optical, mechanical, thermal, and electrochemical capabilities, GNP is an ideal electronic material for electrochemical sensing. It is primarily produced through the micromechanical cleavage of GP, epitaxial growth on SiC, CVD of hydrocarbons on transition metal surfaces, and the dispersion of GP in organic solvents. However, these processes present challenges in obtaining processable GNPs in large quantities, preventing the full exploitation of its intriguing properties. For its usage in EMI shielding and flame retardancy in a variety of application niches, GNP has recently gained a great deal of interest [6–10]. The most pertinent GNP dielectric behavior for EMI shielding applications is its high electrical and thermal conductivity. These are some of the outstanding features of GNP. In the polymer matrix, lamellar GNP can be disseminated and create a "tortuous route" effect that reduces the rate of matrix degradation and thermal diffusion [11]. As a result, it is frequently employed as a flame-retardant carrier. In addition to improving the matrix's flame-retardant properties, it can also improve the mechanical properties of the matrix. In the area of flame-retardants, GNP has gained a lot of interest and investigation recently [8, 11–14]. Due to the powerful stacking interactions between π-π-bonds and the van der Waals force, GNP is prone to becoming irreversibly aggregated, which may reduce its flame-retardant property. GNP or GP is typically used to prepare functionalized GNP with covalent or non-covalent bonds to increase the dispersibility of GRNPs in polymer matrices in order to solve the problem of pore dispersity with polymeric matrices. To achieve the anticipated synergistic effect, GNP is frequently employed in combination with other flame-retardants [15].

GNP and also its composites are being used in exciting new research by research organizations in applications like supercapacitors, batteries, sensors, solar panels, diodes, transistors, tissue regeneration, EMI filters, flame-retardant materials, etc. [16–18].

The oxidized forms of GNP are known as "GO" and "RGO," and they are often created via the exfoliation of GPO. They are more well-liked than pristine GNP because both of them include reactive oxygen groups and can disperse better in common solvents and polymeric systems.

Though we have discussed the comely features of GNP such as its outstanding electrical and thermal conductivity and enhanced control of its functionalization, its demerits are high cost, difficulty encountered in workability, low production rate, and high hydrophobicity. It is apparent that its selection as GNP-based material of choice should be made with a holistic consideration of its merits and demerits.

2.3 GO

GO is often made by chemically treating GP through oxidation, followed by dispersion and exfoliation in water or appropriate organic solvents. It is composed of a single layer of GPO. The Hummers and Offeman approach [19] and its modified form suggested by several studies [20–23] can also be used to generate GO sheets. According to speculation, different functional groups with oxygen are likely to be present in GO as shown in Fig. 2.1. The advantages of GO over graphene and rGO are its excellent polar functionalization, cheapness, good dispersion in most conventional solvents, as well as ease of workability. However, it has its disadvantages, which are low electrical and thermal conductivity, random surface functionalization, and the difficulty encountered in additional functionalization after its preparation. In view of these points, researchers should be able to choose wisely as per their intended application niche, which graphene-based material best matches their desired application.

It has been determined that the oxygen functionalities in GO primarily take the form of hydroxyl but also epoxy groups on its basal plane, with minor levels of lactone, carboxyl, phenol, carbonyl, and quinone at the sheet borders. Its applications in nanocomposites [18], batteries [24], photocatalysis [25], and sensors [21] are made possible by the wide variety of oxygen functional. The oxygenated groups in GO can significantly impact its electrical, mechanical, and electrochemical characteristics. As a result, they account for the variations between GO and pure graphene. Controllable densities of GO monolayers can be deposited onto a wide range of substrates, allowing for the preparation of thin conductive films on rigid and flexible substrates, which improves their suitability for use in nanocomposites. While

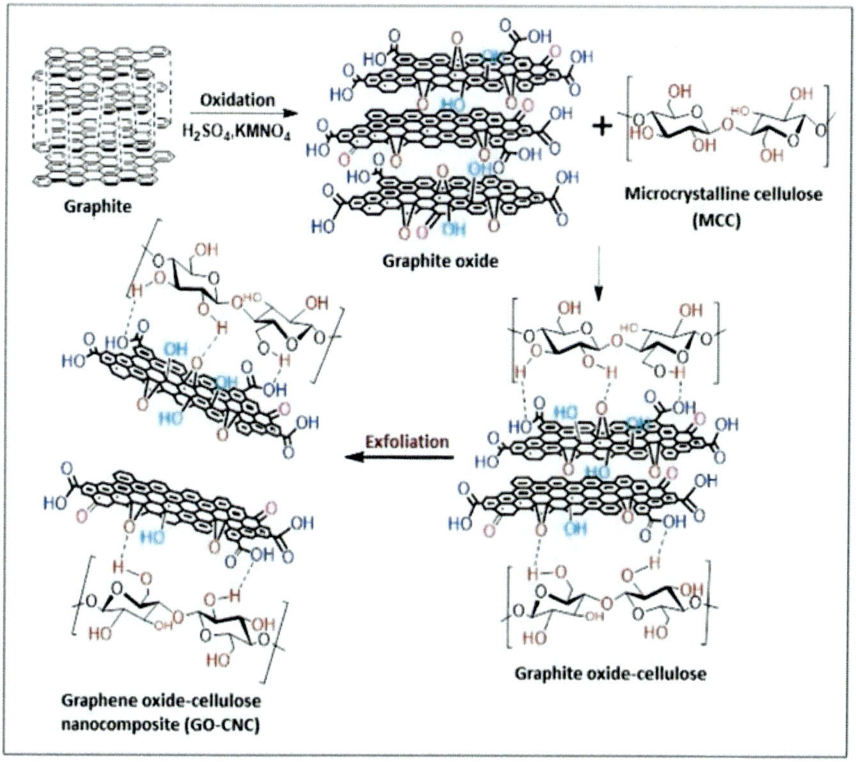

Fig. 2.1 Graphical scheme showing the plausible mechanism for in situ formation of GO-cellulose nanocomposite (GO-CNC). Reproduced with permission from Zaman et al. [23], Copyright 2020, Elsevier Science Ltd

some researchers have investigated using GO with nanocomposites to improve their mechanical, electrical, and thermal properties, many researchers have used GO to create graphene-based nanocomposite materials for other advanced applications.

2.4 rGO

Covalent modification or thermal degradation takes two steps to create rGO from GNP oxide. As previously indicated, GO is initially enhanced with reactive oxygen-containing functional groups using various chemical modification techniques placed on the nanocarbon basal plane. Modified GNP oxide is further treated with reducing agents (organic [25, 26] or synthetic [22, 26]), resulting in rGO (see Fig. 2.2). Thermochemical and electrochemical methods of reducing GO are also options [20, 22, 25]. rGO has advantages such as high electrical and thermal conductivities, better

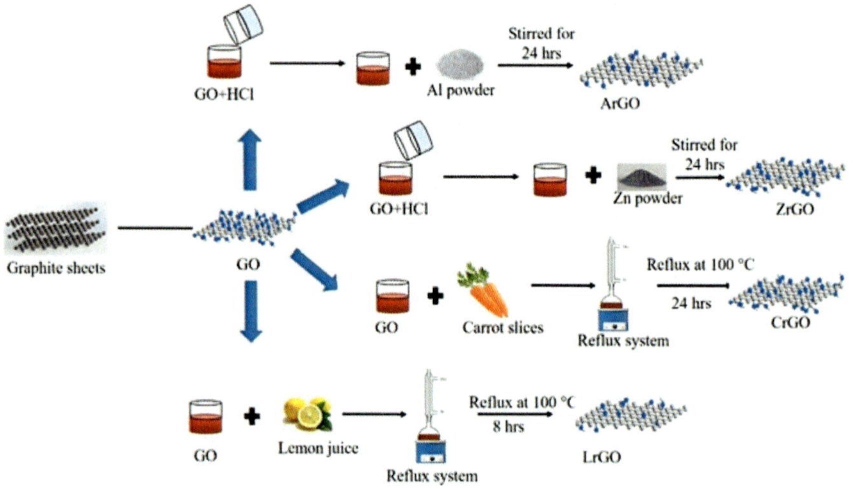

Fig. 2.2 Schematic illustration of the synthesis procedures for reduction of GO, i.e., aluminum reduced GO, zinc reduced GO, carrot reduced GO, and lemon reduced GO. Reproduced with permission from Joshi et al. [26], Copyright 2023, Elsevier Science Ltd

control of functionalization in comparison to GO, and cheaper than neat GNP, though its demerits are its high hydrophobicity and poor workability.

2.5 Importance of Graphene-Bearing Nanocomposites

Biosensors, energy storage devices, photocatalysts, and drug delivery could benefit from incorporating nanofillers like GNP within the polymer host. Numerous processing methods have been reported for dispersing fillers derived from GNPs and GO into polymeric matrices. Although the controlled loading and size of the particle were carefully taken into consideration, the fabrication of GNP-bearing polymer nanocomposite has been made more accessible by the use of ultrasonication process for the nanofiller dispersion. Most of the GNP-bearing polymer composite materials that have been studied were made with GO, CRGO, and/or TrGO as fillers. For a variety of multifunctional applications, 3D GNP-bearing PNC (also known as 3D-GPNCs) are being touted as new-generation materials [27].

Engineered NPs are not new, although they have recently gained popularity. Well-known common materials now referred to as "NMTs" include amorphous silica, zeolites, carbon black, pigments, and smectic layered clays. Long known are the special qualities of metal colloids; for instance, the vivid ruby red hues of stained-glass windows are made by adding either copper or gold to the glass during fabrication, creating colloids. The Lycurgus Cup, made of dichroic glass and dated to the fourth century A.D., is a stunning illustration of this type of glassmaking [28]. When light is transmitted through the glass, it appears ruby red while having an opaque green-yellow color in direct light. This phenomenon is thought to be caused by the presence of NPs of a silver-gold alloy in the glass, which are generally 50–100 nm in diameter. It is thought that these NPTs were generated in situ in the glass during heat treatment after the addition of metal salts to the molten glass.

Polymeric (nano)composites provide enhanced mechanical and tensile strength compared to typical polymer composites, as well as decreased scratch and mark resistance, a higher temperature of thermal distortion, and greater noise dampening. The difficulties typically associated with high reinforcement content in composites, such as decreased toughness, poor optical clarity, and higher melt viscosity, are less of an issue in producing high-performance polymer nanocomposites because nano-reinforcement loading of less than 10 wt.% is sufficient.

Extrusion, blow molding, thermoforming, injection molding, compression but also transfer molding are only a few examples of the well-established conventional techniques that can be utilized to treat nanocomposites [29, 30]. Additionally, these techniques enable high throughput, which is necessary for profitable manufacturing. For instance, polymer or glass fiber-reinforced car parts must be manufactured at a rate of more than one part per minute in order to compete with conventional materials and technologies. Given that processing expenses make up between 2/3 and 3/4 of the price of engineered composite materials, better processing results in lower total costs, which balance out potential increases in material costs due to the use of nano-reinforcing fillers [5].

However, there are certain drawbacks to using nanocomposites as well. Due to the nano-reinforcement, these polymers often have lower durability and less optical clarity when compared to virgin polymers. They also have melt viscosities that are noticeably higher than those of virgin polymers, which complicates processing. However, producing polymer nanocomposites is still less complex than producing high-performance composite materials utilizing autoclave processing or resin transfer molding with typical carbon fiber reinforcements. The material qualities that can be obtained from polymer nanocomposites depend greatly on the nano-reinforcement and matrix that are utilized.

Nanocomposites have become widely used in various industries, including packaging, aerospace, semiconductors, automotive, construction, electronics, energy, cosmetics, and others. The global market for polymer nanocomposites is anticipated to surpass $26.1 billion by 2028, growing at a compound annual growth rate of 15.84% from $10.8 billion in 2022 to that amount [31].

Nanoclay is the NMT that is most frequently utilized to create commercially viable nanocomposites [31]. However, commercial companies are now adopting CNT, GNP, and its derivatives as alternative reinforcement for advanced electronics materials production. The book's discussion will focus on GNP-bearing nanocomposites used in EMI shielding and flame retardancy for the remaining sections.

2.6 History and Progress in Polymer EMI Shields

Discoveries and technological advances are causing an increase in the sophisticated use of plastics. A large portion of the increase is thought to be caused by the displacement of metal components. This is especially true with the trend toward getting parts smaller in the electronics industry. Most plastics are great electrical insulators by nature, allowing electromagnetic waves to travel through them without being damaged. When plastics are used to enclose electronic components, this poses a difficulty. Electromagnetic waves released from an exposed gadget hinder other electronic devices' operations. RFI or EMI are terms used to describe this phenomenon "RFI." Dreger, Hormuth, and Simon provided early accounts of the sources of EMI [32]. Electronic equipment is both an emitter and a target of EMI. Electronic devices must be able to work under EMI from other sources or electromagnetic compatibility, in addition to the need that they do not emit radiation that interferes with other EMC. Electronic equipment must be insulated to allow for the filtering of incoming and outgoing interferences in order to avoid malfunction. Shielding has the dual benefit of lowering unwanted emissions from the shielded product and guarding against any interference from stray external signals on sensitive components. Metal shrouds were practical to utilize in numerous instances in the past. When developing laptop and notebook PCs, for example, inside metal shrouds are disadvantageous in terms of weight. They require inexpensive, light, strong, and simple plastics to construct, which can survive harsh conditions during processing and throughout use. To stop electromagnetic interference, plastics need to undergo a few more alterations.

In 1983, the FCC created a rule restricting EMI emissions from all electrical devices. Electronic equipment was divided into two categories in FCC regulation docket 20,780: class A (remote control devices, telephones, word processors, industrial computers, data processing equipment, programmable controllers, microprocessors, electronic typewriters, digital scales, large calculators, copiers, navigation equipment, sensitive test instruments) for non-consumer products and class B (CB receivers, digital clocks, pocket calculators, electronic games, digital watches, cardiac pacemakers, audio and high-fidelity equipment, radios appliances, video equipment, TV, personal computers, etc.) for consumer products. The same frequency range and testing distance are used for both devices, but the emission levels are measured at different levels [32, 33]. Table 2.2 provides the maximum amount of transmitted field strength for electronic equipment used in the USA. Table 2.2 also provides the comparable regulation for information technology equipment used in the European Union [32].

The breakdown of these and other surrounding electronic systems is brought on by the EM waves produced by machinery, tools, mobile phones, televisions, etc. These waves pose a severe threat to society. Cell phones are prohibited in numerous locations to prevent unwarranted instrument crackdowns, especially in operating rooms and flights. Therefore, there is a pressing need to create materials that can shield these equipment from dangerous EM radiation in order to solve these issues. In this regard, research teams are making great efforts to design and create materials that

Table 2.2 FCC radiation limits and EC emissions requirements for information technology equipment

FCC radiation limits			EC emissions requirements for information technology			
Frequency (MHz)	Distance (m)	Field strength (μ V/M)		Limits (dBμ V/m)		
Non-consumer			*Class A*			
30–88	3	100	Frequency (MHz)	30 m	10 m	3 m
89–216	3	150	30–230	30	40	59
217–960	3	200	231–1000	37	47	57
>960	3	500				
Consumer			*Class B*			
30–88	10	90	Frequency (MHz)	10 m	3 m	
89–216	10	150	30–230	30	40	
217–960	10	210	231–1000	37	47	
>960	10	300				

shield and safeguard both human life and such instruments. Since Schelkunoff [34] originally described the transmission line theory in 1943, it is crucial to comprehend the theory underlying the shielding mechanism. The plane wave shielding theory was later presented by Schultz and his colleagues in 1988 [35]. All shielding materials created to date have their foundation in these two notions. Copper was the first metal utilized for shielding, and metals were discovered to be the greatest EMI shielding materials, according to Joe Weibler's report [35]. Four different metals—copper, aluminum, zinc, and tin—were examined for their RF SE. Copper was discovered to be the best shielding material of all of them. As a result of their outstanding conductivity, stability, and capacity to reflect, absorb, and transmit EM waves, metal-based EMI shields have since been described by several researchers [36, 37]. The dielectric, electric, and magnetic properties of shielding materials have since been enhanced by developing metal-based composite materials. Although they were discovered to be the most promising materials, they lacked stability and serviceability. As a result, lightweight carbon materials, including GNP, CF, MXene, CNTs, GDY, conducting polymers, etc., have been added to create high-performance shielding materials that are lightweight and have improved qualities. These materials are being employed extensively in EMI shielding applications, but more development is ongoing.

2.7 History and Progress in Fire Retarding Polymer Materials

2.7.1 Incidences of Fire

Using data collected from 48 countries, or 1/6 of the world's countries, 1/3 of the world's population, and 40 cities, statistics for 2020 were compiled. According to the International Association of Fire and Rescue Services, out of the 68 million calls for fire and rescue services made in 2020 for a population of 3.3 billion (or 42% of the world's population), 3.8 million (or 5.7% of all calls) were related to fires. These fires resulted in 20.6 thousand fatalities and 69.5 thousand injuries. Accordingly, there were an average of 20 calls per year for every 1000 people in these nations, out of which 1.2 were for fires. In the same year, on average, 0.6 people died and 2.1 people were injured in fires for every 100,000 people, and on average, 0.5 people died and 1.8 people were hurt in each fire. Figure 2.3 depicts the distribution of fire by type and shows that 33.6% of all fires of which 24.2% were in residential buildings and 8.0% were in all other facilities occur in buildings. In comparison, 11.5% happened in vehicles, 2.5% happened in forests, 21.9% occurred in grass and bushes, 15.8% occurred in landfills, and 16.0% happened in other fires [38].

When looking closely at the number of fatalities caused by fire as per Fig. 2.4, it is evident that residential building fires account for 82.7% of the fatalities, followed

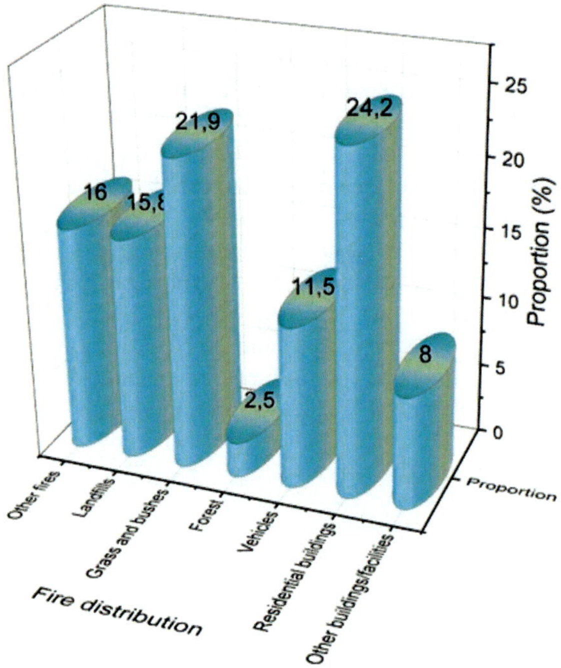

Fig. 2.3 Fire by types with respect to 33.6% related to its distribution for residentials, other buildings/ facilities, vehicles, forest, grass and bushes, landfills, and other fires

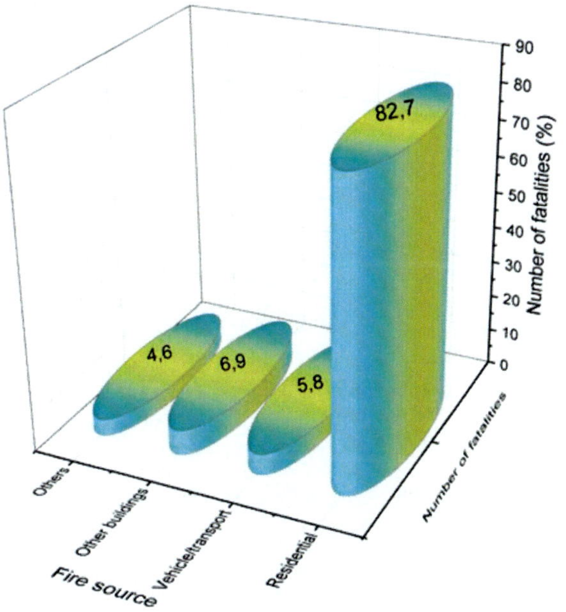

Fig. 2.4 Scheme depicting the number of fatalities with respect to fire sources

by 6.9% in other types of buildings, vehicle fires accounting at 5.8%, and different types of fires account for 4.6% of fatalities.

Moreover, it has been estimated that over 4 million fires every year in the world are either directly or indirectly caused by polymers, resulting in significant loss of life and property. Yet today's society is heavily reliant on synthetic polymeric materials. They pervade our surrounding and may be found in almost every office building, home, and car [39]. Currently, synthetic polymer materials continue to replace more conventional materials like steel and nonferrous metals, and naturally occurring polymeric materials like wood, cotton, natural rubber, and so forth. These synthetic polymer materials are also original, with distinctively valuable physical properties. However, the drawback of synthetic polymer materials over steel and other metals is that, in most circumstances, they can easily catch/support fire [40].

2.7.2 *Behavior of Polymer Materials in the Presence of Fires*

Gaining a thorough understanding of the combustion process of a polymer is essential before addressing the problem of the fire threat posed by polymeric materials. Polymer combustion is typically a multistage process with numerous interconnected chemical and physical processes. To start and maintain the combustion of polymers, three essential elements heat, fuel, and oxygen typically depicted as a fire triangle as depicted in Fig. 2.5 must occur simultaneously [41]. Polymers are flammable due to their predominantly carbon- and hydrogen-based chemical composition [12].

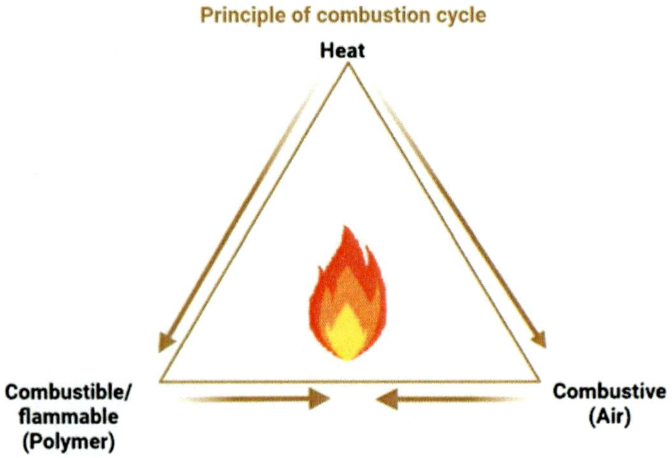

Fig. 2.5 Principle of the combustion cycle

Two components are necessary for the combustion reaction: one or more combustibles (reducing agents) and a combustive (oxidizing agent). Typically, oxygen in the air is what causes combustion. The entire process often begins with an increase in the polymeric material's temperature brought on by a heat source to the point where it causes polymer bond scissions. The resulting polymer fragments' volatile portion diffuses into the atmosphere and produces an explosive gaseous mixture (also called fuel). When the auto-ignition temperature is reached, which is the temperature at which the activation energy of the combustion reaction is attained, this gaseous mixture ignites and releases heat. As an alternative, the fuel can also ignite when exposed to a potent energy source outside of it at a lower temperature (known as the flash point) (spark, flame, etc.). The amount of heat released during fuel combustion determines how long the combustion cycle will last. More combustibles are created when the heat released surpasses a particular threshold because it triggers new breakdown reactions in the solid phase. Thus, the combustion cycle is maintained, and the term "fire triangle" is used as per Fig. 2.5 [12].

The energy needed to start combustion in polymers varies depending on the material's physical properties. For example, the polymer softens, melts, and drips when heating semicrystalline thermoplastics. The amount of energy that the polymer can store throughout these processes is determined by its ability to store heat and its enthalpy of fusion and crystallinity level. Therefore, the temperature difference caused by the exothermicity of the reactions involved, the specific heat, and the thermal conductivity of the semicrystalline thermoplastic all play a significant role in the increase in polymer temperature and the associated rate. However, the heating stage results in polymer breakup in the case of most thermosets and amorphous thermoplastics because these materials lack a melting point [12].

An energy input is necessary for the endothermic process known as covalent bond dissociation, which causes a polymer to decompose thermally. The energy supplied to the system must be more than required to bind the covalently linked atoms together (200–400 kJ/mol for most C–C polymers). The weakest bonds and whether or not oxygen is present in the solid and gas phases are significant determinants of the decomposition mechanism. The actions of heat and oxygen work together to cause thermal breakdown in most cases. Thus, oxidizing thermal degradation and non-oxidizing thermal degradation can be distinguished [42].

2.7.3 *Flame-Retardants*

Breaking the fire triangle necessitates the addition of particular substances known as "flame-retardants" to reduce the risk of ignition in the event of contact with a heat source or deceleration of combustion if the polymer or a neighboring material has already caught fire. The physical mixing of flame-retardant with the polymer during transformation has the benefit of not changing the chemical makeup of the macromolecules. The flame-retardant functionality may also be added before (as monomers) or after polymerization (by chemical grafting), which necessitates a potential alteration to the synthesis process but guarantees that the additive won't migrate. The class of reactive flame-retardants is made up of the flame-retardant employed in this fashion. Numerous different usable compounds are needed due to the diversity of polymers that need to be protected, the various fire safety requirements for various application sectors, and the current regulations and standards. However, to limit the quantities introduced and the deterioration of the material's mechanical and functional properties, optimizing the fire behavior frequently necessitates using two or more flame-retardant compounds that combine various modes of action [43].

History of Flame-Retardants

The Eastern civilizations, notably the Egyptians and Chinese, were the first to use flame-retardants. About 3000 years ago, the ancient Egyptians would immerse grass and reed in seawater before using it for roofing, causing the mineral salts to crystallize and act as a fire-retardant when the grass and reed dried. Alum and vinegar were also used by the Chinese and Egyptians to paint temple timbers in order to protect them from fire [44, 45]. At the same time, the Roman Empire's army made the first efforts to use flame-retardants in Western civilizations when they besieged the city of Piraeus in about 87–86 B.C. The Roman army coated them with an alum, vinegar, and clay concoction to prevent wooden siege towers from catching fire from city defenses. The ancient Egyptian flame-retardant combination was the basis for the Roman formulation [44]. The development of an "incombustible cloth" for Parisian stage curtains came later in the seventeenth century after treating canvas with a mixture of clay and gypsum. In 1735, Wyld received the first patent for a fire-retardant treatment for fabrics and wood [45]. This effort can be viewed as the first known attempt to give polymeric materials flame resistance.

King Louis XVIII asked Gay-Lussac to investigate ways to safeguard the materials used in theaters in 1786 after many fires in French theaters. Gay-Lussac discovered that ammonium salts of phosphoric, sulfuric, and hydrochloric acids were efficient in flame-retarding hemp and linen fabrics. Post 34 years of methodical examination of all available chemicals. He also noticed that Borax alone did not prevent after-low but was very effective and that a blend of ammonium phosphate and ammonium chloride was more effective in giving improved outcomes. The relevance of this work has endured the test of time. Consequentially, the fundamental components of contemporary fire-retardant chemistry were established early in human history and remained cutting-edge until the beginning of the twentieth century. The Groups III, V, and VII components contain the majority of the most efficient treatments for cellulosic materials [45].

William Henry Perkin, a highly regarded chemist, became interested in reducing the high flammability of the then-common fabric known as "flannelette" in 1913. During his work, he first outlined the most crucial criteria for a fire-retardant fabric, incorporating properties like durability, feel, non-poisonous nature, low cost, and printability after treatment. The Perkins treatment involved soaking the fabric in aqueous sodium stannate and ammonium sulfate solutions. After heating, the compounds are transformed into soluble stannic oxide, which was once assumed to be the active retardant. The Perkins method was not well received, and until new developments in synthetic polymers during World War II, there was little significant scientific research on fire retardance [45]. Since the 1930s, technical formulations of PCA known as CPs have been created for use as flame-retardants in plastics and sealants as well as additives in lubricants and cutting fluids [46]. The development of other reactive flame-retardants containing phosphorus and halogens followed quickly after the invention of the reactive flame-retardant chlorendic acid in the 1950s. In 1956, inert substances like aluminum hydroxide started to be employed as flame-retardant fillers for plastics because the aforementioned flame-retardants could not satisfy the flame-retardant standards for PE plastic [47–49]. DCRP, a cyclic compound flame-retardant, was first marketed in 1960 [50]. In the 1980s, a new class of flame-retardant known as additive-type flame-retardant polymers, which utilized bromine-based flame-retardants as organic halogen-based flame-retardants, started to be widely used in a variety of commodities [51, 52]. Due to their great flame-retardant efficacy and inexpensive production costs, consumers preferred these. They held a monopoly in the flame-retardant market, with their consumption making up 85% of the total volume of organic flame-retardants. Since the 1990s, bromine-based flame-retardants have been prohibited everywhere due to their environmental durability, biotoxicity, bioaccumulation, and migratory properties since they were recognized as organic pollutants [53, 54]. At the beginning of the twentieth century, OPFRs were introduced to use. After 1940, the production and use of OPFRs skyrocketed. They did not, however, become widely used flame-retardants until the 1970s and 1980s, when bromine-based flame-retardants rapidly disappeared from the market due to legislative restrictions on their usage and production. As a result of their outstanding flame resistance and plasticizing effects when employed as a replacement, OPFRs eventually evolved into becoming common items. Alone, the manufacturing of flame-retardants made from organophosphate ester rose from 100,000 tons in 1992 to 341,000 tons in 2007 [55, 56]. Currently, in addition to the search for improved flame-retardant efficacy, additional causes are driving the evolution. Improving the fire performance of polymers is a high goal due to the need to strike a balance between minimizing fire danger and environmental concerns as well as the demand to reduce the ecological footprint of materials. Thus, the introduction of a new class of sustainable, natural, and renewable flame-retardant technologies has been prompted [57]. Figure 2.6 reports an evolution of the use of flame-retardants over the years.

2.8 Summary

In summary, the brief exploration of graphene nanoplatelets provides a glimpse into the extraordinary properties and diverse applications of these unique nanomaterials. Their two-dimensional structure, composed of graphene sheets, imparts exceptional mechanical, electrical, and thermal characteristics.

The versatility of graphene nanoplatelets is evident in their applications across various industries, ranging from electronics and energy storage to composites and coatings. Their high surface area, conductivity, and mechanical strength make them promising candidates for enhancing the performance of a wide array of materials and devices.

While the potential of graphene nanoplatelets is vast, ongoing research and development efforts are essential to address challenges related to large-scale production, cost-effectiveness, and optimal integration into different matrices. Collaborative endeavors across scientific disciplines will play a crucial role in unlocking the full potential of graphene nanoplatelets for practical applications, contributing to advancements in materials science and technology.

In conclusion, this brief overview serves as an introduction to the remarkable world of graphene nanoplatelets, inviting further exploration and innovation in harnessing their unique properties for a myriad of technological breakthroughs.

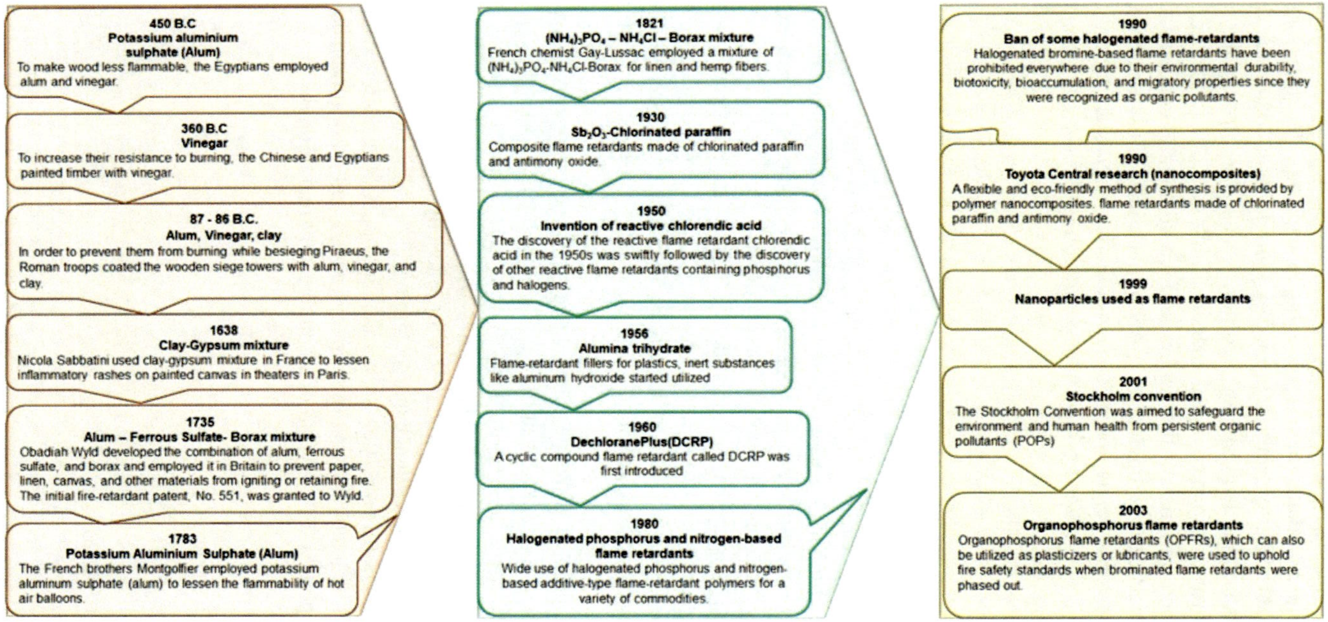

Fig. 2.6 Chronology of the early use of flame-retardants

References

1. J. Colapinto, Material question. Graphene may be the most remarkable substance ever discovered. But what's it for. The New Yorker **22**, 29 (2014)
2. H. Kim, A.A. Abdala, C.W. Macosko, Graphene/polymer nanocomposites. Macromolecules **43**(16), 6515–6530 (2010)
3. J.R. Potts, D.R. Dreyer, C.W. Bielawski, R.S. Ruoff, Graphene-based polymer nanocomposites. Polymer **52**(1), 5–25 (2011)
4. K.R. Paton, E. Varrla, C. Backes, R.J. Smith, U. Khan, A. O'Neill, C. Boland, M. Lotya, O.M. Istrate, P. King, Scalable production of large quantities of defect-free few-layer graphene by shear exfoliation in liquids. Nat. Mater. **13**(6), 624–630 (2014)
5. G. Armstrong, An introduction to polymer nanocomposites. Eur. J. Phys. **36**(6), 063001 (2015)
6. J.T. Orasugh, S.S. Ray, Functional and structural facts of effective electromagnetic interference shielding materials: a review. ACS Omega **8**(9), 8134–8158 (2023)
7. J.T. Orasugh, S.S. Ray, Graphene-based electrospun fibrous materials with enhanced EMI shielding: recent developments and future perspectives. ACS Omega **7**(38), 33699–33718 (2022)
8. Y. Shi, A. Yao, J. Han, H. Wang, Y. Feng, L. Fu, F. Yang, P. Song, Architecting fire safe hierarchical polymer nanocomposite films with excellent electromagnetic interference shielding via interface engineering. J. Colloid Interface Sci. **640**, 179–191 (2023)
9. A. Łapińska, N. Grochowska, K. Filak, P. Michalski, K.R. Szymański, P.A. Zaleski, K. Dydek, A. Daniszewska, K. Żerańska, A. Dużyńska, S. Kowalczyk, A. Plichta, Non-metallic multifunctional PVDF—Graphene nanoplatelets nanocomposites as an effective electromagnetic shield, thermal and electrical conductor. Mater. Today Adv. **18**, 100365 (2023)
10. B. Lee, U. Hwang, J. Kim, S.-H. Kim, K. Choi, I.-K. Park, C. Choi, J. Suhr, J.-D. Nam, Highly dispersed graphene nanoplatelets in polypropylene composites by employing high-shear stress for enhanced dielectric properties and frequency-selective electromagnetic interference shielding capability. Compos. Commun. **37**, 101409 (2023)
11. W. Xu, X. Wang, Y. Wu, W. Li, C. Chen, Functionalized graphene with Co-ZIF adsorbed borate ions as an effective flame retardant and smoke suppression agent for epoxy resin. J. Hazard. Mater. **363**, 138–151 (2019)
12. F. Laoutid, L. Bonnaud, M. Alexandre, J.M. Lopez-Cuesta, P. Dubois, New prospects in flame retardant polymer materials: From fundamentals to nanocomposites. Mater. Sci. Eng. R. Rep. **63**(3), 100–125 (2009)
13. M.N. Uddin, L. Le, R. Nair, R. Asmatulu, Effects of graphene oxide thin films and nanocomposite coatings on flame retardancy and thermal stability of aircraft composites: a comparative study. J. Eng. Mater. Technol. **141**(3), 031004 (2019)
14. S. Ran, F. Fang, Z. Guo, P. Song, Y. Cai, Z. Fang, H. Wang, Synthesis of decorated graphene with P, N-containing compounds and its flame retardancy and smoke suppression effects on polylactic acid. Compos. B Eng. **170**, 41–50 (2019)
15. X. Wang, E.N. Kalali, D.-Y. Wang, Two-dimensional inorganic nanomaterials: a solution to flame retardant polymers. Nano Adv **1**, 155 (2016)
16. S.S. Ray, A. Geberekrstos, T.S. Muzata, J.T. Orasugh, in *Process-Induced Phase Separation in Polymer Blends: Materials, Characterization, Properties, and Applications* (Carl Hanser Verlag GmbH Co KG, 2023)
17. K. Malkappa, J. Bandyopadhyay, V. Ojijo, S.S. Ray, Superior flame retardancy, antidripping, and thermomechanical properties of polyamide nanocomposites with graphene-based hybrid flame retardant. J. Appl. Polym. Sci. **139**(37), e52867 (2022)
18. B.S. Singu, K.R. Yoon, Exfoliated graphene-manganese oxide nanocomposite electrode materials for supercapacitor. J. Alloy. Compd. **770**, 1189–1199 (2019)
19. W.S. Hummers Jr., R.E. Offeman, Preparation of graphitic oxide. J. Am. Chem. Soc. **80**(6), 1339–1339 (1958)

20. L.T. Temane, J.T. Orasugh, S.S. Ray, Adsorptive removal of pollutants using graphene-based materials for water purification, in *Two-Dimensional Materials for Environmental Applications*. ed. by N. Kumar, R. Gusain, S. Sinha Ray (Springer International Publishing, Cham, 2023), pp.179–244
21. J.T. Orasugh, V. Saasa, S.S. Ray, B. Mwakikunga, Supersensitive metal free in-situ synthesized graphene oxide@cellulose nanocrystals acetone sensitive bioderived sensors. Int. J. Biol. Macromol. **241**, 124514 (2023)
22. J.T. Orasugh, S.S. Ray, Nanocellulose-graphene oxide-based nanocomposite for adsorptive water treatment, in *Functional Polymer Nanocomposites for Wastewater Treatment*. ed. by M.J. Hato, S. Sinha Ray (Springer International Publishing, Cham, 2022), pp.1–53
23. A. Zaman, J.T. Orasugh, P. Banerjee, S. Dutta, M.S. Ali, D. Das, A. Bhattacharya, D. Chattopadhyay, Facile one-pot in-situ synthesis of novel graphene oxide-cellulose nanocomposite for enhanced azo dye adsorption at optimized conditions. Carbohyd. Polym. **246**, 116661 (2020)
24. N. Kumar, J.R. Rodriguez, V.G. Pol, A. Sen, Facile synthesis of 2D graphene oxide sheet enveloping ultrafine 1D $LiMn_2O_4$ as interconnected framework to enhance cathodic property for Li-ion battery. Appl. Surf. Sci. **463**, 132–140 (2019)
25. S. Sadhukhan, A. Bhattacharyya, D. Rana, T.K. Ghosh, J.T. Orasugh, S. Khatua, K. Acharya, D. Chattopadhyay, Synthesis of RGO/NiO nanocomposites adopting a green approach and its photocatalytic and antibacterial properties. Mater. Chem. Phys. **247**, 122906 (2020)
26. R. Joshi, A. De Adhikari, A. Dey, I. Lahiri, Green reduction of graphene oxide as a substitute of acidic reducing agents for supercapacitor applications. Mater. Sci. Eng. B **287**, 116128 (2023)
27. A.T. Lawal, Graphene-based nano composites and their applications. A review. Biosens. Bioelectron. **141**, 111384 (2019)
28. I. Freestone, N. Meeks, M. Sax, C. Higgitt, The Lycurgus cup—a roman nanotechnology. Gold bulletin **40**, 270–277 (2007)
29. J. Bandyopadhyay, S.S. Ray, Are nanoclay-containing polymer composites safe for food packaging applications?—An overview. J. Appl. Polym. Sci. **136**(12), 47214 (2019)
30. D.I. Bower, in *An Introduction to Polymer Physics* (American Association of Physics Teachers, 2003)
31. R. Markets, *Polymer Nanocomposites Market: Global Industry Trends, Share, Size, Growth, Opportunity and Forecast 2023–2028—Product Image Polymer Nanocomposites Market: Global Industry Trends, Share, Size, Growth, Opportunity and Forecast 2023–2028*. https://www.researchandmarkets.com/reports/5769248/polymer-nanocomposites-market-global-industry?gclid=CjwKCAjw3ueiBhBmEiwA4BhspAUVtmANDRVMD9KwWgaAgFnhlqNOxAx2jef1UWXabwcTtnepQtnSwhoCfRIQAvD_BwE. Accessed 9 May 2023
32. J.C. Huang, EMI shielding plastics: a review. Adv. Polym. Technol. J. Polym. Process. Inst. **14**(2), 137–150 (1995)
33. A. Polak, J. Margolis, *Conducting Polymers and Plastics* (Chapman & Hall, New York, 1989), p.4
34. S.A. Schelkunoff, *Electromagnetic Waves* (1943)
35. Weibler, J.; Enclosures, L., Properties of Metals used for RF shielding. *EMC Test and Design* **1993**, 100.
36. J.T. Orasugh, C. Pal, M.S. Ali, D. Chattopadhyay, Electromagnetic interference shielding property of polymer-graphene composites, in *Polymer Nanocomposites Containing Graphene*, ed. by M. Rahaman, L. Nayak, I.A. Hussein, N.C. Das (Woodhead Publishing, 2022), pp. 211–243
37. J.T. Orasugh, C. Pal, A.P. Samanta, D. Chattopadhyay, Carbon nanotube and nanofiber reinforced polymer composites, in *Encyclopedia of Materials: Plastics and Polymers*. ed. by M.S.J. Hashmi (Elsevier, Oxford, 2022), pp.837–859
38. N. Brushlinsky, S. Sergei, P. Wagner, B. Messerschmidt, *World Fire Statistics* (2022)
39. T. Kashiwagi, iPolymer combustion and flammability—Role of the condensed phase, in *Symposium (International) on Combustion* (Elsevier, 1994), pp. 1423–1437
40. B. Research, *Global Market for Plastics Compounding 2021–2026*. https://www.researchandmarkets.com/reports/5515323/global-market-for-plastics-compounding-2021-2026#:~:text=The%20global%20market%20volume%20for,the%20period%20of%202021%2D2026. Accessed 10 Jan 2023

41. W. He, P. Song, B. Yu, Z. Fang, H. Wang, Flame retardant polymeric nanocomposites through the combination of nanomaterials and conventional flame retardants. Prog. Mater. Sci. **114**, 100687 (2020)
42. H. Macskásy, G. Palyi, in *Plastics: Their Behaviour in Fires* (Elsevier, 2012)
43. A. Boudenne, L. Ibos, Y. Candau, S. Thomas, *Handbook of Multiphase Polymer Systems* (Wiley & Sons, 2011)
44. A.I. Al-Mosawi, Flame retardants, their beginning, types, and environmental impact: a review. ÉPÍTŐANYAG: A SZILIKÁTIPARI TUDOMÁNYOS EGYESÜLET LAPJA **74**(1), 2–8 (2022)
45. R.R. Hindersinn, *Historical Aspects of Polymer Fire Retardance* (ACS Publications, 1990)
46. S. Bayen, J.P. Obbard, G.O. Thomas, Chlorinated paraffins: a review of analysis and environmental occurrence. Environ. Int. **32**(7), 915–929 (2006)
47. A. Dey, S. Mandal, S. Bhandari, C. Pal, J.T. Orasugh, D. Chattopadhyay, Characterization methods, in *Fiber-Reinforced Nanocomposites: Fundamentals and Applications* (Elsevier, 2020), pp. 7–67
48. L.S. Birnbaum, D.F. Staskal, Brominated flame retardants: cause for concern? Environ. Health Perspect. **112**(1), 9–17 (2004)
49. M. Alaee, P. Arias, A. Sjödin, Å. Bergman, An overview of commercially used brominated flame retardants, their applications, their use patterns in different countries/regions and possible modes of release. Environ. Int. **29**(6), 683–689 (2003)
50. E. Hoh, Zhu, R.A. Hites, Dechlorane plus, a chlorinated flame retardant, in the Great Lakes. Environ. Sci. Technol. **40**(4), 1184–1189 (2006)
51. R. Hou, Y. Xu, Z. Wang, Review of OPFRs in animals and humans: absorption, bioaccumulation, metabolism, and internal exposure research. Chemosphere **153**, 78–90 (2016)
52. A. Covaci, S. Harrad, M.A.-E. Abdallah, N. Ali, R.J. Law, D. Herzke, C.A. de Wit, Novel brominated flame retardants: a review of their analysis, environmental fate and behaviour. Environ. Int. **37**(2), 532–556 (2011)
53. R.J. Law, C.R. Allchin, J. De Boer, A. Covaci, D. Herzke, P. Lepom, S. Morris, J. Tronczynski, C.A. De Wit, Levels and trends of brominated flame retardants in the European environment. Chemosphere **64**(2), 187–208 (2006)
54. F. Tao, M.A.-E. Abdallah, D.C. Ashworth, P. Douglas, M.B. Toledano, S. Harrad, Emerging and legacy flame retardants in UK human milk and food suggest slow response to restrictions on use of PBDEs and HBCDD. Environ. Int. **105**, 95–104 (2017)
55. S. Lee, H.-J. Cho, W. Choi, H.-B. Moon, Organophosphate flame retardants (OPFRs) in water and sediment: occurrence, distribution, and hotspots of contamination of Lake Shihwa, Korea. Marine Pollut0 Bull. **130**, 105–112 (2018)
56. G. Chen, Y. Jin, Y. Wu, L. Liu, Z. Fu, Exposure of male mice to two kinds of organophosphate flame retardants (OPFRs) induced oxidative stress and endocrine disruption. Environ. Toxicol. Pharmacol. **40**(1), 310–318 (2015)
57. H. Vahabi, F. Laoutid, M. Mehrpouya, M.R. Saeb, P. Dubois, Flame retardant polymer materials: an update and the future for 3D printing developments. Mater. Sci. Eng. R. Rep. **144**, 100604 (2021)

Chapter 3
Electromagnetic Interference Shielding Materials

3.1 Principle of Polymeric Electromagnetic Interference Shielding Materials

By utilizing the phrase "SE," the effectiveness of a shielding material against EM radiation is commonly expressed in dB. With EMI shielding materials that reduce EMW by 20 dB, commercial requirements can be satisfied [1]. The ratio of the power intensity of the incident wave(s) to that of the transmitted EM wave(s) (measured on a logarithmic scale) is known as the total SE (SE_T) [2].

$$SE_T = 10 \log_{10} \frac{E_t}{E_{in}} = 20 \log_{10} \frac{H_t}{H_{in}}. \tag{3.1}$$

P_t, P_{in}, E_t, E_{in}, H_t, and $H_{in,}$ respectively, denote the transmitted power intensity, incident power intensity, transmitted electric field intensity, incident electric field intensity, transmitted magnetic field intensity, and incident magnetic field intensity.

$$SE_T = SE_R + SE_A + SE_M. \tag{3.2}$$

The above equation was developed in order to create SE_T, which may be written as Schelkunoff proposed combining attenuation via reflection (SE_R "approximately 377 ohms for air"), attenuation via absorption (SE_A), plus attenuation by multiple reflections (SE_M) [1].

3.1.1 Attenuation via Reflection

When shielding materials are exposed to radiation, reflection is the easiest event to see. The shielding material's electrical conductivity dramatically affects how much

S. S. Ray et al., *Graphene-Bearing Polymer Composites*, Springer Series in Materials Science 340, https://doi.org/10.1007/978-3-031-51924-6_3

R_L there is. The electrical conductivity of the shield(s) will determine how high the R_L will be. An EMW must interact with free electrons in order for it to be reflected by the substance. The following can be done to calculate the reflection loss (SE_R) using Fresnel's equations [1]:

$$SE_R = 10 \log (Z + Z_0) / (4ZZ_0) \tag{3.3}$$

$$SE_R = 39.5 + 10 \log(\sigma / \mu\omega). \tag{3.4}$$

Z, Z_0, σ, μ, and ω denote the shield impedance, free space impedance, which is known to be 376.73 Ω, the shields' electrical conductivity, its magnetic permeability, as well as the incident EM waves angular frequency.

According to Eqs. 2.3 and 2.4, SE_R is a ratio of the conductivity to permeability along with angular frequency "σμω" of the shield. As a result, SE_R drops as frequency rises for a given shielding material. Composites with high and low SE_R values are characterized by high SE_R values. As a result, the first air/shield interface in conducting metal-based shielding materials typically has a high SE_R. . Mobile charge carriers in the shield should interact with EM waves to create a counter-field (referred to as induced field/scattered field).

3.1.2 Absorption-Based Attenuation

Absorption happens as the EM waves pass through the shielding substance. The absorption process is frequently related to the thickness (d) of the shielding material, and better shielding is seen when the material has electric and magnetic dipoles that interact with the electromagnetic waves. The surface's electrical conductivity, which in turn depends on the interaction between incident waves and charge carriers, determines how much nano/micro capacitors form and how much permeability increases as a result of eddy current losses. Therefore, the development of absorption events involves the interaction of magnetic loss, polarization loss, and ohmic loss.

Shielding materials' definition of the SE_A of EM waves is

$$SE_A = 20\left(\frac{d}{\delta}\right) \log_{10}^{e} \tag{3.5}$$

$$SE_A = 8.686\left(\frac{d}{\delta}\right) \tag{3.6}$$

$$SE_A = 8.686d\sqrt{\frac{\sigma\mu\omega}{2}}. \tag{3.7}$$

σ, δ, ω and μ, are the electrical conductivity, skin depth, angular frequency, and magnetic permeability. According to Eq. 3.7, effective SE_A contributions require a shielding material with good electrical conductivity. Additionally, these phrases demonstrate that SE_A rises as σ rises. This shows that for a given EMI shield, SE_A surges with the rising frequency and that the EMI shields with high values of μ result in a larger SE_A. A good absorbing material, according to the statement, should have high μ, good σ, and enough "d" to achieve the required δ. The dielectric characteristics of the shielding materials determine how well EM waves absorb and reflect in the absence of μ.

Skin depth (δ) is an important factor in EM wave shielding. With increasing depth, the EM wave penetration amplitude progressively decreases. As shown in Fig. 3.1, skin depth is defined as the expanse at which the EM field intensity is reduced to 1/e (37%) of the initial EM field amplitude value. The following is a universal formula for the skin's depth:

$$\delta = \sqrt{\frac{2}{\sigma\mu\omega}}\sqrt{\sqrt{1+\left(\frac{\omega\varepsilon}{\sigma}\right)^2}+\frac{\omega\varepsilon}{\sigma}}. \tag{3.8}$$

$\frac{\omega\varepsilon}{\sigma} > > 1$ for excellent conducting materials, which transforms Eq. 2.8 into

$$\delta = \sqrt{\frac{2}{\sigma\mu\omega}}. \tag{3.9}$$

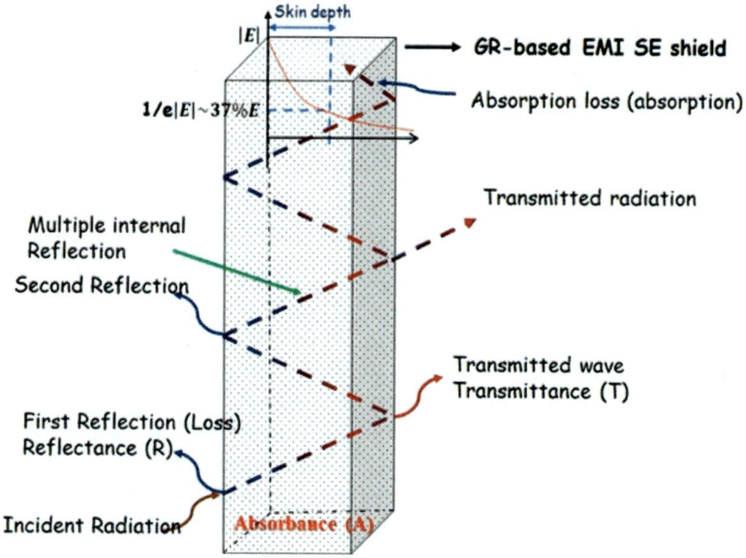

Fig. 3.1 Scheme explaining the mechanism of EM wave shielding

The incident EM waves' wavelength (ω), as well as the shielding materials σ and μ, are all important factors in determining the delta, according to Eq. 3.9. As a result, only a small portion of high ω EM waves can get through the materials. δ can be decreased by raising the conductivity, permeability, and frequency of EMI shields. At a given frequency of 1 GHz, Cu has a δ of 2.09 μ m, and Ni has a δ of 0.47 μ m ($\mu_r = 100$, $\sigma = 1.15$ 107/m, respectively). Due to Ni's ferromagnetic properties, δ for Ni is relatively low compared to Cu [1].

3.1.3 Attenuation via Multiple Reflections

EM waves experience multiple reflections at the interfaces as they travel through the shielding materials (Fig. 3.1), but these reflections do not attenuate EMWs as much as absorption or reflection. One way to write the SE_M is as.

$$SE_M = 20\log_{10}\left(1 - \left(e^{-\frac{2d}{\delta}}\right)\right) \tag{3.10}$$

$$SE_M = 20\log_{10}\left(1 - \left(1 - 10^{-\frac{SE_A}{10}}\right)\right). \tag{3.11}$$

A direct correlation between SE_A and the attenuation of EM waves as a result of numerous reflections may be found in Eqs. (3.10 and 3.11). A shielding material has an extremely tiny SE_M because of its great absorption. SE_M is often ignored when calculating SE_T for thick shielding materials with high SE_R and SE_A values (also ignored if $SE_T > 15$ dB). Since the waves that are internally reflected several times are eventually absorbed or lost as heat, repeated internal reflections and absorption can be combined. According to Schelkunoff's shielding theory, numerous reflections should not be considered if the shielding material's thickness (d) exceeds δ. Typically, this mechanism works betwixt the front and rear surfaces of EMI shields.

3.1.4 Evaluation of the SE

EM waves' incident power intensity (Pin), which is affected by shielding materials, is split into three components: (a) reflected power intensity (P_r); (b) absorbed power intensity (P_a); and (c) transmitted power intensity (P_t). To calculate the shielding in relation to the reduction of the electric field intensity (E) and the EMW power, three coefficients are used: transmittance (T), reflectance (R), and absorbance (A) of EMWs (P). Both their power and the strength of the electric field can be expressed as.

$$R = P_r / P_{\text{in}} = \left|\frac{E_r}{E_{\text{in}}}\right|^2 = |S_{11}|^2 = |S_{22}|^2 \tag{3.12}$$

$$T = P_r / P_{in} = \left| \frac{E_t}{E_{in}} \right|^2 = |S_{12}|^2 = |S_{21}|^2 \tag{3.13}$$

$$A = P_r / P_{in} = \left| \frac{E_a}{E_{in}} \right|^2 = 1 - |S_{11}|^2 - |S_{22}|^2 \tag{3.14}$$

$$A = P_r / P_{in} = \left| \frac{E_a}{E_{in}} \right|^2 = 1 - R - T. \tag{3.15}$$

Herewith E_t, denotes the transmitted wave, E_{in} is the incident wave, E_a represents absorbed waves, and the E_r reflected waves, respectively.

The SE of EM waves was assessed using vector network analyzers for both the coaxial transmission line approach and the waveguide method. As complex scattering parameters made up of S_{11}, S_{12}, S_{21}, but also S_{22} where S_{11} or S_{22} stands for reflection and S_{12} or S_{21} for transmission, the aforementioned coefficients can be discovered utilizing two-port vector network analyzers.

If there are not many internal reflections of the shielding material, then $1 - R$ is used to indicate the relative intensity of the EM waves inside the material after reflection. In light of this, the effective absorbance A_{eff} could be expressed as:

$$A_{eff} = \left(\frac{A}{1 - R} \right) = \left(\frac{1 - R - T}{1 - R} \right) \tag{3.16}$$

SE_R and SE_A could be expressed in relations to R, T, $|S_{11}|$ and $|S_{21}|$ as:

$$SE_R = 10 \log\left(\frac{1}{1 - R} \right) = 10 \log\left(\frac{1}{(1 - |S_{11}|^2)} \right) \tag{3.17}$$

$$SE_A = 10 \log\left(\frac{1}{1 - A_{eff}} \right) = 10 \log\left(\frac{1 - R}{T} \right) = 10 \log\left(\frac{(1 - |S_{11}|^2)}{|S_{21}|^2} \right). \tag{3.18}$$

With the aid of Eqs. 3.17 and 3.18, total SE (SE_T) can be represented in terms of S_{21} or T ($SE_T > 15$ dB, the item SE_M can be disregarded).

$$SE_T = SE_R + SE_A = -20 \log(S_{21}) = -10 \log T. \tag{3.19}$$

3.1.5 Mitigation of Electromagnetic Wave Reflection

Impedance Matching

The σ and μ have the biggest impact on the SE_R and SE_A values, according to Eqs. 3.3–3.7, for EM waves shielding materials with constant thickness (with fixed EM waves frequency), and the relevant preliminary relationships are as follows:

$$SE_R \sim (\sigma/\mu) \tag{3.20}$$

$$SE_A \sim (\mu\sigma). \tag{3.21}$$

It demonstrates that the SE_R and SE_A values get better as the σ value rises, indicating that this feature has a positive effect on EM wave absorption and reflection. In contrast, the value of SE_A would increase with an increase in μ while the SE_R would decrease, showing that a rise in magnetism would aid in reducing EM reflection and increase EM absorption.

In terms of the shielding materials' ability to absorb and reflect EM waves, impedance matching provides important information. Frequently, reflection coefficient (Γ) and reflection loss are used to describe EMI shielding materials (RL). Different impedances or refractive indices between two mediums lead to RL, which results in EM reflection on the interface or surface. In terms of Z and Z_0, Γ can be represented as:

$$\Gamma = \frac{(Z - Z_0)}{(Z + Z_0)}, \tag{3.22}$$

where Z and Z_0 (376.73 Ohms) stand for the corresponding impedances of the shielding materials and empty space. The following equation explains how Γ and RL are related.

$$\Gamma = 10\left(\frac{-\mathrm{RL}}{20}\right). \tag{3.23}$$

Z and Z_0 terms for RL are as follows:

$$\mathrm{RL} = 20\log_{10}|\Gamma| = 20\log_{10}\left|\frac{(Z - Z_0)}{(Z + Z_0)}\right|. \tag{3.24}$$

μ_r and ε_r are expressed with respect to Z as follows:

$$Z = \sqrt{\mu_r/\varepsilon_r}\tan h\left(j\frac{2\pi \mathrm{fd}}{c}\sqrt{\varepsilon_r\mu_r}\right). \tag{3.25}$$

The complex permittivity and also permeability terms are denoted as ε_r and μ_r. More than 90% and 99%, respectively, of the energy of EM waves can be absorbed when the value of RL is less than −10 and −20 dB.

When $Z = Z_0$ is used in Eq. 3.24 to show that impendence must be matched ($Z = Z_0$), it is revealed that all incidence EMW will pass through the shielding material without being reflected if the impedances of the shielding material and the incident EMW are closely matched to each other (Γ=0). Z is thus expressed in terms of μ and ε as:

$$Z = \sqrt{\frac{\mu}{\varepsilon}},\ Z_0 = \sqrt{\frac{\mu_0}{\varepsilon_0}} \tag{3.26}$$

$$Z = Z_0 \Rightarrow \sqrt{\frac{\mu}{\varepsilon}} = \sqrt{\frac{\mu_0}{\varepsilon_0}} \tag{3.27}$$

$$\frac{\mu}{\mu_0} = \frac{\varepsilon}{\varepsilon_0} \Rightarrow \varepsilon_r = \mu_r. \tag{3.28}$$

When the relative permeability (μ_r) and relative permittivity (ε_r) of the shielding material are equal as obtained by Eq. 3.28 and none of the incident EM waves are reflected at the surface, the principle presented in Eq. 3.28 is applicable. The incident EMWs penetrate the material to be absorbed, ε_r *and* μ_r affects how well an EMI shielding shields the incident EM waves and the SE of the shield. In other words, by narrowing the gap between the relative permeability and permittivity values of the shielding material, it will be easier for the shielding surface and the air to match impedance, which will lessen reflection and boost absorptivity. The magnetic and dielectric synergetic effects are important components and elements for EMW shielding. A method for merging the dielectric and magnetic characteristics in composite materials is impedance matching. Surface reflection is reduced by EMW absorption with significant incident wave attenuation, which improves absorption effectiveness.

When EMWs come into contact with the shielding material under the situation of mismatch ($Z \neq Z_0$), some of the wave is reflected. Because there will be a significant disparity between Z and Z_0 if the shielding is constructed of highly conducting materials with very low impedance, Γ will be quite high. Maximum incident EM waves will reflect with the least amount of absorption under this mismatch circumstance.

Electromagsnetic Wave Attenuation

Dielectric Loss

Dielectric loss is the dissipation of EM waves caused by the electric field's interaction with shielding materials. The dielectric loss factor is commonly used to describe the ability of an electrical system to attenuate due to polarization loss and conduction loss (dissipation factor or electric loss tangent)

$$\tan \delta_\varepsilon = \frac{\varepsilon''}{\varepsilon'}. \tag{3.29}$$

δ_ε represents the dielectric loss angle; $\varepsilon = \varepsilon' = j\varepsilon''$ denotes the complex permittivity while ε'"the real permittivity" stands for the dielectric constant (stored charge) and the imaginary permittivity "ε''" represents the dielectric loss (dissipation). The real permittivity can, therefore, be determined thus:

$$\varepsilon' = \varepsilon_\infty + \frac{(\varepsilon_s - \varepsilon_\infty)}{\left(1 + \omega^2 r^2\right)}. \tag{3.30}$$

As per Debye's theory, ε'' denotes EM wave energy dissipation which is calculated as per Eq. (3.31).

$$\varepsilon'' = \frac{(\varepsilon_s - \varepsilon_\infty)}{\left(1 + \omega^2 r^2\right)} \omega\tau + \frac{\sigma}{\omega\varepsilon_0}\varepsilon_0. \tag{3.31}$$

The static dielectric constant, relative dielectric constant (high frequency), angular frequency, electrical conductivity, and polarization relaxation time are denoted by ε_s, ε_∞, ω, σ, and τ, respectively.

With regards to Eq. 3.31, it can be seen that EMI filters having high electrical conductivity produce filters with high ε'' which signifies high dielectric loss. It is vital to note that the dielectric loss is dominated by the conduction loss in materials having high electrical conductivity.

All polarization types, plus electron polarization, interfacial polarization, ion polarization, and dipole polarization, are often explained by polarization loss. Instead of the MW region (0.3–300 GHz), electron and ion polarization often take place in the high-frequency range (103–106 GHz). In the MW area, defects, functional groups, heteroatoms, and interfaces produce dipole polarization. The polarization loss in the dipole direction that results from a dipole rotation under an alternating electric field being unable to keep up with the shift in the applied electric field causes the EM energy to be lost as heat. Interfacial polarization also makes a major contribution to relaxation loss. At the contact surfaces, charges are redistributed and accumulated, resulting in a capacitance-like orientation that causes polarization loss.

Magnetic Loss

The permeability (magnetic energy component) of an EMI shield(s) also contributes to its shielding effectiveness in addition to the dielectric characteristics. The magnetic field strength of EM waves may weaken when magnetic materials are used. Magnetic losses occur when an EM field is applied to a magnetic substance. The relaxation phenomenon that takes place during magnetization and contributes to the magnetic effect includes hysteresis loss, domain wall resonance loss, natural resonance, exchange resonance, and eddy current loss [1]. In contrast to hysteresis loss, which is minimal in a weak magnetic field, domain wall resonance loss often occurs in the MHz frequency range. Eddy current loss, exchange resonance, and natural resonance are thus the three most significant types of magnetic loss in the MW range (GHz frequency range) [1]. An energy gap between the two spins is produced when a magnetic vector is subjected to an alternating electric field. This energy gap is what results in resonance loss [1]. The magnetic loss factor (also known as the magnetic loss tangent) could be expressed as given in the equation to quantify the magnetic attenuation as in expression 3.32.

$$\tan \delta_\mu = \frac{\mu''}{\mu'}. \tag{3.32}$$

δ_μ, μ'', and μ' denote magnetic loss angle, imaginary permeability, as well as real permeability. From these parameters, the complex permeability is calculated thus:

$$\mu = \mu' - j\mu''. \tag{3.33}$$

This representation is common in electromagnetics, particularly when dealing with materials that exhibit both magnetic permeability and loss. The real part μ' is associated with the ability of the material to support magnetic fields, while the imaginary part μ'' represents the energy loss in the material due to magnetic effects.

Understanding and controlling both magnetic and dielectric losses are essential in various technological fields. Engineers strive to optimize material properties and design structures to minimize unwanted losses and enhance the performance of devices and systems that rely on the transmission and reception of electromagnetic waves.

3.2 Summary

In conclusion, the exploration of graphene-bearing polymer composites for EMI shielding materials represents a remarkable convergence of advanced materials science and practical engineering solutions. The integration of graphene, a two-dimensional wonder material, into polymer matrices has unleashed a new paradigm in the development of efficient and multifunctional EMI shielding materials.

Graphene's exceptional electrical conductivity, combined with the flexibility and processability of polymers, has resulted in composites that exhibit outstanding EMI shielding performance. These materials not only effectively redirect and absorb electromagnetic waves but also bring additional benefits such as mechanical strength, lightweight characteristics, and thermal stability.

The tunability of graphene-bearing polymer composites allows for the tailoring of EMI shielding properties to meet specific application requirements. Whether in telecommunications, aerospace, or automotive industries, these composites offer a versatile and promising solution for safeguarding electronic devices from unwanted interference.

As this field matures, challenges related to scalability, cost-effectiveness, and standardization of testing protocols must be addressed to facilitate widespread adoption. Collaborative efforts among researchers, industry professionals, and regulatory bodies are pivotal in bridging the gap between scientific innovation and practical implementation.

The significance of graphene-bearing polymer composites as EMI shielding materials extends beyond mere technological advancements. These materials contribute to the development of more energy-efficient and reliable electronic systems, fostering progress in various industries. The synergy between graphene's unique properties and the versatility of polymer matrices positions these composites as key players in the ongoing quest for robust and efficient EMI shielding solutions.

In essence, the journey into graphene-bearing polymer composites for EMI shielding materials marks a transformative chapter in materials science, where the pursuit of cutting-edge technology aligns seamlessly with the demands of an interconnected and rapidly evolving electronic landscape.

References

1. R. Kumar, S. Sahoo, E. Joanni, Composites based on layered materials for absorption of microwaves and electromagnetic shielding. Carbon 118072 (2023)
2. R. Banerjee, A. Gebrekrstos, J.T. Orasugh, S.S. Ray, Nanocarbon-containing polymer composite foams: a review of systems for applications in electromagnetic interference shielding, energy storage, and piezoresistive sensors. Ind. Eng. Chem. Res. **62**(18), 6807–6842 (2023)

Chapter 4
Principle of Fire/Flame-Retarding Polymer Materials

4.1 Introduction

The principle of fire/flame-retarding polymer materials is rooted in the imperative to enhance the safety and fire resistance of polymers, ubiquitous in diverse industries. Polymers, while valued for their versatility, are often susceptible to combustion, posing a significant hazard in various applications. The fundamental objective of fire retardancy is achieved through the strategic incorporation of additives or modifications to the polymer's chemical structure. This includes introducing flame-inhibiting compounds, creating intumescent layers, employing halogenation, incorporating phosphorus compounds, utilizing inert fillers, and exploring nanotechnology for enhanced dispersion of flame-retardants. By disrupting the combustion process or forming protective layers, these principles aim to mitigate flammability, slow down the spread of fire, and contribute to the development of safer materials in fields such as construction, automotive, textiles, and electronics. The ongoing evolution of fire-retardant polymer technologies reflects a commitment to advancing safety standards and minimizing the potential risks associated with polymer-based materials.

4.2 General Principle

As illuminated, polymers go through various stages of pyrolysis and combustion. Through the pyrolysis process, combustible fuel is produced from the polymeric substrate that an external heat source has heated. When this fuel is combined with the stoichiometric amount of air oxygen in the flame, only a portion is typically totally combusted. The remaining portion can be burned off using strong catalysts and abundant oxygen, among other extreme methods. The combustion cycle is maintained by returning some of the generated heat to the substrate, which in turn causes it to

S. S. Ray et al., *Graphene-Bearing Polymer Composites*, Springer Series in Materials Science 340, https://doi.org/10.1007/978-3-031-51924-6_4

continue to pyrolyze. The environment takes away another part of the heat [1]. Flame-retardant systems can either act physically (by cooling, creating a protective barrier, or diluting fuel) or chemically, depending on their nature (reaction in the condensed or gas phase). They could obstruct the various polymer combustion processes (heating, pyrolysis, ignition, propagation of thermal degradation) [2].

4.3 Physical Action

Some flame-retardant additives undergo endothermic breakdown, which results in heat consumption and a lowering of temperature. This calls for a small amount of reaction cooling media to be below the temperature of polymer combustion. Hydrated tri-alumina and magnesium hydroxide both fall within this group and begin to release water vapor at temperatures of about 200 and 300 °C, respectively. This type of pronounced endothermic reaction is said to serve as a "heat sink." The flammable gas mixture is diluted as a result of the flame-retardants' breakdown, which produces inert gases such as water, carbon dioxide, and others, thus reducing the concentration of reagents and the likelihood of ignition. A barrier layer, either solid or gaseous, may also form between the solid phase, where thermal deterioration occurs, and the gaseous phase, where combustion happens, as a result of various flame-retardant additives. Such a barrier restricts the movement of oxygen and flammable, volatile vapors. The quantity of decomposed gases generated as a result is significantly reduced. Additionally, it is possible to physically separate the fuel gases from the oxygen, which stops the combustion process from continuing [2].

4.4 Chemical Action

The flammability of the substrate is dependent on the energy required to heat the polymer to the pyrolysis temperature, decompose, gasify, or volatilize the combustibles, as well as the quantity and nature of the gaseous products. By favoring the creation of carbonaceous char and water, a flame retardant that works through a condensed phase chemical process can change the substrate's pyrolytic route and significantly lower the amount of gaseous combustibles. In this instance, as the amount of the flame-retarding chemical increases, the heat emitted during combustion is reduced [1, 3].

The amount of combustible material in the gas-phase mechanism remains constant, but the heat emitted during combustion typically decreases as the amount of the flame-retarding chemical increases. As the surface temperature drops, less heat is returned to the polymer surface, which causes the pyrolysis to slow down or stop altogether. The flame-retarding component must be gaseous when it reaches the flame and must be volatile. As an alternative, it must break down and supply the gaseous phase with the active portion of its molecule. There will be a less active agent in the char that still remains in the substrate. In the worst-case scenario, the polymer's pyrolysis should proceed as if there had been no flame-retarding chemical added to it. Additionally, the volatiles' composition as they approach the flame shouldn't be impacted by the presence of the gas-phase active agent [1, 3].

4.4.1 The Gas-Phase Mechanisms

The active flame-retardant's gas-phase activity is based on its ability to obstruct the polymer's combustion process. This is realized by stopping free radicals and diluting combustible gas. Free radical inhibitors are produced when the FR is heated or burned, like $HPO^{\cdot}$, which can absorb the $HO^{\cdot}$, $H^{\cdot}$, and $O^{\cdot}$ released during the burning of the matrix and so prevent the chain combustion reactions. Additionally, FRs emit many inert gases during burning, including N_2, NH_3, HCl, CO_2, and H_2O, which can lower the oxygen and small molecule fuel concentration and stop combustion. For instance, in the polymer matrix, some FRs that contain P/N have impacts on both free radical capture and flammable gas dilution [4–6]. Similar to other fuels, polymers also undergo pyrolysis to form species capable of reacting with oxygen from the air, resulting in the H_2–O_2 scheme that fuel combustion spreads through in a branching process [1, 7]:

$$H^{\circ} + O_2 = OH^{\circ} + O^{\circ}, \tag{4.1}$$

$$O^{\circ} + H_2 = OH^{\circ} + H^{\circ}. \tag{4.2}$$

The primary exothermic reaction that supplies the majority of the energy needed to keep the flame is

$$OH^{\circ} + CO = CO_2 + H^{\circ}. \tag{4.3}$$

Inhibiting the chain branching reactions (4.1) and (4.2) is essential for slowing or prohibiting the combustion.

4.4.2 The Condensed-Phase Mechanism

The condensed-phase process assumes that the flame-retardant ingredient, which is often provided in significant proportions, and the polymer interact chemically. Flame-retardants have the potential to cause two different types of chemical reactions during the condensation phase. First, they may hasten the breakdown of polymer chains. In this instance, the polymer drips and displaces the flame action zone. Alternately, the flame-retardant may, through chemical alteration of the deteriorating polymer chains, result in the production of a carbonized (and possibly expanded) or vitreous layer at the surface of the polymer. The vitrified or char layer is a physical barrier between the gas and condensed phases [2]. Cross-linking and dehydration have also been proposed as the two main forms of this interaction for various polymers, including synthetic and cellulosic [1].

4.5 Summary

In conclusion, the principles of fire/flame-retarding polymer materials represent a multifaceted approach aimed at enhancing the fire resistance of inherently combustible polymers. The imperative to mitigate the risks associated with polymer flammability has driven extensive research into various strategies and additives designed to impede the ignition, combustion, and spread of flames.

The addition of flame-retardant additives, such as halogenated or phosphorous compounds, plays a pivotal role by disrupting the combustion process. Halogens release free radicals, breaking the chain reaction of combustion, while phosphorous-based compounds promote the formation of a protective char layer on the polymer surface, acting as a barrier against oxygen and heat.

The concept of intumescence, where flame-retardants expand and form an insulating layer during exposure to heat or flames, provides another effective means of delaying the propagation of fire. This approach, along with the use of nanocomposites and engineered polymer structures, showcases the versatility of flame retardation methods.

The development of environmentally friendly and less toxic flame-retardants is increasingly becoming a focal point, considering the regulatory landscape and heightened awareness of ecological impacts. As research progresses, synergistic approaches combining multiple flame-retardant mechanisms are emerging as effective strategies for achieving enhanced fire resistance.

The evolving field of fire-retardant polymer materials underscores the importance of balancing fire safety requirements with environmental considerations. It is a dynamic arena where innovative solutions continue to be explored to meet the demands of diverse industries while adhering to stringent safety standards.

In essence, the principles of fire/flame-retarding polymer materials encapsulate a commitment to advancing materials science, engineering, and safety standards. As technologies evolve, so too does our ability to engineer polymers that not only fulfill functional requirements but also contribute to a safer and more sustainable future.

References

1. M. Lewin, E.D. Weil, Mechanisms and modes of action in flame retardancy of polymers. Fire Retard. Mater. **1**, 31–68 (2001)

2. F. Laoutid, L. Bonnaud, M. Alexandre, J.M. Lopez-Cuesta, P. Dubois, New prospects in flame retardant polymer materials: From fundamentals to nanocomposites. Mater. Sci. Eng. R. Rep. **63**(3), 100–125 (2009)
3. D. Van Krevelen, Some basic aspects of flame resistance of polymeric materials. Polymer **16**(8), 615–620 (1975)
4. Y. Xue, J. Feng, Z. Ma, L. Liu, Y. Zhang, J. Dai, Z. Xu, S. Bourbigot, H. Wang, P. Song, Advances and challenges in eco-benign fire-retardant polylactide. Mater. Today Phys. **21**, 100568 (2021)
5. L. Liu, Y. Xu, Y. Pan, M. Xu, Y. Di, B. Li, Facile synthesis of an efficient phosphonamide flame retardant for simultaneous enhancement of fire safety and crystallization rate of poly (lactic acid). Chem. Eng. J. **421**, 127761 (2021)
6. X. Hu, J. Sun, X. Li, L. Qian, J. Li, Effect of phosphorus–nitrogen compound on flame retardancy and mechanical properties of polylactic acid. J. Appl. Polym. Sci. **138**(7), 49829 (2021)
7. G.J. Minkoff, C.F.H. Tipper, *Chemistry of Combustion Reactions*. Butterworths (1962)

Chapter 5
Techniques for Polymer-Based EMI Shielding and Fire Retarding Characteristics Measurement

5.1 Brief Introduction

Generally speaking, nanocomposites are engineered materials that are an assemblage of two or more components with very different characteristics that improve one or more properties of the polymer matrix. GNPs are added to the polymer matrix to create a type of composite known as a GNP-reinforced composite. The composite is referred to as a nanocomposite if any one of the dispersed GNP dimensions is <100 nm. Several characterization methods are presented in various works of literature in order to investigate the impact of GNP integration. This section aims to bring together the methods used to evaluate the physical properties, rheological behavior, mechanical properties, and fracture analysis of GNP-based nanocomposites used as EMI filters or flame-retardants. A few specific characterization technologies and advanced characterization tools are also presented to further analyze the microstructural characteristics of composites. The focus of modern researchers is on discovering new characterization and evaluation techniques for analyzing the products for their diverse applications, ranging from simple constructions like leisure, sports items, interior furnishing, etc., to advanced engineering constructions like EMI shields, advanced armory systems, fire retardants, aerospace, etc.

5.1.1 Techniques for GNP-Based Polymer EMI Shields Characterization

The predetermined application area of GNP-integrated polymer-based EMI shielding nanocomposite determines the characterization techniques to be utilized. For instance, EMI shields designed for biological or pharmaceutical purposes are researched and compared differently from those designed for electronics, aircraft, packaging, vehicles, furniture, or other uses. Researchers have used a variety of

S. S. Ray et al., *Graphene-Bearing Polymer Composites*, Springer Series in Materials Science 340, https://doi.org/10.1007/978-3-031-51924-6_5

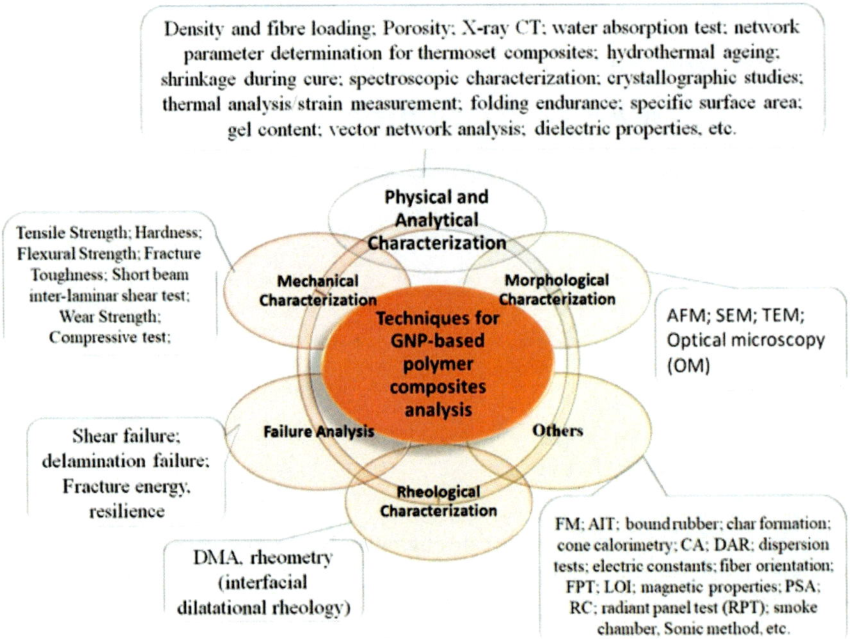

Fig. 5.1 Characterization methods for polymer@GNP-based EMI shields are briefly summarized

techniques to study the structural as well as performance characteristics of GNP and its nanocomposites, including SEM, TEM, XRD, XPS, RS, NMR, EC, TGA, dielectronic properties, FM; AIT; bound rubber; char formation; CC; CA; DAR; dispersion tests; electric constants; fiber orientation; FPT; LOI; magnetic properties; PSA; RC; RPT; smoke chamber, sonic method, DSC, EIS, DMA, and flame retardancy. FTIR, EDX, and other techniques have also been used to study the chemical characteristics of these materials. In Fig. 5.1, characterization methods for GNP- and GNP-based (nano)composites are briefly outlined. Tensile, thermal conductivity, electrical conductivity, morphological, optical properties, equilibrium swelling by organic vapors, chemical reactivity, and other features of GNP-reinforced (nano)composites are a few of the few but crucial ones.

Morphology and Crystallographic

The crystal pattern of the nanocomposite and the effect of the integration of GNPs on crystallite size, which is one of the keys influencing factors for the degradation of mechanical behavior of the composite materials, are determined using crystallographic analyses. Crystallographic studies are performed to identify the nanocomposite's crystal structure and the effect of GNP inclusion on crystallite size, which is one of the key factors controlling how quickly composite materials degrade mechanically. One example is the measurement of the intensity of the diffracted CuKα radiation using the Bruker AXS X-ray diffractometer advanced system at a generator current and voltage of 40 mA and 40 kV. It ranges in measurement from 5 or 10 to 80 or 90° at a 2θ angle. Crystallinity and crystallite size analysis are other tasks for which the XRD diffractogram is utilized. The peak fitting program "Peakfit" is typically used for peak deconvolution. The Scherrer equation is used to determine crystallite size [1].

$$\tau = K\lambda(\beta \cos\theta). \tag{5.1}$$

Herewith, θ denotes incident radiation angle, K denotes shape factor (varying for various polymers), λ stands for incident ray wavelength in Å. At the same time, β represents the Full Width at Half maxima (FWHM). The percentage crystallinity could be analyzed by means of the usual technique illustrated by Dey et al. [1]. The following equation can be used to calculate the crystalline index (I_c) [1]:

$$I_c = \frac{I_{\text{Crystalline}} - I_{\text{Amorphous}}}{I_{\text{Crystalline}}} \times 100. \tag{5.2}$$

$I_{\text{crystalline}}$, in this case, is the counter reading at peak intensity at a 2θ angle that corresponds to the distinctive (Sharp) peak of a crystalline material, and $I_{\text{amorphous}}$ is the counter reading at peak intensity at a 2θ angle for the peak that represents the amorphous material in the composite.

Effective morphological characterization is crucial to establishing structure–property connections for these materials since property enhancements substantially correspond with nanocomposite microstructure. For example, single-layer platelets may be challenging to directly observe by TEM, whereas microtome thin sections of the composite can enable direct observation of dispersed multilayer GNPs. Such thicker platelets typically show adequate contrast against the polymer matrix to be imaged without staining. WAXS, in comparison with TEM, can more quickly shed light on the state of dispersion over a larger volume of composite; however, since the scattering intensity varies with the concentration of the scattering feature, some morphological information may be overlooked [1]. GP and GO, which serve as the building blocks for many GNP-based materials, have layered structures, as do some silicates (such as montmorillonite), which have received much attention as composite fillers [2–4]. Both nanoclays and GNPs do display comparable states of dispersion when distributed into a polymer matrix, depending on factors including the processing method and phase affinities.

Furthermore, GNPs and nanoclays frequently have similar aspect ratios (reaching 1000) [2]. Nevertheless, reinforcements like TrGO frequently have a localized appearance that is wavier than nanoclays [5]. In earlier research on nanoclay-based composites, it was hypothesized that platelet dispersion might occur in one of three general states on, short-length scales: stacked, intercalated, or exfoliated. Therefore, we propose extending this terminology to GO-derived and GNP-bearing polymer nanocomposites since similar morphologies have been noted in the literature on these systems.

The two methods most frequently used to evaluate the condition of dispersion of GNPs within a polymeric (nano)composite blend are probably TEM and WAXS experiments. When detected by TEM, immiscibility of the phases and/or insufficient exfoliation of the filler made of GP or GO can lead to huge agglomerates made of stacked platelets. The appearance of a diffraction peak matching the interlayer spacing of GP or GO may also suggest this [6, 7]. A shifted diffraction peak from that of unaltered GP or GO demonstrates that intercalated platelets still have a stacked structure but have increased interlayer spacing (on the order of a few nanometers) [8]. High aspect ratio platelets are frequently discovered to be advantageous to the mechanical, electrical, and thermal properties of composite material, as will be covered in more detail in the following sections. In order to create better aspect ratio platelets than stacked or intercalated platelets, an exfoliated morphology of GO or GICs is typically needed [9]. Although multilayer intercalated platelets may appear to be dispersed (as typically seen by TEM) despite the lack of a diffraction peak, this condition of dispersion may be implied by a scattering profile similar to that of the neat matrix polymer.

On the other hand, using TEM to assess platelet exfoliation and shape quantitatively has its own set of difficulties (such as sampling a sufficient number of filler particles at such high magnification and the possible influence of TEM sample preparation on the apparent level of dispersion). While individual platelet dispersion can be evaluated using WAXS or TEM, neither technique is capable of spotting more significant morphological details. The presence of fractal-like aggregate particles of filler at length scales beyond that of individual particles has been detected using SAXS and USAXS measurements on various nanocomposite systems. However, there is only a small amount of information available on GO-bearing polymer composites, possibly because such techniques are not widely available [10]. However, a recent study using small-angle neutron scattering looks at thermally reduced GO (TrGO)/polycarbonate nanocomposites. Based on a hypothetical platelet model created for montmorillonite, these measurements were utilized to quantify the platelet dispersion, which predicted that the effective aspect ratio would decrease as TrGO loading rose and that filler would aggregate more as loading reached greater levels [11].

Last but not least, cross-sectional SEM analysis has been utilized to assess filler dispersion made of GNP as well as to check the surface for pull-out of filler, potentially providing information about the strength of interfacial adhesion. However, caution must be taken when recognizing the distributed filler. In addition, SEM typically cannot clarify the extent of platelet exfoliation and is, therefore, best used as a supplemental approach.

When disseminated in a polymer matrix, exfoliated GNP-based materials are often not seen as rigid disks but rather as bent or crumpled platelets because they are frequently compliant. Additionally, it has been demonstrated that GNP would permanently "scroll up" when its polymer host is heated above its glass transition temperature (T_g) [12]. According to certain reports, the conformation of the chemically modified GNPs can also be impacted by the polymer matrix's compatibility

[13]. The particles may take on a more extended shape if the platelets have a strong affinity for the matrix. But as the affinities between the constituents diminish, the platelets can eventually take on a more crinkled structure. The method used to process the composites can have an impact on the microstructure: melt-processed composites may favor a more oriented, intercalated/stacked structure due to platelet restacking, while solution- or in situ-processed composites may favor randomly oriented, exfoliated platelets. The processing method can also cause the dispersed platelets to become oriented, which can be advantageous for reinforcing [14] but may increase the percolation threshold. Platelets may be aligned parallel to the surface and more randomly positioned near the interior of composites treated by injection molding [15]. Comparatively, aligned platelets may be present throughout the whole cross-section of sufficiently thin, compression-molded specimens [15] or solution-cast films. The orientation of the distributed platelets may also be impacted by the filler type.

The addition of a nanofiller to semicrystalline polymers can change their degree of crystallinity, crystallite size, and spherulite structure and even cause otherwise amorphous polymers to crystallize [2–4, 6, 7, 16]. The degree of crystallinity of semicrystalline polymer matrices has been reported to increase [17], decline [18], or not change [19] upon the incorporation of graphene-based fillers, depending on the identity of the polymer. Alterations in the melting temperature of the matrix material also has been disclosed [20]. By acting as a heterogeneous nucleation site for crystal formation, graphene-based fillers may also have an impact on the crystallization rate. Additionally, GNPs have been found to have a less dramatic effect than CNTs in speeding up the crystallite development kinetics of poly(l-lactide) [21]. Other changes in the morphology of the polymer matrix or composite structure, beyond crystallization, can result from the introduction of graphene-based filler. It was claimed that the presence of TrGO filler has an impact on the morphology of the self-assembling triblock copolymer poly(styrene-block-isoprene-block-styrene): The TrGO was preferentially disseminated in the PS blocks where they took on a folded shape, and AFM and EFM investigations on roughly 300 nm-thick films made by spin coating demonstrated loss of long-range order in the domains [22].

In light of the material that is referenced above, this section discusses the various characterization methodologies. The prognosis of some outcomes anticipated for various characterization approaches is also provided below.

AFM

In comparison with the complex sample preparation processes necessary for electron microscopy, this characterization technique is thought to be more convenient from the perspective of sample preparation. Using this technique, a sample without any surface alterations can be examined, and the atmospheric pressure allows it to be done. Over a piezo scanner, the sample is placed. A raster scan of the sample's surface is carried out by a cantilever with a very small probe and sharp tip measuring <100 Å.

To create a two- or three-dimensional image of the sample's surface topology, a laser beam that has been reflected from the cantilever is detected. We can perform the scanning in touch, non-contact, or tapping (intermittent) mode. The tip touches the sample surface while the probe is in contact mode, which increases the risk of probe damage. When operating in non-contact mode, the probe tip is kept 50–150 above the surface and relies on a weak Van der Waal force to interact with the sample surface. High-frequency oscillation is delivered to the cantilever in tapping mode in addition to the non-contact type feature. The oscillation's amplitude decreases when the oscillating cantilever intermittently contacts the sample surface.

The following is how the roughness parameters are calculated:

$$\text{Mean square roughness (RMS) } R_q = \sqrt{\frac{\sum (Z_i - Z_a)^2}{N}}, \tag{5.3}$$

Maximum peak to valley height R_{max}, where N is the number of points inside the given region, Z_i is the height value, and Z_a is the average height values within the given area.

$$\text{Mean roughness } R_a = \frac{1}{L_x L_y} \int_0^{L_x} \int_0^{L_y} |f(x,y)| dx dy, \tag{5.4}$$

$f(x,y)$, L_x and L_y are the sample surface relative to the center plane and the surface dimensions [1].

For a better understanding of the distribution of roughness, it can also be plotted as a histogram. Additionally, bearing area analysis can be used to determine the %age of surface area on the sample surface that is above a specific height.

With the height distribution in topography, it is also possible to understand the distribution of resonance in the composites along a particular direction on the sample surface along a line on the AFM topology image. The distributions of height and the spatial distribution of the polymeric matrix's fibers can be examined from the generated graph.

In contrast to AFM topology, the FMM mode of AFM produces a significantly crisper image based on the differential between local stiffness values at various spots on the sample surface. The cantilever's deflection is kept constant in this mode, but a driving signal is also applied to it. The ensuing DC and AC components of the electric signal display distinct characteristics. The surface roughness picture is influenced by the DC component, which is connected to the constant force acting on it. A link exists between the sample's elastic properties and the phase difference and amplitude variation of the AC component. Thus, a fiber dispersed as the hard phase in a softer matrix may be well defined by the FMM approach, especially for some areas of the sample where a substantial height difference between the softer matrix and harder fibers does not exist.

With the aid of AFM, Das and his colleagues studied the 2D as well as the 3D topography of PANI and its nanocomposite systems [23]. PANI emerges as a spherical structure in the topographic photos, which is compatible with SEM analyses. The average diameter of PANI particles is determined from the height profile to be 5.65 nm. The increase in particle size of PANI/S-RGO NPs is shown in Fig. 5.2d and e which demonstrates that PANI nanofibers coat the surfaces of S-RGO nanosheets during in situ polymerization. The PANI/S-RGO NPs' estimated diameter was raised to 56.1 nm, further confirming the successful polymerization of aniline on the surface of S-RGO nanosheets as seen in Fig. 5.2f [23].

Optical Microscopy

The simplest microscopic technique is optical microscopy (OM), which uses optical lenses to magnify objects up to 1500 times. This technique provides an image of living creatures in natural environments without requiring any special sample preparation steps. OM images are frequently employed for macrosized graphene-based reinforcing fillers and their mixtures [24]. This characterization approach can also be utilized to analyze the crystallization behavior of the polymers/blends with respect to the inclusion of the graphene filler materials used in its polarized OM form.

Physical and Analytical Characteristics

The shelf life as well as application range of the GNP-based composites are determined by physical characterization, which also examines the dimensional stability. The filler loading, swelling, void content, water absorption, hardness, hydrothermal aging, network characteristics (special to thermoset composites), and cyclic tests are a few examples of this characterization.

Density and Graphene-Loading

In researching dimensional stability, density is a key factor. Based on their density, composites can be divided into low (<500 g/cc), medium (500–800 g/cc), and high (> 800 g/cc) categories. Comparing low-density composites to high-density composites, voids, porosity, and intercellular gaps are more prevalent in the former. The mass of a substance per unit volume is its density. Weighing a material with a specific dimension allows us to calculate mass. By measuring the dimension, volume is determined (Length X Width X Height). A composite's density is discovered to rise as certain qualities, such as its propensity to retain water and trap moisture, alter with density [1].

Although the density of GNP-based polymer nanocomposites has apparently been evaluated using ASTM D792-13, reinforced polymeric composites are tested according to standard ASTM D792-75 or ASTM D792-91. Weighing the composite

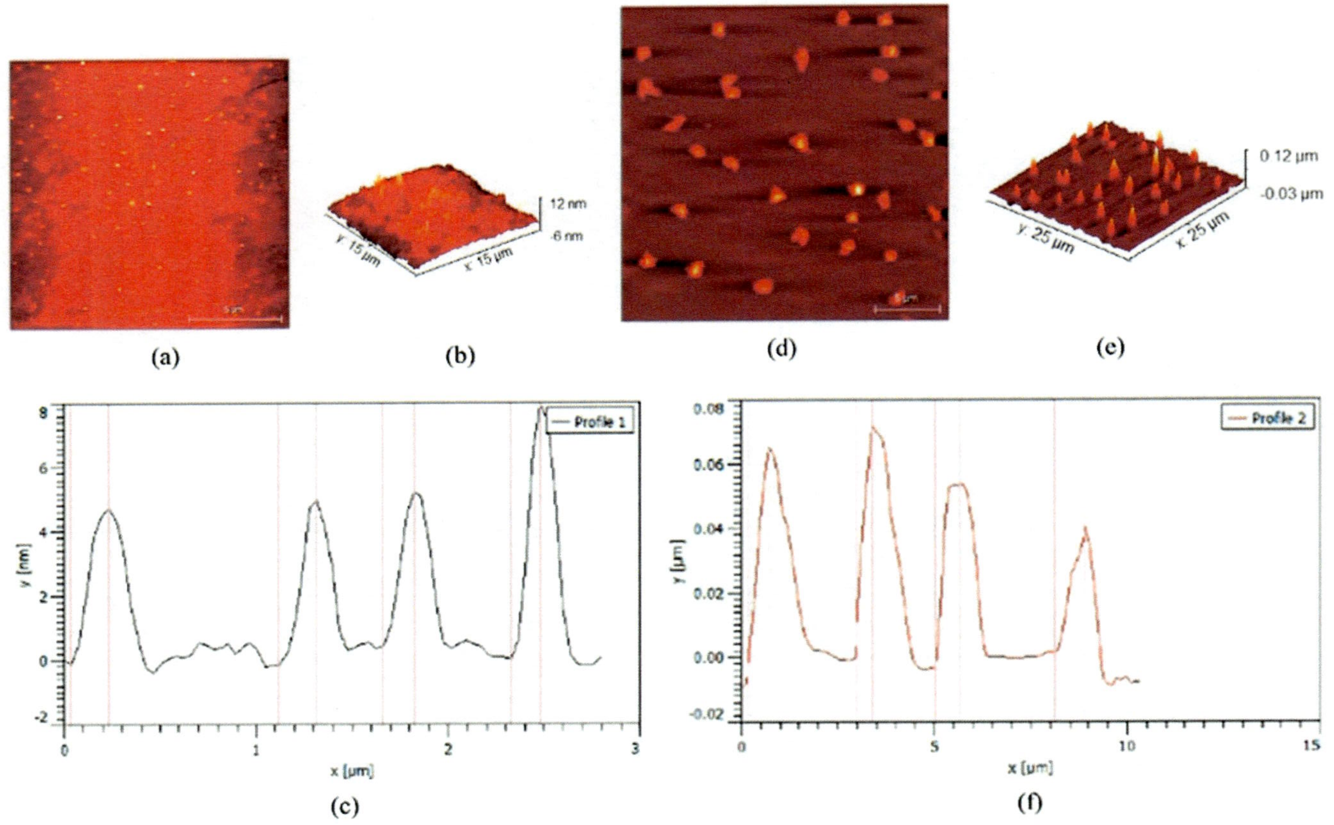

Fig. 5.2 AFM images of **a** 2D PANI topography, **b** 3D PANI topography, **c** height profile of PANI, **d** 2D PANI/S-RGO topography, **e** 3D PANI/S-RGO topography, **f** height profile of PANI/S-RGO composites. Reproduced with permission from Das et al. [23]. Copyright 2022, Elsevier Science Ltd

in air, suspending it on a wire, and submerging it in water allows us to calculate the density of the nanocomposite. It is noticed that the weights differ. A sinker of a specific weight is affixed to the composite specimen being tested when the composite has a lower density than water. The following equation is then used to compute the density:

$$\rho = \frac{(0.9975)a}{(a + w - b)}. \tag{5.5}$$

Here, a represents the weight of the specimen in air, b represents the combined weight of the specimen and sinker when submerged, and w represents the weight of a sinker that is entirely submerged but a wire that is only partially submerged.

In order to determine the graphene-based filler loading, V_f in accordance with ASTM D3171-15, the matrix containing graphene-based filler is heated linearly at a rate of 20 K/min between 298 and 773 K in order to remove the matrix material. The weight of the residue that is left over is then utilized to calculate V_f.

Then, using the following equation, the ASTM D2734 approach is utilized to determine the void content from the theoretical and experimental density of the composites [1].

$$\text{Void content}(\%) = \frac{\rho_{\text{theoritical}} - \rho_{\text{experimental}}}{\rho_{\text{theoritical}}}, \tag{5.6}$$

where

$$\rho_{\text{theoritical}} = 1/\left[\left(\frac{W_f}{\rho_f}\right) + \left(\frac{W_m}{\rho_m}\right)\right]. \tag{5.7}$$

The terms, W_f, W_m, ρ_f and ρ_m, denote the weight fractions of the fibers, the matrix, the fibers, and or the matrix, correspondingly.

The density of GNP-based polymeric EMI shield can also be directly calculated from the geometric-mass volume relationship thus:

$$\rho = \frac{m}{V}. \tag{5.8}$$

And the porosity of the material is then estimated as follows:

$$\text{Porosity} = 100\left(1 - \rho_f / \rho_s\right). \tag{5.9}$$

Here, the density of the foam and the solid samples are denoted by ρ_f and ρ_s.

Porosity

Different types of porosity exist in composite materials, including polymeric composites based on graphene-based filler. Pore Type 1 is known as a "gas evolution pore," and it is primarily created when making feedstock filament during the extrusion process. Physical gaps at layer/layer interfaces, which are caused by voids at the boundaries of several layers, are what are referred to as pores of type 2. Pore type 3 is typically produced as a result of the specimen's shattered interface's pulled-out reinforcing fillers. However, because it results from the post-processing fracture of the material, it cannot be regarded as the composite's inherent porosity. Therefore, while analyzing overall porosity, porosity types 1 and 2 can be considered (*P*). The following equation represents *P*:

$$P = \frac{V_t - V_a}{V_t} \tag{5.10}$$

Thus, the theoretical volume of the specimen following FDM fabrication is denoted by, V_t., while the actual volume of the specimen is denoted by, V_a. A designed solid object in CAD software has a volume of 8.45 cm^3 [1].

However, practically speaking, shrinkage changes the shape; hence, the value cannot be directly applied to, V_t. The standard approach is to weigh the specimen after cutting it into a normal morphological volume to estimate its actual measured mass (*M*) and determine its average density "ρ." The following equation can be used to compute *M* and ρ, V_a based on average values:

$$V_a = \frac{M \times W(\%)_{\text{Polymer}}}{\rho_{\text{Polymer}}} + \frac{M \times W(\%)_{\text{Fibre}}}{\rho_{\text{Fibre}}}. \tag{5.11}$$

Another method for figuring out the porosity of the graphene-based filler-reinforced foam is mercury porosimetry(s). Foam cylinders with diameters between 6.1 and 6.5 mm are used to create triplicate samples that weigh between 0.07 and 0.43 g. A porosimeter (such as the MicroActive AutoPore V 9600) is utilized with a filling pressure range of 0.5–50 psi. Porosity denoted as ε is determined through measurements of the overall volume of mercury intrusion represented as V_i thus:

$$\varepsilon = \frac{V_i}{V_i + \frac{1}{\rho}}. \tag{5.12}$$

Thus, ρ, being the GNP-based nanocomposite density, is determined via the equation below:

$$\rho = \frac{1}{\frac{W_p}{\rho_p} + \frac{W_f}{\rho_f} + \frac{W_{\text{rp}}}{\rho_{\text{rp}}}}. \tag{5.13}$$

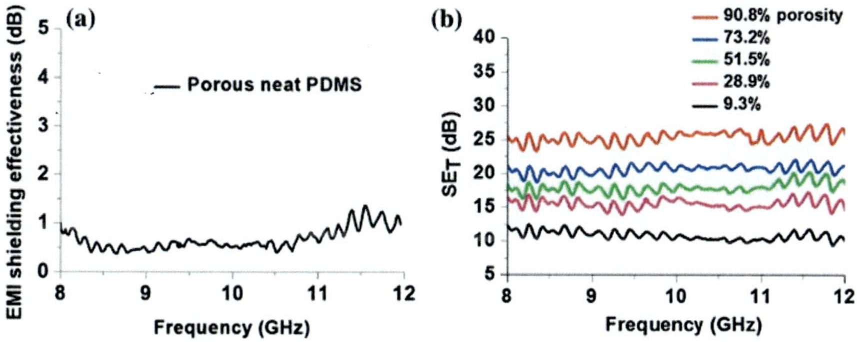

Fig. 5.3 **a** EMI shielding effectiveness of neat PDMS with 90.8% porosity and **b** SET of GF/PDMS composites for different porosities as a function of frequency. Reproduced with permission from Sun et al. [26]. Copyright 2016, Elsevier Science Ltd.

The quantities ρ_p, ρ_f, and ρ_{rp} denote the density of the polymer, fiber, and porogen, correspondingly. W_p, W_f, and W_{rp} are the polymer, fiber, and porogen weight fractions, respectively. Using the Washburn equation and the equivalent cylindrical pores, it is also possible to estimate the pore diameter d:

$$d = \frac{-4\gamma \cos\theta}{P}. \tag{5.14}$$

γ, P, and θ stand for mercury surface tension (480 dyn cm^{-1}), pressure above the mercury, and advanced contact angle betwixt the pore wall and the mercury. α "surface/volume ratio," is afterward computed with the aid of

$$\alpha = A_i \rho (1 - \varepsilon), \tag{5.15}$$

where A_i denotes total surface area per unit mass.

The need for determining the porosity of the EMI shield is thus higher porosity has been said to cause enhanced multangular EM wave reflectance on the surface along with increased SE_T: The relationship is a direct one.

A good example of the use of mercury porosimetry is the determination of the porosity of skin-structured Ag@rGO/AF@rGO/F/PDMS nanocomposites aimed at application as efficient porous structured EMI shields [25]. The authors used a MicroActive AutoPore V 9600, China mercury porosimetry for this purpose, where their results revealed a porosity of 89.1% for the rGO/AF foam.

Another instance of the consideration of porosity to study the property performance of an EMI shield is from the report by Sun et al. [26]. These authors observed that their composite sample "GF/PDMS" was reinforced with 2.7 wt.% GNP displayed a porosity of 90.8%, and it produced the highest EMI SE of ~25 dB. They also noted that the virgin PDMS specimen, having a porosity of 90.8%, produced a highly negligible EMI SE of $\leq$ 1.0 dB (Fig. 5.3) [26].

It can, therefore, be seen that porosity is an important property of graphene-based filler containing polymer composite porous EMI shields as it affects the EM wave transmittance as well as reflectance, which eventually affects the EMI SE of the shield.

Porosity has also been reportedly demonstrated to enhance the absorption of the EM waves via a highly conductive porous graphene-based filler-bearing film [27].

X-Ray CT Scan

The mesostructure of porous material can be seen in three dimensions using the non-destructive X-ray CT scan. This method is also possible to assess the porosity of fiber-reinforced composites. A typical example of an instrument used in this regard is the "microCT, VTomex m300, GE." In their report, Morales-Rodriguez et al. [28] described a method for precisely characterizing the volumetric porosity of the composites. The sample's internal structure, which specifies the porosity, is built using piled-up 2D reconstructed tomographic slices. By applying 3D median filtering to a nearby one voxel, these images are further processed to lower the noise level. While cracks and pores are visible as white phases, the fiber/matrix is visible as black. The white voxel ratio (= number of white voxels/total block voxel size) quantifies porosity. The path inside an open pore is straight, which is the tortuousness. The ratio of the distance between two places in the pore phase divided by their separation along a straight line can be used to calculate it, according to the source measuring porosity step by step.

In this regard, Ho and Kontopoulou [29] demonstrated the efficacy of this technique to study the cross-sectional morphology of 2D as well as 3D PP/GNP nanocomposites with the aim of estimating the porosity of the samples. They reported that their fabricated foams demonstrated a highly porous morphology where the specimen with 3 wt.% GNP filler revealed a clearly defined cell structure, as presented in Fig. 5.4 [29].

Water/Moisture Absorption Analysis

According to ASTM D570, water absorption tests are conducted. The outcomes are utilized to shed light on the performance of composites in humid settings. Using the following equation, one may determine the %age of water absorption [1]:

$$\text{Water absorption } (\%) = \frac{W_t - W_d}{W_d} \times 100. \tag{5.16}$$

W_t and W_d are the composite weight after its immersion in water and dry composite weight.

Another important test related to this is the swelling ratio, which is determined thus: after being dried in an oven until equilibrium weight is reached, the rectangular

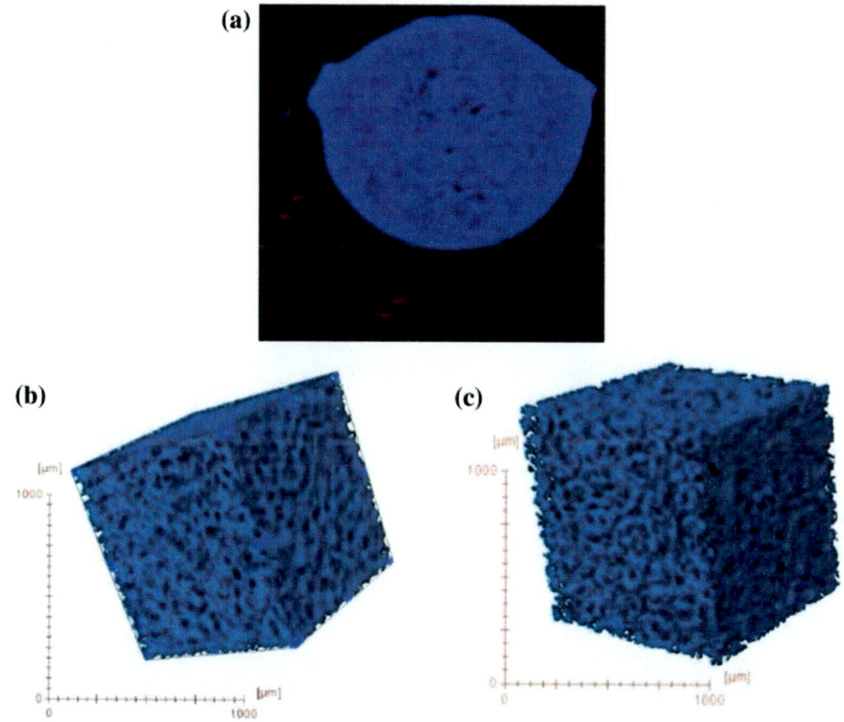

Fig. 5.4 **a**, **b** 2D microCT scan and a 3D rendering of a PP-GNP 3 wt.% foam; **c** 3D rendering of a PP-EG 3 wt.% foam. Reproduced with permission from Ho and Kontopoulou [29]. Copyright 2020, Elsevier Science Ltd.

pieces of composite material are put in a desiccator to cool. After that, the samples are submerged in water at 298 K for 24 h or until equilibrium is reached. After being removed, the wet samples are dried using a lint-free cloth and precisely weighed. The samples were then withdrawn from the deionized water at regular intervals. The excess water was blotted off with filter paper. The samples were weighed again to calculate the swelling ratio using the equation below.

$$\text{Swelling ratio} = \frac{\text{weight of the swollen sample}}{\text{weight of the dry sample}}. \quad (5.17)$$

A polymer-encapsulated IC package will swell when exposed to a humid environment because moisture diffuses through the plastic's pores. However, neither the silicon chip nor the metallic lead frame is permeable to moisture; therefore, their dimensions are unaffected. Hygrostress develops as a result of the plastic expanding differently from the other materials.

$$\text{Hygrostrain}\ \varepsilon_h = \text{SC} \times C, \quad (5.18)$$

where, as illustrated below, C is the moisture concentration (wt.%) and SC is the swelling coefficient or the coefficient of hygro-expansion.

$$\sigma_h = E\varepsilon_h. \tag{5.19}$$

Moisture concentration C is precisely defined in the governing Eqs. (5.20) and (5.21) for moisture diffusion.

$$\frac{\partial}{\partial x_i}\mathrm{DM}_{ij}\left(\frac{\partial C}{\partial x_j}\right) = \frac{\partial C}{\partial t}, \tag{5.20}$$

and for the resin's (polymer matrix) isotropic area,

$$\nabla \cdot \mathrm{DM}\nabla C = \frac{\partial C}{\partial t}. \tag{5.21}$$

x is the Cartesian location, DM is the moisture diffusivity, t is the time, and C is the moisture concentration. The comparable equations regulating thermal diffusion are as follows for processes of moisture diffusion occurring simultaneously with those of heat conduction in the same area:

$$\frac{\partial}{\partial x_i}\left(k_{ij}\frac{\partial T}{\partial x_j}\right) = \rho c_P\frac{\partial T}{\partial t}, \tag{5.22}$$

$$\nabla \cdot k\nabla T = \rho c_P\frac{\partial T}{\partial t}, \tag{5.23}$$

T, k, ρ, Cp denote the temperature, thermal conductivity, density, and specific heat, respectively. It is believed that DM and k are temperature-dependent variables. Solubility and diffusivity are both involved in the resin's water absorption process. The aforementioned diffusion Fick's laws give an explanation of diffusion. In the linear regime, water molecule solubility in resin abides by Henry's law. The following Arrhenius equation yields the diffusivity DM:

$$\mathrm{DM} = \mathrm{DM}_0 \exp\left(-\frac{Q_{\mathrm{act}}}{R_u T}\right). \tag{5.24}$$

T, R_u, and Q_{act} are the absolute temperature, universal gas constant, and activation energy (4.84×10^4 J/mol), while $\mathrm{DM}_0 = 47.2\ \mathrm{mm}^2/\mathrm{s}$ [30]. The Arrhenius equation can also be used to describe the solubility "s."

$$s = s_0 \exp\left(-\frac{Q_s}{R_u T}\right), \tag{5.25}$$

$s_0 = 4.96 \times 10^{-7}\ \mathrm{mg\ mm}^{-3},\ \mathrm{Mpa}^{-1};\ Q_s = 3.87 \times 10^4$ J/mol [30].

For instance, in polymeric materials aimed at application in electronics, the use of moisture absorption and desorption, two different types of moisture preconditioning, is common. In an experiment by Kitano et al. [30], plastic SOJ packages that had been moistened were placed in a VPR oven for a period of time, and afterward, the packages were inspected for signs of package cracking. Completely dried IC packages were subjected to a temperature and humidity environment of 85 °C/85% RH for varied amounts of time during the moisture absorption preconditioning procedure. The IC packages are subjected to an 85 °C/85% RH atmosphere until saturated as part of the moisture desorption preconditioning process, and then they are baked in dry air at 80 °C. According to research, packages that have either been preconditioned by moisture absorption for longer than 34 h. or by moisture desorption for <26 h shatter [30]. These investigations have shown that the moisture concentration at the crucial pad-resin interface, rather than the absolute moisture content in the package, controls package breaking. The critical moisture concentration level at the chip pad-resin interface, which corresponded to the critical 34 h of absorption, is found to be 63%, and compared with the critical 26 h of desorption, it is found to be 60% after performing a one-dimensional analysis of the moisture distribution within the layer of plastic resin adjacent to the bottom of the die pad during the preconditioning moisture absorption and desorption processes. The center of the die pad can be solved using Kitano et al. [30] 1D's solution. However, it is unclear how moisture concentration and diffusion rates change along the die pad's full width and length. The moisture there may prove to be more crucial than that in the middle since the thermomechanical stress is highest near the border of the die pad. This is especially true for small-outline packages with very thin resin (polymer matrix) layers at the edges.

Considering the fact that polymeric materials intended to be used in electronic are required to possess very low moisture absorption, it is apparent that graphene-based filler-bearing polymer composite EMI shield should possess very low moisture absorption in order to avoid the development of hysteresis within the electronic device shielded from EM waves.

Network Parameter Determination for Thermoset Composites

For graphene-based filler-bearing thermoset composites, determining network parameters such as molecular weight between cross-links and mesh size is crucial to assessing their qualities and maximizing their mechanical integrity. The "rubber elasticity theory of random Gaussian networks" has been suggested by Halake and Lee [31], as a method for determining the molecular weight between cross-links. Based on the relationship between G' and swelling ratio, storage modulus (G') is determined (SR). Here, G' is calculated using a frequency sweep experiment with a parallel plate oscillation rheometer under conditions of constant temperature and strain. The following equation provides an estimate of the molecular weight between two cross-links:

$$M_c = \frac{cRT}{G'}. \tag{5.26}$$

C denotes the concentration of the polymer (g/m^3, SR = ρ/c = 1, ρ represents the hydrogel density).

The frequency sweep experiment employing parallel plate rheology is another experiment that is published for the determination of the network characteristics using a rheometer. The plateau modulus is computed for which $G_i\prime$ denotes the entire storage modulus value within 1–10 rad/s region as per Eq. 5.27 Using Eqs. 5.28 and 5.29, the cross-linking density and mesh size are computed [32].

$$G =_i^n \sum \mathrm{G}_i'/n, \tag{5.27}$$

$$\rho = G/\mathrm{RT}, \tag{5.28}$$

$$\xi = (\mathrm{RT}/G \cdot N_A)^{1/3}, \tag{5.29}$$

Here, ξ: Debye length, *R*: Gas constant (8.314 J/(mol K)), *T*: Absolute temperature (in Kelvin), *G*: Permittivity of free space (8.854 × 10^(−12) C^2/(N m^2)), NA: Avogadro's number (~0.022 × 10^(23) mol^(−1)).

This equation is often used in the context of plasma physics or electrolyte solutions to characterize the distance over which electric fields can effectively screen out the influence of a charged particle.

Hydrothermal Aging

It is common knowledge that graphene-based filler-reinforced materials must be durable against thermal and hydrothermal aging in order to prevent catastrophic failure in an aqueous environment. Since one of the key elements that affect graphene-based filler-reinforced materials' strength and stiffness is temperature. All polymer-based materials' elastic modulus and strength decrease with rising temperature under sustained load conditions due to thermal softening. Additionally, it is also necessary to study the performance of graphene-based filler-reinforced composites in thermal and hydrothermal environments for particular applications, such as those involving electrical and electronic materials. Only the effect of temperature on a material's properties is examined in thermal aging. However, the effects of both temperature and water molecules are investigated in hydrothermal aging.

As per European aerospace standard NF EN 2489 [AFNOR NF EN 2489: Aerospace series, fiber-reinforced plastics, determination of the action of test fluids], hydrothermal aging properties are assessed by submerging composite and resin plates at 70 °C in deionized water for up to 16 weeks. Following that, the water absorption is presented as a function of time or as the square root of time.

$$\text{Hydrothermal aging} = M(t)(\%) = \frac{m(t) - m(0)}{m(o)} \times 100. \tag{5.30}$$

Here, the masses at the initial state and time t, respectively, are denoted by $m(0)$ and $m(t)$. Two analytical models are primarily employed to characterize the relationships: (i) Fick and (ii) Langmuir models, according to Joliff et al. [33]. Below is provided the condensed equation for each model.

Fick model:

$$\frac{M_t}{M_s} = \frac{2}{h}\sqrt{\frac{D_t}{\pi}}. \tag{5.31}$$

Herewith, M_t and M_s denote absorbed water mass at time t as well as mass of saturated water. D and h stand for the diffusion coefficient and thickness of the sample.

The Carter and Kibler [34] two-phase diffusion model is frequently referred to as "Langmuir-type" diffusion. It takes into account the presence of both free-floating and bound water molecules inside the polymer network. The formula is provided below:

For all t values, $\gamma \cdot n_\infty = \beta N_\infty$ with the boundaries conditions: $n(x, 0) = 0$ and $N(x, 0) = 0$ for $x < h/2, n(h/2, t) = n_\infty$ and $n(-h/2, t) = n_\infty$.

At short periods of time ($kt \leq 0.7$),

$$m_t = \frac{4}{\pi^{3/2}}\left(\frac{\beta}{\gamma + \beta}m_\infty\right)\sqrt{k.t.} \tag{5.32}$$

For kt ~ 5, a pseudo saturation level is acquired, as recounted by Joliff et al. [33], thereby transforming the equation thus:

$$m_t = \left(\frac{\beta}{\gamma + \beta}\right)m_\infty. \tag{5.33}$$

For extended times ($kt >> 1$),

$$m_t = m_\infty\left(1 - \frac{\gamma}{\gamma + \beta}e^{-\beta t}\right). \tag{5.34}$$

The ASTM D 570-98 can also be used to carry out this technique: Tensile, flexural, impact, and fracture toughness samples are typically subjected to hygrothermal aging in accordance with ASTM D 570-98, where the samples are immersed in distilled water at 25 and 50 °C and removed from the water after a set amount of time, wiped with a clean, dry cloth to remove the surface water, weighed in a high precision balance to assess the weight change, and then re-immersed for continued aging. When additional weight gain was determined to be minimal, the samples' weighting was discontinued. Testing was done on the aged samples for tensile, flexural, impact, and fracture toughness.

Shrinkage During Cure

Resin chemical shrinkage is particularly important because it explains surface failure and aids in the production of products with high-quality surfaces. It has been shown that the addition of GO to PAA led to restriction in the PPA chain mobility, resulting in reduced shrinkage of the nanocomposite in the direction of the nanocomposite thickness [35]. Shrinkage of graphene-based filler-bearing polymer nanocomposite shields is mostly attributed to the curing of the polymeric resin or the direct evaporation of solvents due to surface tension. It is important to understand the shrinkage percentage of the final products/shields in order to design the desired size of the product accurately. The linear shrinkage of polymeric nanocomposites can be estimated using a rheometer and has been reported by a group of researchers [36]. In order to provide shear strain to the resin sample positioned between the plates, a compressive normal force Fz and torque or rotary frequency f can be applied to the top plate, which can move vertically in contrast to the base plate, which is immovable. In order to measure the shrinkage, they also used a helium-based pycnometer. The thermoset composite material is oscillated at a single frequency of 30 Hz during the post-gelation phase. The stress-controlled test is conducted in an isothermal environment with a torque of 500 m Nm and a constant compressive normal force of 0.1 N. To estimate the linear dimensional change throughout the cure, the actual distance between the two plates is tracked and plotted against time. Because of the gelled glue and the low normal force used, Shah et al. asserted that "no material is driven out from between the plates, producing a false measurement."

The same material is submitted to a post-cure study once the gap becomes consistent. In order to prevent thermal expansion, the material is heated to a high temperature (55 °C) for 6 h. before the gap is measured at room temperature. The chemical shrinkage ε_v is then calculated using the expression below, where h_0 and h are the starting gap and the actual gap, accordingly.

$$\varepsilon_v = \left(1 + \frac{1}{3}\left(\frac{h - h_0}{h_0}\right)\right)^3 - 1. \tag{5.35}$$

The shrinkage behavior of graphene-based filler-bearing polymer nanocomposites and/or hybrids can be effectively studied via rheometry technique. With reference to the work by Haider et al. [37], in oscillation mode, disposable metal parallel plates with a 40 mm diameter are employed. Figure 5.5 depicts the rheology experimental setup. A known amount of torque was given to the top plate while the bottom plate was fixed in this setup. In order to quantify the viscosity under both isothermal and non-isothermal circumstances, the space between the plates is maintained constant. The top plate's force on the resin sample is represented by the normal force F_z in Fig. 5.5 as a percentage. When the resin sample is compressed, this force is considered positive. These tests are conducted at a constant angular velocity (ω) while being torque- or strain-controlled [37].

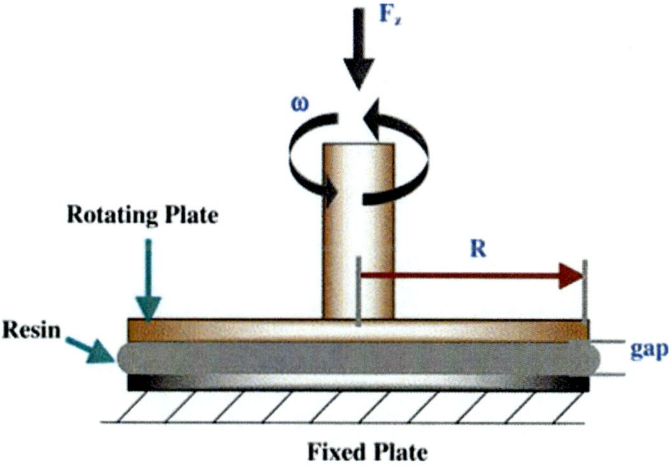

Fig. 5.5 Parallel plate geometry setup. Reproduced with permission from Haider et al. [37], Copyright 2007, Elsevier Science Ltd.

The important point to note is that graphene-based filler-bearing polymer nanocomposite systems have been demonstrated to suppress shrinkage in their host polymer during curing or carbonization [38].

Spectroscopic Characterization

For graphene-based filler-bearing polymer nanocomposites, a variety of instruments, including FTIR, Raman, and CP-MAS, are used to identify molecular structure, estimate molecular weight, determine grafting percentage ages, and pinpoint GNP/polymer interactions. The majority of the time, an FTIR spectrum is captured in ATR mode utilizing a diamond plate or zirconium plate in the 4000–650 cm^{-1} (Fig. 5.6) range or by preparing the samples with KBr and analyzing them in transmittance or absorbance mode [39].

On the other side, the Raman spectra are captured using a RENISHAW via Raman microscope coupled with four common Leica microscope objective lenses at room temperature in a backscattering geometry. It uses an argon ion laser source with an excitation wavelength of 514 nm. Typically, a 2400 lines per mm grating is used to gather spectra for the 150–800 cm^{-1} range, with a 20 s data point acquisition period.

The Raman microscope of graphene-based filler-bearing polymer composite EMI shields is mostly studied to understand whether the graphene-reinforcing filler undergoes structural transformations. An example is the study of the structure of PP-reinforced GNPs in a study by Ho and Kontopoulou [29]. These researchers observed in their study of 3 wt.% GNPs reinforced PP that the I_G/I_{2D} ratio of the nanocomposite remained the same (2.39 ± 0.4), the value practically the same as that obtained for the pristine GNPs, as depicted in Fig. 5.7.

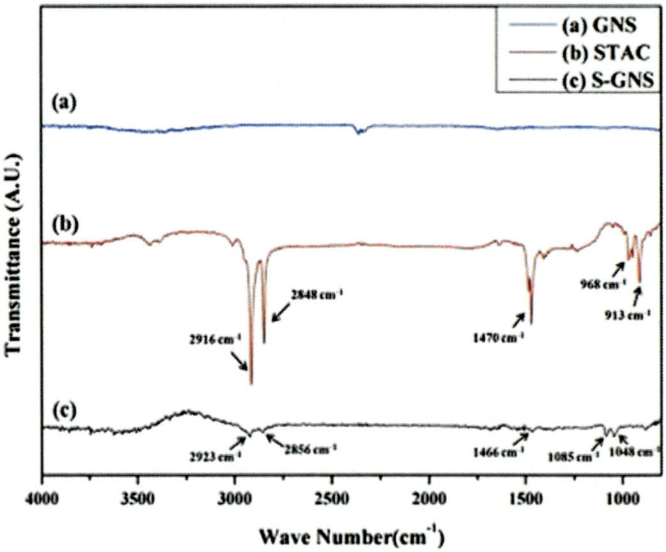

Fig. 5.6 ATR-FTIR spectra of **a** graphene nanosheet (GNS), **b** stearyl trimethyl ammonium chloride (STAC), and **c** S-GNS. Reproduced with permission from Hsiao et al. [39]. Copyright 2013, Elsevier Science Ltd.

Spectroscopy analysis is, therefore, a vital tool for researchers to adopt with the aim of studying the structural properties of GNP-based polymeric nanocomposites as well as the potential changes in the structural properties of the systems.

Thermal Strain (CTE) Measurement

Thermal strain measurements are carried out using high-temperature linear strain gauges that are attached to the specimens using a high-temperature strain gauge adhesive, M-bond 600, in order to look into the hybrid composites' linear coefficient of thermal expansion. Jang et al. [40] provided an explanation of the process. This test, we believe, is essential for graphene-based filler-bearing polymer composite EMI shields intended for high-temperature applications, and it is also important in understanding the buildup of minute stresses in the course of the material preparation such as 3D printing of polymeric composite systems.

Heat will diffuse into a plastic IC package when placed in a solder reflow oven because the ambient temperature is higher than the package's surface temperature. Previous equations, including the one below, regulate and describe the temperature distribution in the plastic IC package [41].

$$\frac{\partial}{\partial x_i}\left(k_{ij}\frac{\partial T}{\partial x_j}\right) = \rho c_P \frac{\partial T}{\partial t}. \tag{5.36}$$

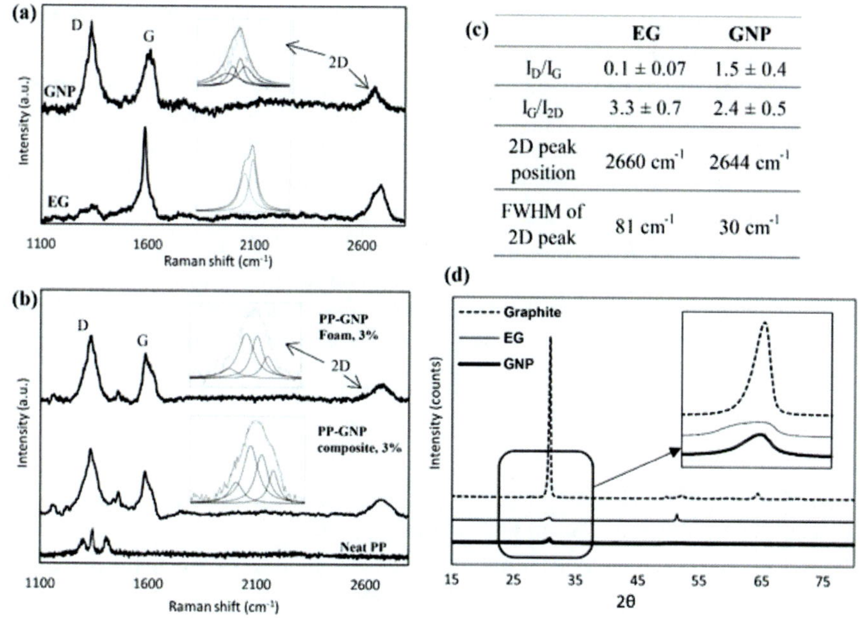

Fig. 5.7 Raman spectra of **a** EG and GNPs; **b** PP-GNP composite and foam. Inset pictures show the deconvoluted peaks of the 2D peak. **c** Intensity ratios, 2D peak positions, and FWHM obtained from Raman spectra; **d** powder XRD graphs of EG, GNPs, compared to natural flake GP. The inset shows the low intensity and broadening of the (002) peaks. Reproduced with permission from Ho and Kontopoulou [29]. Copyright 2020, Elsevier Science Ltd.

The surface temperature of the package is considered to be the same as the ambient temperature for the boundary condition. This is especially relevant to VPR operations because of how much heat is transferred from the surface. In this portion, the molding compound was dried at 175 °C with no stress on the package. The thermal strain is described as follows:

$$\varepsilon_t = \int_{T_{\text{cure}}}^{T} \alpha \mathrm{d}T. \tag{5.37}$$

α, and T denote the CTE and temperature. The thermal stress "σ_t" is presented thus:

$$\sigma_t = E\varepsilon_t. \tag{5.38}$$

E represents the modulus, while 175 °C is the mold compound curing temperature and the package is in a zero-stress state.

In integrated circuit packaging with an EMI shield, a better packaged IC is sought that contains a semiconductor die affixed to a substrate, an EMI shield that wraps the semiconductor die, and a package body that, in most electronics systems, also encapsulates the EMI shield [42]. In computer systems' chips, the chip failure is explained by diverse cracking at varying points of the package and die like the die lateral side, its corner, its backside, wire bond pad low k, solder bump, under bump metallization "UBM," etc. Cracking with underfill, substrate, and bump near substrate are the key characteristics of package failures brought on by thermal–mechanical stress. During assembly and field applications in various environmental situations, thermal–mechanical stress causes CPI problems [43].

GNP-based EMI shields are not left out of these challenges, seeing the ultimate target for their fabrication is real-life application in these systems. Therefore, understanding the behavior of graphene-based filler-bearing polymeric nanocomposites under diverse thermal stress and strain will help, in the long run, its preferred application in electrical systems.

Folding Endurance

(Nano)composite films for applications such as flexible membranes and packaging are subjected to this test. The number of folds that the sample film could withstand before breaking is known as the folding endurance value. According to the literature, the standard sample size for this test is 2.5 cm by 2.5 cm.

A folding endurance meter (folding index) (YQ-Z-31, Sichuan, China) is a common tool that can be used to assess the folding resistance of polymeric composite materials, according to a report [44].

It is expected that the property performance of GNP-based polymer nanocomposites should remain the same even after thousands of cycles (at least 10,000 folding cycles), especially with the aim of their application in electrical/electronic devices, which is also so required of robust EMI shield [27, 45].

Mechanical Characterization

Graphene-based filler-reinforced polymer composites are now widely employed in a variety of high-performance applications, including those in the domains of aerospace, sporting goods, construction, automotive, etc. These composite parts have replaced metallic ones in the automotive, electronics, and aerospace industries. By creating most of the key structural components in contrast to their carbon fiber-reinforced counterparts, the most recent innovations in commercial aircraft offered a weight reduction of 17% [46]. This report also had it that, as a result, its use makes it possible to increase the range of action and transportable load. Takeoff is also less energy-consuming. As a bonus, graphene-based filler has another advantage of avoiding the accumulation of ice on the wing at altitude. It can also deal with thunderbolts without flinching because of its high conductivity. These have a great degree of strength while being relatively light.

The class of composite materials known as nanocomposites includes those with a matrix material reinforced by a variety of NPs or nanofillers. The NMTs' dimensional properties, such as grain size, cylinder cross-section diameter, etc., and layer thickness, are all <100 nm. One can utilize a matrix made of the most popular polymeric materials, including epoxy, nylon, polyetherimide, and others. Using two different types of fillers, namely fillers at the microscale level and fillers at the nanoscale, nanocomposites, an appealing alternative material, was produced in recent decades. It has been demonstrated that the mechanical, thermal, electrical, and electromechanical properties of the nanocomposites have significantly improved.

A crucial necessity for product design, analysis, and life prediction is the mechanical characterization of the composites. With only a very small amount of filler loading (~ <5 wt.%), the mechanical properties of polymeric nanocomposites made with various types of NPTs, including nanoclays, GNP, GO, rGO, cellulose NPTs, MXene, cellulose nanofibrils, nanowhiskers, graphdiyne, and chitin nanofibrils, have significantly improved. These nanofillers primarily control the mechanical characteristics of polymeric nanocomposites. The improvement in mechanical properties, including tensile strength, hardness, Young's modulus, etc., of the nanocomposites can be attributed to the high rigidity of graphene-based fillers, their high surface area, as well as their affinity for the majority of polymer matrices.

Comparing graphene-based filler-reinforced polymer composites to the traditional materials typically utilized in structural applications, they exhibit superior characteristics. Future development in this area of materials is possible, nevertheless. Therefore, researchers have concentrated on the streamlined production process, process cost reduction, application, and improved material qualities.

Tensile Strength

The flat dog bone-shaped specimens with dimensions of 150, 30, and 3 mm are subjected to tensile tests in accordance with the standard test method ASTM D638-14 [47], though ISO 527-1 covers the concern. The standard addresses how to determine the in-plane tensile properties of high modulus fiber-reinforced polymer composite materials. The qualities of the generated composite materials, which might contain continuous or discontinuous fibers, depending on the direction of the test. Instron H10KS, Instron 5567, Hounsfield Tensometer, Hounsfield H25KS, Zwick 1445, and RSA3 (TA Instruments, USA) are a few examples of the various types of UTM used to conduct the testing [1]. The recommended sample geometry/dimension for this standard (ASTM D638-14) is also enshrined in the test standard. The specimens are prepared in accordance with the appropriate moisture profile before being evaluated at the same degree of preparation. Tests carried out in altered environments should be documented. The specimens must also be kept until the test is finished in the climate-controlled environment. The specimens are positioned in the tensile testing machine's grips with their axes pointing in the test direction. The grips are tightened, and the amount of pressure placed on them is noted: For more details, see the test standard.

The tensile properties of graphene-based filler-bearing polymer nanocomposite systems have been widely studied by researchers, with most of the researchers revealing from their study that the addition of graphene-based reinforcing fillers to polymer matrices results in dramatic enhancement in the mechanical properties of the resulting nanocomposites even at very low volume fractions [19]. The important parameter to be considered when estimating the tensile strength of graphene-based filler-bearing polymer composite materials is the ultimate tensile strength [UTS (Most frequently, this is expressed as the maximum applied load divided by the initial specimen cross-sectional area.)]. The following equation can be used to determine the UTS. Three relevant figures can be informed of the findings.

$$\text{Ultimate Tensile strength } F_u = \frac{P_{\max}}{A}. \tag{5.39}$$

F_u denotes the UTS, MPa.

$P_{\max}$ signifies the maximum load before failure.

A stands for average cross-sectional area, mm^2.

The following equation describes how to determine the tensile stress at each data point in order to determine the tensile modulus:

$$\text{Tensile modulus } \sigma_i = \frac{P_i}{A}. \tag{5.40}$$

σ_i, and P_i are the *i*th tensile stress (in Mpa) and the *i*th load (in *N*).

There are several known theoretical approaches for calculating tensile modulus, including the rule of mixture, the Halpin–Tsai method, Hirsch's model, and the Einstein-Guth model [1]. According to Dey et al. [1], several models are employed to determine a relationship between the mechanical properties of the composites and the reinforcing fillers.

The tensile strength of graphene-based filler-bearing nanocomposite EMI shields has been demonstrated to increase with the inclusion of the graphene-based filler material up to its percolation threshold [48]. For instance, Joseph and his colleagues studied the mechanical properties of PVC/graphene and PMMA/graphene nanocomposites. They observed that the behavior of the two nanocomposite systems was much different. The UTS of the composite having the highest filler loading (20 wt.%) for PMMA composite was less than the neat PMMA. On the other hand, the PVC composite system having 20 wt.% of the graphene-based filler material presented an UTS of 42.1 Mpa, which was greater than the value for the neat PVC revealed to be 31.8 Mpa [48].

Compressive Test

According to ATSM standard D6641M [49], compressive testing for composites is conducted. The DIC method, which is becoming more and more important in recent years for tracking the deformation of the surface of composites, has been reportedly utilized as per literature [1]. The compressive speed is held constant at 1.3 mm per minute. According to ASTM D6641M, the experimental apparatus and the specifications of the specimen with the CLC test fixture were completed. The specimen has a gauge length of 13 mm, a width of 13 mm, and a thickness of 2 mm. Along the testing apparatus, the specimens were lined up.

In an excellent instance, the compressive strength of Epoxy/AgNW/GNP nanocomposite aerogels was studied with the help of a universal testing machine (CMT5105) [50]. These researchers reported a compressive strength of 173.54 MPa, showing the robustness of the novel aerogel, which possessed an excellent EMI SE of 84.01 dB as well as an outstanding electrical conductivity of 26.5 S/m. These material systems were postulated to serve well as excellent EMI shields in niches such as electronics, construction, and aerospace sector [50].

Another report of compressive strength study [using a compression measurement instrument (SANS CMT5105, China)] for graphene-bearing polymer composite EMI shield is the report on Epoxy nanocomposite foams containing 32.26 wt.% GNP [51]. The compressive strength of the nanocomposite systems, which demonstrated an EMI SE and electrical conductivity of 86.6 dB and 3.14×10^2 S/m, was revealed to be 27.4 MPa, which was not affected by the increased or reduced inclusion of the rGO [51]. These authors calculated the compressive strength of their fabricated system from the compressive stress vs strain data collected from the experimental report.

Compressive strength study is mostly performed for GNP-polymer aerogel systems aerogels, though reports have shown its vitality for foam systems [52]. They conducted their experiment as follows: Using a universal material testing equipment (Zwick/Roell Z020) equipped with a 2.5 kN load at a compression rate of 2 mm/min, compression stress–strain curves of neat PDMS, and its nanocomposites were obtained. The authors of this study examined the compressive strength of flexible PDMS/rGO/SWCNT EMI shields with an EMI SE and electrical conductivity of 31 dB and 1.2 S/m, and they discovered that the resulting composites contained just 0.28 wt.% of rGO and SWCNTs. Additionally, the PTGCAs reinforced with the strong 3D skeleton had exceptional mechanical characteristics; improvements of 233 and 108% in compression strength and modulus, respectively, were made in comparison with the plain PDMS [52].

The compression experiment demonstrates how the graphene-based filler-bearing polymer composite EMI shielding material will respond under compression. Compression testing is capable of determining how these materials behave or responds to crushing loads and measuring the plastic flow behavior and ductile fracture limits of a material, which is a particularly important attribute for materials used in electronic packaging. Additionally, compressive strength tests can be adopted to evaluate batch quality and manufacturing consistency to obtain basic mechanical property data useful in designing industrial or laboratory processes for the systems fabrication.

Failure Analysis

Due to its significance as a research topic today, failure analysis of graphene-based filler-reinforced nanocomposites has drawn the interest of numerous experts [1]. The transverse matrix cracking, shear matrix cracking, and fracture of GNP sheets are the primary reasons for failure in nanocomposite materials.

Most of the matrix in graphene-based filler-reinforced GNP-reinforced composites experiences plane strain tension during the pure shear test. In contrast, the graphene sheet(s) experience uniaxial tension along the direction of the applied force, which impacts the mechanical stability of the nanocomposite system.

Rheological Characterization

Rheometry

Three distinct graphene-based filler-reinforced nanocomposites can be created: (a) films (solid), (b) sheets, and (c) bars. The creation of a graphene-based nanofiller is necessary for the early development of such composites. These components are then combined with either a monomer, prepolymer, or polymer and allowed to cure to create the composite matrix. In the modern day, transfer molding is the primary method used to create composites, and one of the most crucial factors in the successful development of the composite is flow behavior. Rheology not only influences how we comprehend and utilize these materials but also makes it possible to grasp how the nanocomposites react to electrical, magnetic, and thermomechanical forces. The study of fluid movement under isothermal or isostress conditions is known as rheology. These nanocomposites are characterized rheologically in two primary steps:

(a) Rheology of the graphene-based fillers: Measurements in oscillatory linear shear flow, constant shear flow, plus transient shear flow are typically used to characterize the shear rheology of graphene-based filler-bearing nanofillers such as GO [53]. Very important findings about the rheological properties of GO are depicted in Fig. 5.8 [53].

(b) Rheology of its (nano)composites

The graphene-based filler-based fillers' surface area is another crucial factor affecting the graphene-based filler-bearing nanocomposites' mechanical integrity, stiffness, toughness, and impact strength. Both the distribution of the graphene-based filler in the polymer matrix and the flow behavior of the nanocomposite are influenced by surface area.

Nowadays, rheological studies are common and generally contain all the variables required to determine a material's properties and service life. There are various ways to conduct a rheological investigation, including (i) shear, (ii) tension, (iii) compression, and (iv) torsion.

In the past few years, there has been a significant increase in interest in using graphene-based nanofillers in the polymer matrix to create composites. These materials can potentially improve composites' strength, impact resistance, and thermal barrier, making them suitable for various industrial and engineering applications.

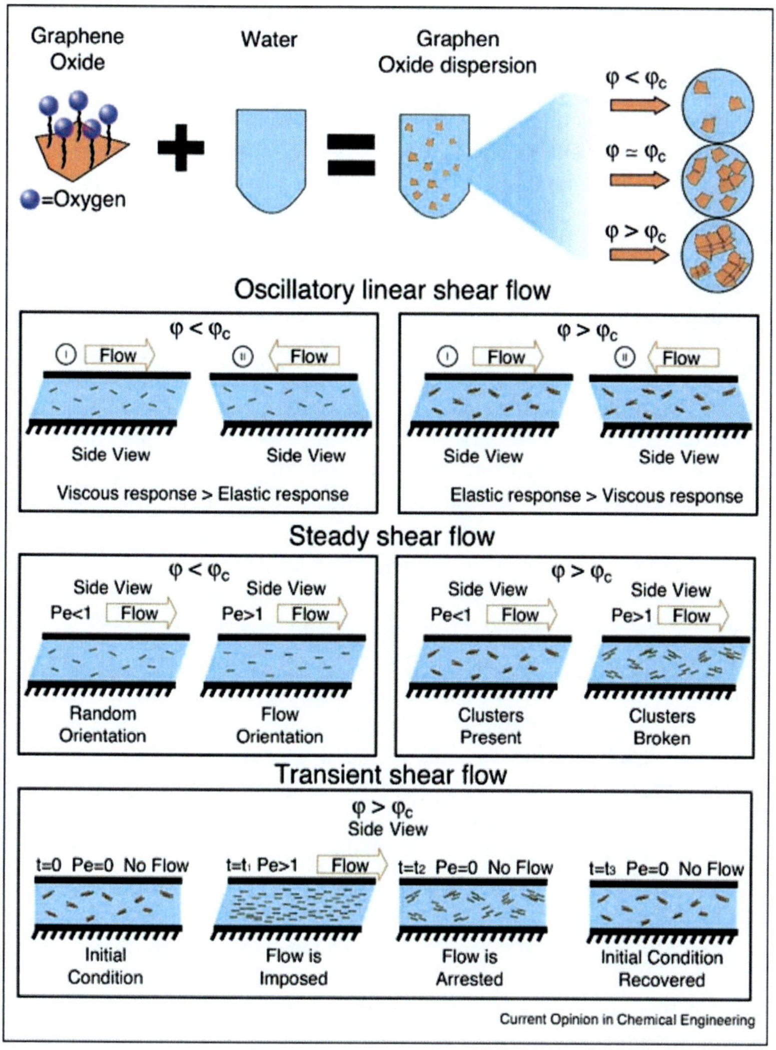

Fig. 5.8 GO aqueous dispersions at different volume concentrations ϕ are prepared by adding GO in water. Oscillatory linear shear flow: When $\phi < \phi_c$ (ϕ_c is the critical volume or mass concentration), viscous response due to the drag of the GO sheets prevails over the elastic response. When $\phi > \phi_c$, elastic response due to the GO self-aggregation prevails over the viscous response due to the drag of the clusters. Steady shear flow: When $\phi < \phi_c$ and at Peclet number $Pe = \dot{\gamma}a^2/D_0 < 1$, where $\dot{\gamma}$ is the applied shear rate, a is the average sheet radius, and D_0 is the sheet diffusivity, GO sheets are randomly oriented; while at $Pe > 1$, GO sheets are oriented along the flow direction. When $\phi > \phi_c$ and at Peclet number $Pe < 1$, GO sheets are arranged in randomly oriented clusters, while at $Pe > 1$, clusters are broken down. Transient shear flow: Only the case $\phi > \phi_c$ is considered. Initially, the dispersion is arranged in randomly oriented clusters. After applying a flow at $Pe > 1$, clusters are broken down, and GO sheets are oriented along the flow direction. When the flow is arrested, $Pe = 0$, GO sheets start to self-arrange. After sufficient resting time, GO sheets recover the initial cluster configuration. Reproduced with permission from Del Giudice and Shen [53]. Copyright 2017, Elsevier Science Ltd.

Rheology becomes a crucial technique for manufacturing composites as well as for characterizing and assessing the performance of the material.

Melt rheology is typically used to examine the flow characteristics of polymeric materials based on thermoplastics. For various loads of the graphene-based filler, it is explored for composite materials, the frequency dependence of the shear storage modulus G', the loss modulus G'', the inverse of the loss tangent G'/G'', and the complex viscosity $\eta*$. The steps for doing rheological characterization are shown below.

Procedure 1, Procedure 2, Procedure 3, Procedure 4, and Procedure 5 can be found in the report by Dey et al. [1].

Diverse researchers have reportedly used the rheological characterization to ascertain the flow properties of graphene-based reinforcing filler–polymer nanocomposites where the property performance of the nanocomposite systems such as storage modulus, loss modulus, complex viscosity, and loss factor is obtained [54]. The rheological properties of graphene-based filler-bearing PMMA nanocomposites have been shown to increase with the increase in the graphene-based filler concentration, as depicted in Fig. 5.9c and d [54]. These authors demonstrated in their study that GNPs revealed better rheological property performance in comparison with their counterparts reinforced with CNTs (Fig. 5.9a–d). With the addition of nanofillers, they saw that the PMMA's flow behavior changed, and finally, the viscosity of the complex rose. According to this effect's proponents, it was more evident at lower frequencies but gradually diminished as frequency rose as a result of shear-thinning behavior. The behavior of the filled composites changes as the amount of nanofiller increases, especially at 5 wt.%. This nonlinear behavior shows the creation of an extended, interconnected network of particles inside the polymer matrix [54].

Dynamic Mechanical Analysis

For GNP-reinforced nanocomposites, dynamic mechanical analysis has established itself as a reliable method of examining the relationship between GNP and polymer.

To evaluate the contribution of each component property, such as matrix and reinforcement modulus, volume fraction of reinforcement, reinforcement aspect ratio, reinforcement orientation, etc., theoretical prediction of the properties of composite materials is crucial. Several models are covered below:

(i) Models by Einstein: For filler-incorporated polymeric materials, this is the most basic model that Einstein has offered. The use of this model is restricted to polymers reinforced with spherulite particles that are present in lower concentrations. The formula comes from:

$$\frac{E_C}{E_P} = \left(1 + 2.5V_f\right). \tag{5.41}$$

V_f is the volume % of the reinforcing filler, while E_p and E_c are the moduli of the matrix polymer and reinforced composite, respectively.

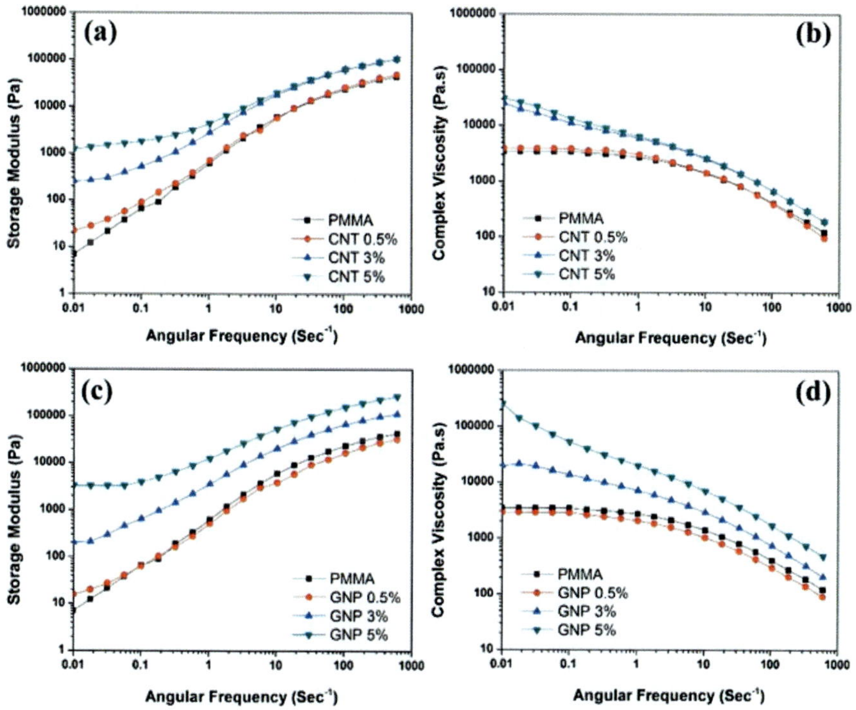

Fig. 5.9 Viscoelastic response of GNP-contating PMMA nanocomposites. Reproduced with permission from Zakiyan et al. [54]. Copyright 2017, Elsevier Science Ltd.

(ii) Guth models: Even at greater filler loading concentrations, the Guth equation is utilized for composites and is given as

$$\frac{E_C}{E_P} = \left(1 + 2.5V_f + 14.1V_f^2\right). \tag{5.42}$$

(iii) Quemada models: Given below is the Quemada equation.

$$\frac{E_C}{E_P} = 1/\left(1 - 0.5K_\delta V_f\right). \tag{5.43}$$

In this equation, a variable term K_δ is used to represent the particle interaction and geometric properties of the GNP nanoplatelets. This number for GNP-loaded composites is typically 2.5.

Different rheological flow models are used to analyze rheology-based data like strain sweep. These models can alternatively be categorized as two- and three-parameter models. The thinning and thickening behavior of the fluid is described by the power-law model (also known as Ostwald's model). Here, the relationship between log (Shear Stress) and log (Shear Strain) is displayed, and the slope (n) is inferred from the linearly fitted equation. When shear rates are low, the experiment is accurate, but when shear rates are high, it diverges from reality. The appropriate equation is provided by

$$\log\sigma = \log K + n\log\gamma(48) \text{ or } \log\eta = \log K + (n-1)\log\gamma(49). \quad (5.44)$$

The viscosity is denoted by "η". For $n < 1$, the fluid exhibits shear-thinning behavior; if $n > 1$, shear-thickening behavior.

It is also common practice to assess the flow characteristics using the Bingham rheological model. Given by is the worry equation.

$$\sigma = \sigma_0 + \mu_p\gamma. \quad (5.45)$$

σ and σ_0 represent the shear stress and yield stress correspondingly, while the plastic viscosity is denoted by μ_p.

The flow pattern of viscoelastic fluids is also described using the Casson model. The given equation is

$$\sigma^{0.5} = K_{oc}^{0.5} + K_c^{0.5}\gamma^{0.5}. \quad (5.46)$$

σ and γ denote the shear stress and yield stress, while K_c stands for the Casson plastic viscosity. Casson yield stress is denoted by K_{oc}.

Regarding examples from the literature, Fig. 5.10 represents temperature-dependent viscoelastic behavior and shows the fluctuation of storage modulus with temperature for GNP-based unfilled and 5 wt.% filled nanoplatelets. The size of the loss modulus and the T_g, which is introduced as a distinguishing peak, are the diagram's two key characteristics. The characteristic peak in this graph, which is caused by the coordinated segmental motion of the polymer, is connected to alpha relaxation (T_g). As can be seen, the presence of NPs causes the loss modulus magnitude to increase and the T_g to move upward. In reality, this is proof that interphase exists and is significant since it restricts chain mobility and sharply raises the system's glass transition temperature [54].

In comparison, a nanocomposite that contains 5 wt.% CNT, one that has 5 wt.% GNP has a greater T_g. Given that the surface energies of GNPs and CNTs are nearly the same, it can be concluded that the two-dimensional GNPs have a significant impact on the PMMA chain dynamics reduction for CNTs. As a result, compared to CNTs, GNPs are thought to be more successful at increasing the interactions between filler and polymer and the entrapment of polymer chains [54].

The EMI shielding performance of diverse graphene-based polymeric composite systems is presented in Tables 5.1 and 5.2.

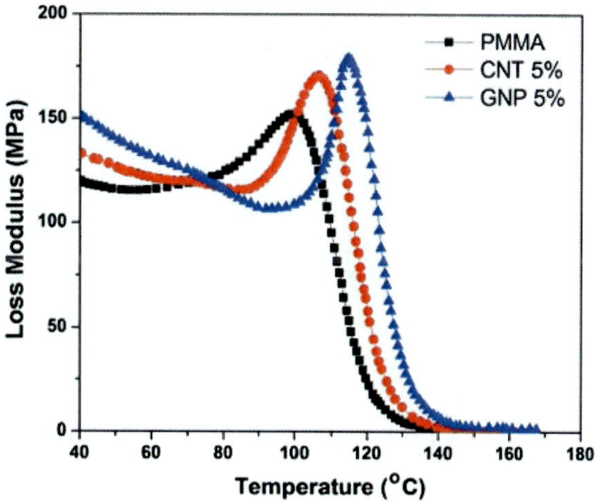

Fig. 5.10 Dynamic mechanical and thermal behavior of neat PMMA and its nanocomposites with GNP and CNT. Reproduced with permission from Zakiyan et al. [54]. Copyright 2017, Elsevier Science Ltd.

Electrical Conductivity

The electrical conductivity of graphene-based filler-bearing-polymer (nano)composite materials applied as EMI shields is a very important parameter that governs the SE of the material [61].

The electrical conductivity and percolation threshold of the different graphene-based filler-bearing polymer composites can be measured by collecting the current–voltage and or electrical resistance data from instruments such as the percolation threshold is given by the following scaling law [24]:

$$\sigma = \sigma_0(\Phi - \Phi_C)^t. \tag{5.47}$$

σ, σ_0, Φ, Φ_C, and t are the electrical conductivity of the composite material, reinforcing electrical conductivity, the filler content, percolation threshold, and the critical exponent. The critical exponent "t" is a constant which also determines whether the filler is dispersed in the composite in 2D or 3D, such that $1 < t < 1.3$ for 2D and $1.6 < t < 2.0$ for 3D [24].

For instance, in a specific study where it was found that the few layers graphene-based MLG formed a 3D structure, the computed value of t was found to be 2.48 in a HDPE/MLG. For the HDPE/MLG nanocomposite with a 3D percolation network, it was reported that the electrical conductivity increased quickly with the addition of filler, reaching about 1.3 S/cm at 20 wt.% MLG concentration. The crucial MLG content, or percolation threshold, was discovered to be as low as 0.25 wt.%, or 0.1 vol.%. The diameter of the polymer particles has a big impact on the composite's percolation threshold. The percolation threshold lowers as polymer particle size increases. As a result, the 3D percolation structure and mm-scale HDPE particles in this work account for the low percolation threshold observed. However, compared to the HDPE/MLG composite with a broad percolation network, the

Table. 5.1 Reported property parameters of various graphene-based filler-bearing polymer composite EMI shields

Formulation	Filler loading (wt.%)	Electrical conductivity (S/m)	Thermal conductivity "W/mK"	Dielectric constant "*e*"	EMI SE in dB	Structure	Thickness	References
PVC/graphene	20	4.8×10^{-3} S cm^{-1}	–	–	31	Multilayered	2 mm	[48]
PMMA/GNP	20	4×10^{-4} S cm^{-1}	–	–	21	Multilayered	2 mm	[48]
Epoxy/rGO	1.2	0.03			6.0			[55]
Epoxy/rGH	1.2	40.2			38.8			[55]
PANI/MWCNT/thermally annealed graphene aerogel/epoxy (PANI/MWCNT/TAGA/epoxy)	–	–	–	–	42	Aerogel	800 μm	[56]
GNPs/rGO/Eoxy	0.1 (rGO) and 20.4 (GNP)	179.2	1.56	–	51	Foam	3 mm	[57]
rGO/Epoxy	0.1	2.8×10^{-3}	–	–	13	Foam	3 mm	[57]
GNP/Epoxy	20.4		1.01	–	15	Foam	3 mm	[57]
PVP@GNP/poly(ethylene-co-methyl acrylate)	15	10^{-2}	–	–	30	Solid	1 mm	[58]
Graphene/NBR	4	–	–	4.6	~77	Solid	2 mm	[59]
Graphene/CNT/WPU/cotton (WPU10-3/Textile (CNT + Graphene))	3	6.4×10^{-1}	0.379	–	30	Layered	0.35 mm	[60]
Graphene/PMMA	3	–	–	–		Solid	2 mm	[54]

Table. 5.2 Reported property parameters of graphene-based filler-bearing polymer composite EMI shields considered their best performance

Formulation	GNP conc. (wt.%)	Tensile strength (TS) (Mpa)	Compressive modulus (TS)	Young's modulus	Magnetic	References
PVC/Graphene	20	42.8	–	2858 MPA	–	[48]
PMMA/Graphene	20	9.61	–	1144 Mpa	–	[48]
Epoxy/rGO	–	–	–	4.30 GPa	–	[55]
Epoxy/reduced graphene oxide with honeycomb structure (rGH)	1.2	–	–	4.99 GPa	–	[55]
PANI/MWCNT/TAGA/epoxy	–	–	204.4 kPa	204.4 kPa	–	[56]

HDPE/MLG composite with a randomly dispersed filler structure (referred to as R-composite) displayed reduced electrical conductivity. At 20 wt.% MLG content, both composites have electrical conductivities that are quite similar; however, the HDPE/MLG composite with the 3D mm-scale percolation structure (referred to as mmP composite) has better electrical conductivity than the R-composite. This is due to the fact that, regardless of structure, the polymer composite exhibits significant electrical conductivity if the MLG content is high enough to allow a conductive network to form between the MLG particles over the percolation threshold. Though the MLG network is easily produced in the mmP-nanocomposite, the electrical conductivity of the mmP-nanocomposite is stronger at low MLG levels than the R-nanocomposite [24].

Impedance matching is one element that impacts microwave absorption. Figure 5.11c shows that the DC electrical conductivity of the PEI/G@Fe_3O_4 (Fe_3O_4 NPs decorated graphene) foams was less than that of the PEI/graphene foams reported in our earlier investigation. This behavior implied that the addition of Fe_3O_4 NPs had a tendency to reduce the electrical conductivity of the composite foams, which would enhance the equality of the electromagnetic parameters, enhance the degree of impedance matching, and reduce surface reflection. The attenuation of electromagnetic waves caused by magnetic and dielectric losses is another significant element that affects microwave absorption. As is well known, a material's overall dielectric property is determined by its electronic, ionic, orientational, as well as space charge polarization. Interfacial polarization, sometimes referred to as Maxwell–Wagner polarization, is the result of virtual charges building up at the interface of two media with varying conductivities and dielectric constants in a heterogeneous system. Pure graphene platelet is not magnetic and mostly contributes to microwave absorption through dielectric loss. Due to the formation of a heterogeneous system and increased interfaces, as well as the stronger coupling at the gaps between the neighboring Fe_3O_4 NPs, the introduction of Fe_3O_4 NPs on graphene sheets would not only increase the magnetic losses but also increase the dielectric losses in a wide frequency range. Additionally, the decreased graphene oxygen groups and flaws may be able to enhance the composite foams' ability to absorb microwaves [61].

Due to its inherent ability to absorb electromagnetic radiation, electrical conductivity is crucial for the effectiveness of EMI shielding. In the majority of graphene-based filler-bearing nanocomposites that are composed of non-conducting polymers, the electrical conductivity is mostly connected to the intrinsic properties of the filler because the non-conducting polymer matrix is essentially an insulating material. An intriguing example is shown in Fig. 5.12, where the DC conductivity of the nanocomposites is plotted as a function of the concentrations of CNT and GNP [54]. It has been found that the electrical conductivity rises as the filler content does as well. It appears that when the nanofiller is present in an equivalent amount in both the MWCNT and GNP nanocomposites, the electrical conductivity of the corresponding nanocomposites is higher. For both types of fillers, 5 wt.% was found to be the maximum electrical conductivity. Reaching the maximum number of electrical contacts and the creation of the saturated conductive network in the polymer matrix may be the cause of this behavior. In reality, the network creation, or the development of conductive routes, is more readily possible in the MWCNT nanocomposites due to the one-dimensional structure and longer length than the GNP. The morphological research in Fig. 5.12 illustrates a similar circumstance [54].

Magnetic Properties

Another important factor that influences microwave absorption is the attenuation of electromagnetic waves brought on by magnetic and dielectric losses. The total dielectric property of a material is well known to depend on its electronic, ionic, orientational, as well as space charge polarization. In a heterogeneous system, virtual charges accumulate at the interface of two mediums with different conductivities and dielectric constants, leading to interfacial polarization, also known as Maxwell–Wagner polarization. Since pristine GNP and its various derivatives, such as GO and rGO, lack magnetic properties, they primarily contribute to microwave absorption through dielectric loss. However, its composite systems can be given magnetic characteristics by adding magnetic NPs like Fe_3O_4. The addition of Fe_3O_4 to GNPs would result in an increase in both magnetic and dielectric losses over a broad frequency range because of the establishment of a heterogeneous system, greater interfaces, and stronger coupling at the spaces between surrounding Fe_3O_4 atoms.

A Model-9 PPSM (Quantum Design) was used in one investigation to examine the magnetic characteristics of G@Fe_3O_4, as illustrated in Fig. 5.11c and d [61]. The magnetization curves between -20 and 20 kOe of the magnetic field were captured at room temperature. The magnetic saturation value (Ms) of G@Fe_3O_4 fell from 75.9 to

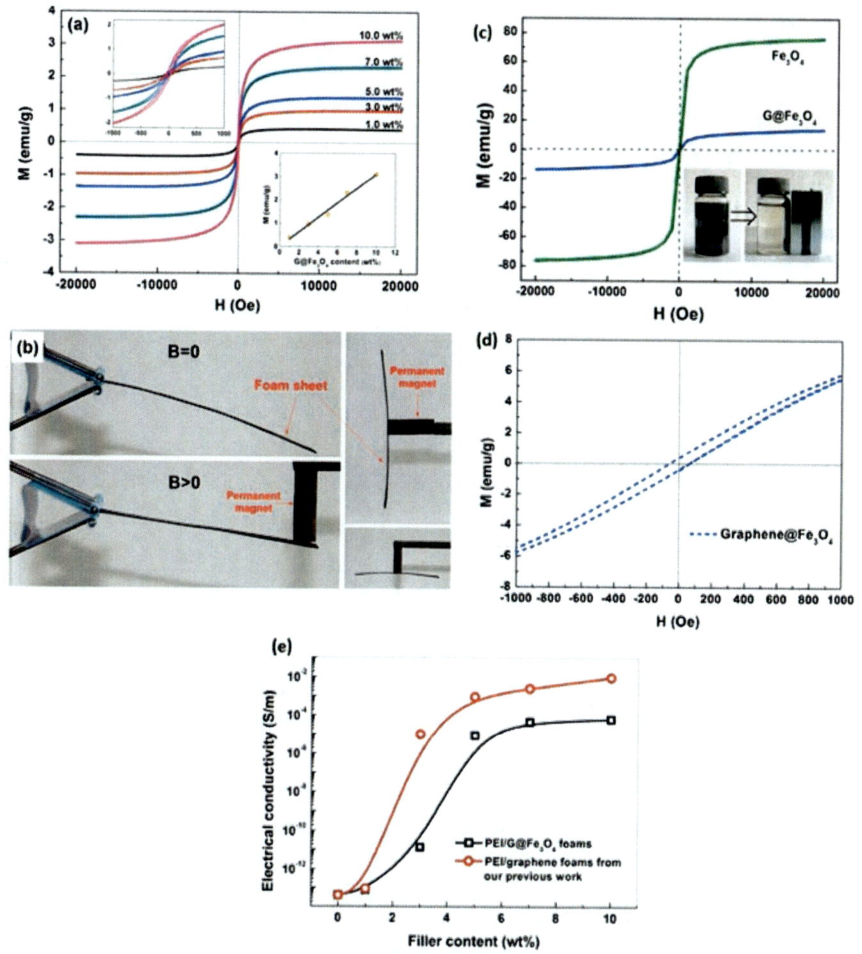

Fig. 5.11 Superparamagnetic microcellular PEI/G@Fe_3O_4 foam sheets: **a** magnetization of the composite foams with different G@Fe_3O_4 loading, **b** this free-standing composite foam sheet can be actuated by an external magnetic field, **c** magnetization hysteresis loops of G@Fe_3O_4 at room temperature. The bottom right inset shows the photographs of G@Fe_3O_4 in water and their response to an external magnetic field, **d** the magnetization hysteresis loops of G@Fe_3O_4 at room temperature, and **e** the electrical conductivity of PEI composite foams. Reproduced with permission from Shen et al. [61]. Copyright 2013, American Chemical Society

14.0 emu/g when compared to pure bulk Fe_3O_4. The smaller Fe_3O_4 NPs and the presence of GNPs may cause the lower value. After the applied magnetic field is removed, the magnetization curves exhibit reversible, nonlinear characteristics with negligible coercivity (~45 Oe, as seen in Fig. 5.11b), suggesting that the G@Fe_3O_4 was superparamagnetic. Additionally, our G@Fe_3O_4 hybrid was simple to separate using a magnet (see Fig. 5.11c insert) [61]. These scientists also said that the PEI/G@Fe_3O_4 foams' magnetic performance was examined by tracking their magnetization under

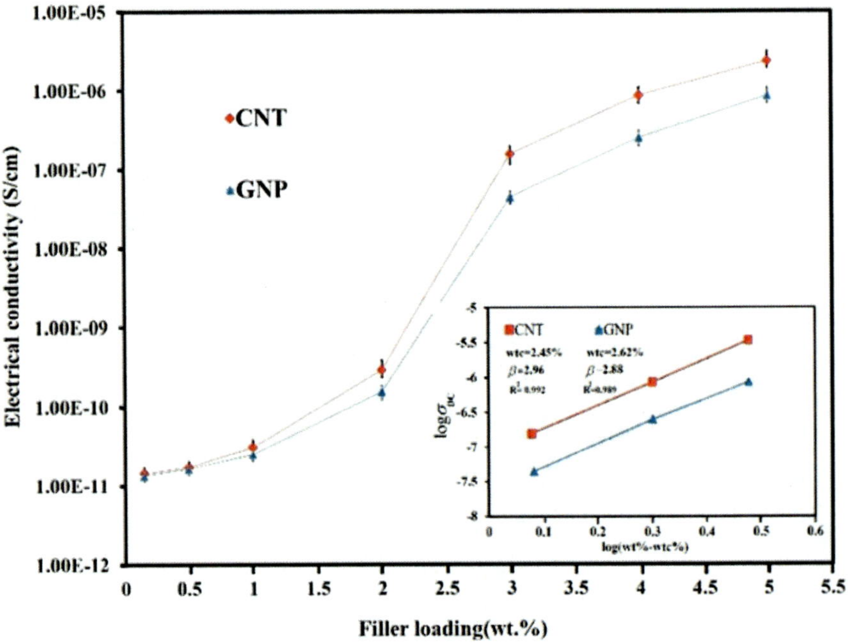

Fig. 5.12 DC conductivity as a function of filler loading for nanocomposites. Inset figure represents the power-law fitting of classical percolation theory for the same composites. Reproduced with permission from Zakiyan et al. [54]. Copyright 2017, Elsevier Science Ltd.

an applied magnetic field of −20 to 20 kOe at ambient temperature. The weight of the composite foam (PEI + G@Fe_3O_4) was used to normalize the magnetization for purposes of comparison. The M_s of the foams was in the range of 0.38–3.09 emu/g, as depicted in Fig. 5.11a, and it tended to rise linearly with the G@Fe_3O_4 content (inset at the lower right). Due to the lower concentration of G@Fe_3O_4 in foams than in G@Fe_3O_4 hybrids, the magnetization of foams dramatically decreased. Furthermore, it is intriguing to learn that the produced PEI/G@Fe_3O_4 foams were superparamagnetic since the Ms exhibited no appreciable hysteresis (Fig. 5.11a, inset at top left). All of these foams had coercivities in the range of 35–5 Oe, which was comparable to the coercivity of the G@Fe_3O_4 hybrid. This occurrence proved that despite being mixed with a polymer matrix, the G@Fe_3O_4 had maintained its superparamagnetism. Consequently, a magnetic field might activate PEI/G@Fe_3O_4 microcellular foams. In Fig. 5.11b, the magnetic response of PEI/G@Fe_3O_4 foam loaded to 10 wt.% in the absence ($B = 0$) and presence ($B > 0$) of a magnetic field is depicted (left). This microcellular foam cantilever was obviously actuated with a quick response time once a permanent magnet was approached. Similar to other foam sheets, this one demonstrated a potent attraction to a static magnetic field, as demonstrated in Fig. 5.11b (right) [61].

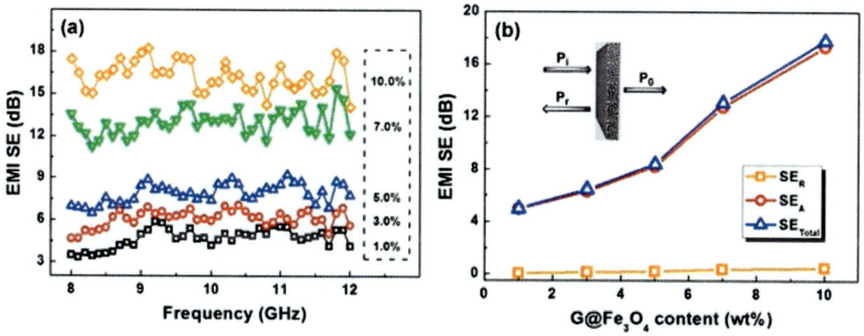

Fig. 5.13 **a** EMI shielding effectiveness as a function of frequency measured in the 8–12 GHz range of the PEI/G@Fe_3O_4 foams with various G@Fe_3O_4 concentrations; **b** SE_{total}, SE_R, and SE_A of microcellular foams at 9.6 GHz. Reproduced with permission from Shen et al. [61]. Copyright 2013, American Chemical Society

An example of a common instrument used for analyzing the magnetic property performance of graphene-based filler-bearing polymer nanocomposite EMI shield(s) is the "model-9 PPSM (Quantum Design)" [61].

EMI Shielding Performance

As the main performance character of study/interest for graphene-based polymeric EMI shields, the EMI shielding performances of these systems have been reportedly characterized using diverse instruments such as WILTRON 54169A scalar measurement system (at room temperature), Vector Network Analyzer (N9926A, Agilent Technologies) [62], PNA Network Analyzer N5222A [63], vector network analyzer (Keysight N5222B) [24], etc. This test is performed mostly within X-band (8–12 GHz); however, many researchers have analyzed these systems outside the X-band, seeing the present world demand for EMI shields operating well even at higher frequencies, even up to 50 GHz [64]. The standard test method utilized for measuring EMI SE of planer systems like graphene-bearing polymer (nano)composites is ASTM D4935-18 [65]. The diverse approaches followed for analyzing the EMI shielding of graphene-polymer nanocomposite systems have been discussed earlier: We refer readers to our earlier discussion in this regard.

A typical example of EMI shielding performance of GNP-based polymer nanocomposite with respect to frequency as well as filler concentration is depicted in Fig. 5.13 [61], where the authors demonstrated from their studies on PEI/G@Fe_3O_4 foams that enhanced inclusion of the graphene-based reinforcing filler "Fe_3O_4 foams" within the PEI matrix in the nanocomposite resulted in great improvement in the EMI SE of the systems.

It has been established from literature reports that the EMI shielding performance of graphene-based filler-reinforced polymeric systems improves with the

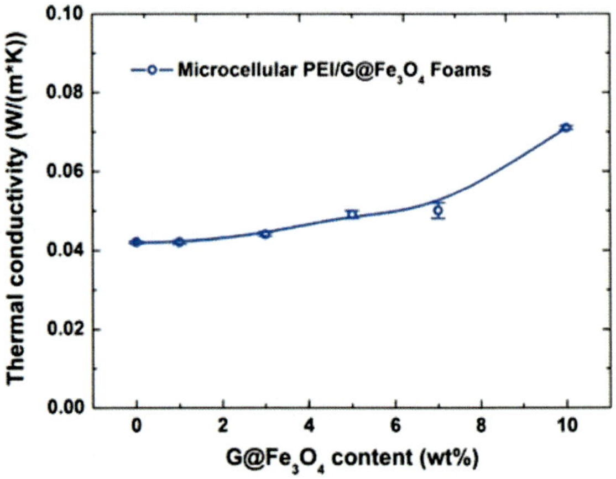

Fig. 5.14 Thermal conductivity of microcellular PEI/G@Fe_3O_4 foams at room temperature. Reproduced with permission from Shen et al. [61]. Copyright 2013, American Chemical Society

increasing inclusion of graphene-based fillers up to the percolation threshold [10, 18, 62, 63, 66–70].

Thermal Conductivity

For aircraft and spacecraft, the thermal insulation of EMI shielding materials is essential to reducing the effect of temperature on the operation of electronic devices and occupants.

According to reports, a Laser Flash System was used to assess the thermal conductivity of both pure PEI foam and the PEI/G@Fe_3O_4 foams [61]. The thermal conductivity of pure PEI foam, as displayed in Fig. 5.14, is 0.042 W/m (mK). The composite foam's thermal conductivity was only raised to 0.071 W/kg by the addition of 10.0 wt.% G@Fe_3O_4 (mK). In general, the modification of cell size and the inclusion of nanofillers have an impact on the thermal conductivity of the composite foam. It is widely acknowledged that the thermal conductivity would typically decrease as the average cell size increases.

In contrast, adding nanofillers with exceptional thermal conductivity would greatly improve the thermal conductivity: The thermal conductivity was postulated to decrease when cell size was decreased. Therefore, the superior thermal conductivity of GNP sheets should be credited for the increased thermal conductivity in the composite foams. However, the foam's thermal conductivity increase is very small, indicating that the performance of PEI/G@Fe_3O_4 foams as thermal insulators would not be affected by this loading of G@Fe_3O_4 [61].

In a different case, a team of researchers revealed that the benefits of mmP-structured MLG-based nanocomposites were seen in relation to thermal conductivity. As simultaneous electron and phonon conduction occur, as illustrated in Fig. 5.15h, thermal conductivities, regardless of structure, exhibit a linear increase. The R-structure and mmP-thermal structure conductivities were 1.56 and 0.81 W/mK, respectively, at 30 wt.% of each structure; revealing a difference of almost

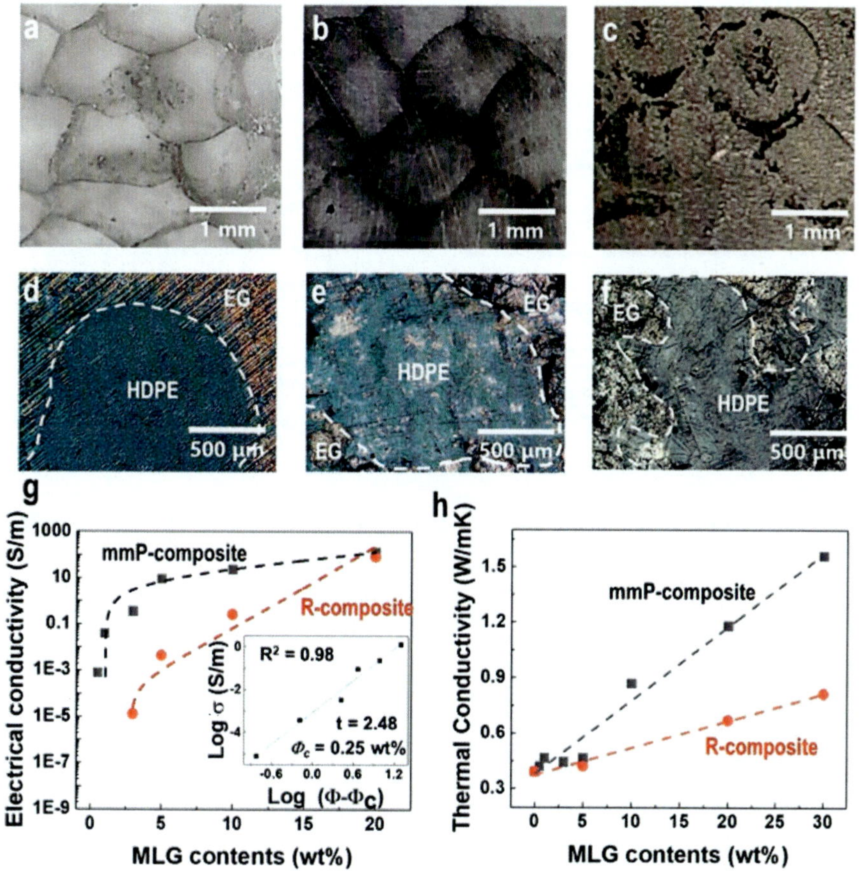

Fig. 5.15 Images of the mmP-structured HDPE/MLG composite with **a** 0.1, **b** 3, and **c** 10 wt.% MLG loading and **d–f** high-magnification OM images. **g** Electrical and **h** thermal conductivities of the mmP-structure and R-structure HDPE/MLG composites with different MLG contents. Reproduced with permission from Ryu et al. [24]. Copyright 2022, American Chemical Society

twice as much as the thermal conductivity as the MLG content increased. The mmP-nanocomposite was proposed to significantly aid in the thermal energy dissipation by EM absorption as the shielding via absorption transforms to heat energy [24].

It is vital that an excellent graphene-based filler-bearing EMI shield should possess low or high thermal conductivity depending on the specified niche of application. For instance, to reduce the effect of temperature on the operation of electronic devices and occupants, it is essential for aeroplanes and spacecraft to use thermal insulation of EMI shielding materials. Therefore, appropriate graphene-based reinforcing fillers should be chosen for specific applications, carefully considering their overall properties as discussed earlier.

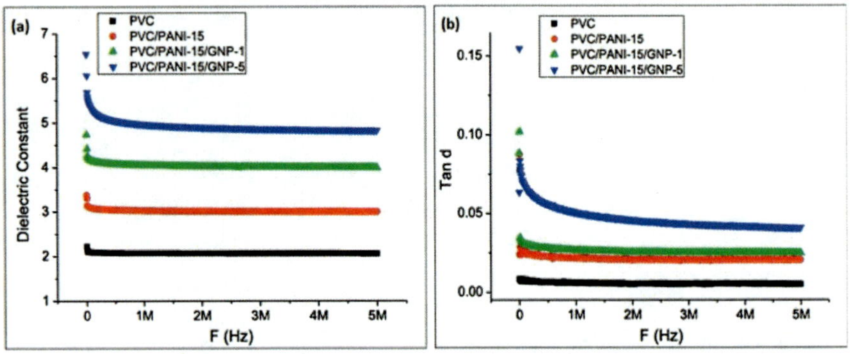

Fig. 5.16 **a** Dielectric constant and **b** Tan δ of PVC, PVC-PANI blend and hybrid polymer composites up to 5 MHz frequency. Reproduced with permission from Shakir et al. [64]. Copyright 2019, Elsevier Science Ltd.

Table 5.3 Various dielectric characteristics of PVC, PVC/PANI blends, and PVC/PANI/GNP composites at 5 MHz

Sample representation	Dielectric constant	Tan δ	AC conductivity (S/cm)
PVC	2.06	0.009	2.8×10^{-6}
PVC/PANI-15	2.98	0.014	1.6×10^{-5}
PVC/PANI-15/GNP-1	3.98	0.025	2.85×10^{-5}
PVC/PANI-15/GNP-5	4.82	0.040	5.6×10^{-5}

Reproduced with permission from Shakir et al. [64]. Copyright 2019, Elsevier Science Ltd.

Dielectric Property Performance

The dielectric constant and dielectric loss are additional significant factors when it comes to polymer composites. The dielectric properties of graphene-based filler-bearing polymer nanocomposite systems have been reportedly characterized by means of instruments such as the Novacontrol Broadband dielectric spectrometer [63].

This quantity has been studied by several researchers for graphene-based filler-bearing polymer nanocomposites intended for EMI shield fabrication [64]. For hybrid composites (12 mm in diameter), the dissipation factor "D" and capacitance "C" are also obtained. At 100 Hz, it was discovered that the dielectric constant of the polymer composites increased from 2 (PVC) to 7 (PVC and PVC-PANI-15-GNP5). As seen in Fig. 5.16a, PVC/PANI-15 and PVC/PANI-15/GNP-1 both showed an improvement in dielectric constant when compared to polymer alone. The polymer composite's dielectric constant (Fig. 5.16a) and dielectric loss (Fig. 5.16b) are measures of how much electromagnetic energy has been dissipated there. Table 5.3 summarizes various dielectric properties that were obtained at a 5 MHz frequency. The mathematical expression can be used to determine the generalized response to AC conductivity.

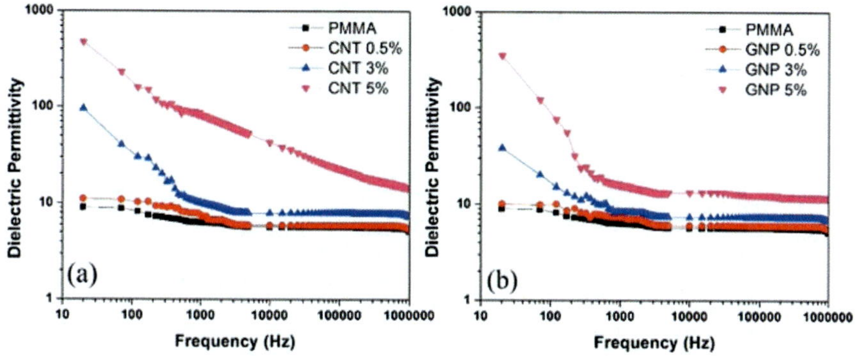

Fig. 5.17 Dielectric permittivity of neat PMMA and its composites in a wide frequency range. Reproduced with permission from Dey et al. [1]. Copyright 2017, Elsevier Science Ltd.

The dielectric permittivity of pure PMMA and its nanocomposites as a function of frequency at room temperature was examined in another report and results are reported in Fig. 5.17. The scientists found that the dielectric characteristics of the nanocomposites compared to the pure polymer did not change noticeably with a 0.5 wt.% nanofiller concentration. The dielectric permittivity, particularly in the sample filled with CNTs, significantly increased with a 5 wt.% loading, and this increase was much higher in the CNT-containing composite than GPN-containing polymer composite [54]. Two mechanisms were used to interpret these findings. The first one is the volumetric effect of filler having a higher dielectric permittivity than the polymer matrix. This indicates that the total effective dielectric constant of the composite is increased by the addition of CNTs with a higher dielectric constant than the polymer matrix. Additionally, as particle concentration rises, interparticle interactions between dielectric particles cause the effective dielectric constant to grow even more. The improvement of the local electric field, which is connected to the effective dielectric constant through polarization, can be attributed to this. Therefore, notwithstanding filler form, size, and surface area, the effective dielectric permittivity of the nanocomposite is increased with an increase in CNT loading over all frequency ranges. The second mechanism relates to the dynamics of polymer chains in the interphase and the polymer matrix, as well as the surface effects of CNTs that are regulated by interfacial polarization. The interfacial polarization is only present at low frequencies and disappears at higher ones. Interfacial polarization, or the buildup of charge on particle surfaces, is minimal at high frequencies [54].

UV–Spectrophotometric Study

The UV–vis spectroscopy can be adopted to study the transparency of graphene-based filler-bearing polymer nanocomposites as well as the effect of the graphene-based filler inclusion in the polymer matrix. Additionally, a PerkinElmer Lambda 35

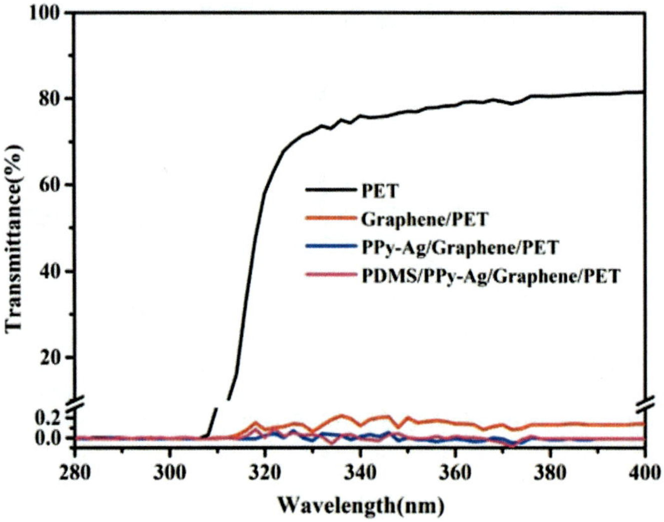

Fig. 5.18 UV transmittance spectra of pristine PET, graphene/PET, PPy-Ag/graphene/PET and PDMS/PPy-Ag/graphene/PET film. Reproduced with permission from Gong et al. [72]. Copyright 2013, Elsevier Science Ltd.

(Singapore) spectrophotometer has reportedly been used to correlate the graphene-based filler dispersion using an absorbance (A), UV–vis spectrophotometric investigation. By applying the Lambert–Beer rule, the molar extinction coefficient of PVP-stabilized graphene was calculated [71].

$$A = \alpha lc. \tag{5.48}$$

A, *c*, and *l* are the molar extinction coefficient, the graphene-based filler dispersion concentration, and path length, correspondingly. Every absorbance test was run against a toluene reference cell that was empty. During the formulation of EMI shields, especially transparent EMI shields, the shielding performance can be predicted from the UV-shielding ability of the materials using the UV–visible spectra.

A great example is found somewhere [72]; the authors studied the UV-shielding performance of the nanocomposite systems (graphene/PET; PDMS/PPy-Ag@Graphene@PET; and PPy-Ag/graphene/PET films) using a Lambda 750 S UV–Visible–Near-infrared Spectrometer (UV–vis Spectrometer, PerkinElmer Corp, USA). By measuring the coating's transmittance between 280 and 400 nm, the UV-shielding effectiveness of the coating was assessed, Fig. 5.18 [72].

The UV transmittance curve of the unblemished PET film revealed that the transmittance in the UV-A region was extremely high, exceeding 60%, that the UV-B region's anti-ultraviolet performance was greater than that of the UV-A region, and that the average transmittance was also 7.95%. The pure PET film's increased transmittance rate indicates how poorly it blocks UV rays. The graphene/PET film transmits about 0.1% of UV-A and UV-B light. The PPy-Ag/graphene/PET film achieves higher UV-shielding performance once the PPy-Ag layer is loaded, with an average transmittance close to 0%, showing that the film has outstanding UV radiation shielding capabilities throughout the UV-A and UV-B regions. Comparing PDMS/PPy-Ag/graphene/PET film to PPy/Ag/graphene/PET film, the UV-shielding performance practically remains unaltered, demonstrating that the PDMS alteration has no discernible impact on the UV-shielding behavior. The authors proposed two explanations for the superior UV-shielding properties of PPy@Ag/graphene/PET and PDMS/PPy-Ag/graphene/PET films. One explanation is the PPy layer's naturally high UV absorption capabilities, which are also amplified by the addition of the PPy/Ag layer, which turns the UV absorption from the dark brown of the graphene layer to black. Another reason suggested was that silver NPTs' high refractive index results in more effective UV scattering [72]. The trend observed by these researchers in their study of the UV-shielding performance of the nanocomposites was similar to that of the EMI SE. The EMI SE of PDMS/PPy-Ag/graphene/PET was reported to be 31.5 dB [72], though they observed that the reaction time had some influence on the EMI SE.

5.2 Techniques for Polymer-Based Fire Retarding Materials Characterization

No single fire scenario or fire test will ever be able to account for all known fire types and their associated fire behaviors since every fire is unique. As a result, the concepts of fire scenarios and fire attributes presented in this section serve as the foundation. The applied heat flux, temperature, length scales, and ventilation are all important aspects that affect the heat and mass transfer characteristics of fires (Fig. 5.19). Their behavior can be divided into three phases [73]:

- Piloted ignition: This is also known as the beginning of blazing combustion, is characterized by an ignition source (such as a flame, cigarette, and glow wire), a tiny length scale (cm), ambient temperatures in the ignition temperature range (600–700 K), and greater ventilation.
- Developing fire: This stage of the fire's development is marked by the continuance of flame combustion, which is defined by an external heat flow around 20–60 kWm^{-2}, larger length scales (dm to m), conditions outside the ignition temperature (700–900 K), and persistently high ventilation.
- Fully developed fire: The ultimate development stage of a fire is marked by a high external heat flux (> 50 kWm^{-2}), vast length scales (> m), ambient temperatures above the point of auto-ignition (> 900 K), and poor ventilation.

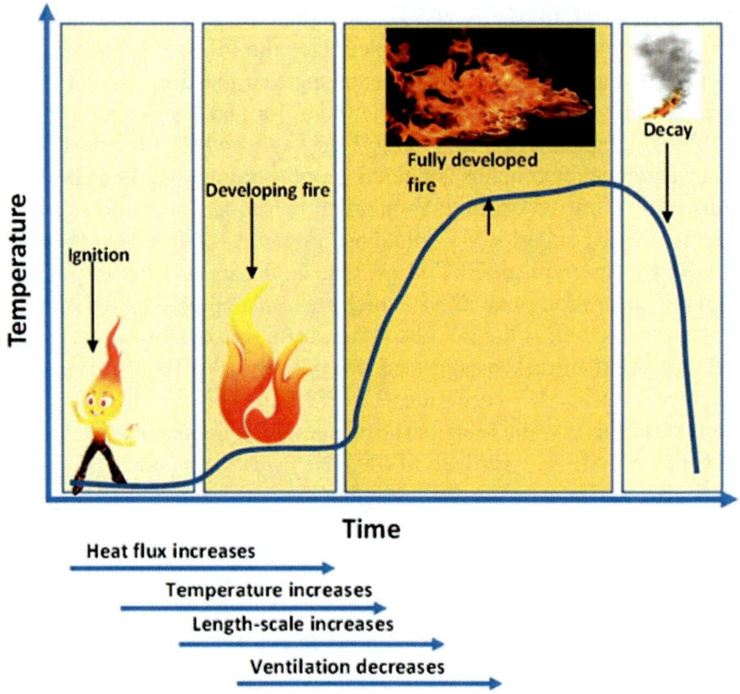

Fig. 5.19 Several fire scenarios that a cone calorimeter may measure, the stages of a fire and its core characteristics

5.2.1 Cone Calorimeter

Cone calorimeter testing, a bench scale process developed by NIST, quickly gained popularity in the academic community and was standardized by ASTM E1354 and ISO 5660-1. The test involves exposing a specimen to a certain heat flux and is regarded as the most efficient and successful medium-scale fire behavior test for polymers. It is also used as a method for engineering fire protection since it allows for the prediction of the results of some large-scale tests, particularly the flash-over time, which is essential for the length of time needed for an escape. The rate of heat and smoke emission as well as variations in mass loss and ignitibility are tracked in the cone calorimeter under a variety of radiant heat exposure conditions [6]. There have been many CACC, some of which are schematically shown in Fig. 5.20. At the start of the test, a 100 mm × 100 mm × 4 mm square specimen is mounted on a load cell and exposed to a predetermined incidence heat flux from a truncated cone radiant heater. The flux may be set up to mimic any level of fire, from a small flame to a full-fledged inferno. An electric spark ignition source powers the guided ignition. The combustion byproducts and entrained air are collected in a hood, and a blower

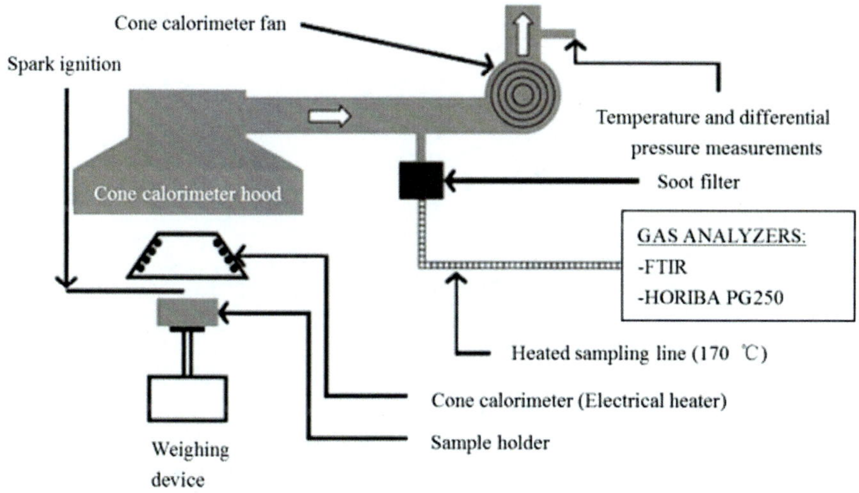

Fig. 5.20 Cone calorimeter experimental setup Reproduced with permission from Fateh et al. [74]. Copyright 2018, the Authors

removes them through a conduit. The heat release rate is calculated from oxygen concentration measurements using the oxygen consumption concept [6].

The idea behind cone calorimetry is that the quantity of heat created during combustion is related to the quantity of oxygen consumed, and this relationship may be measured by observing the decline in oxygen content of combustion gases. The proportionality constant for polymers, which are composed of organic matter, is 13.1 kJ per kilogram of oxygen used [75]. Calculating the heat release rate (heat output in terms of time and surface area-HRR) requires measuring the oxygen concentration and the gas flow (using a paramagnetic analyzer). Curves of HRR vs time that are representative of several sorts of normal burning behavior can be seen. Figure 5.21 provides examples of some [73]:

(a) Thermally thick non-charring (and non-residue generating) samples show a notable initial increase after ignition up to a quasi-static HRR value, which is equivalent to the "averaged or stable HRR." Up until the PHRR, which occurs at the end of the exam, this plateau persists. As the pyrolysis zone approaches, the glass fiber supporting the sample prevents heat transfer to the sample holder, which causes a fall in q_{cond} and the appearance of this peak.
(b) The plateau disappears for non-charring samples of intermediate thickness. There is merely a shoulder to indicate the averaged or steady HRR. Since its origin lies halfway between the thermally thick non-charring and thermally thin behavior, which is covered later, the PHRR rises in comparison with samples that are thermally thick but do not char.
(c) Prior to the development of an actual char layer, HRR shows an early rise in samples of thermally thick charring (residue generating). The thickening of the

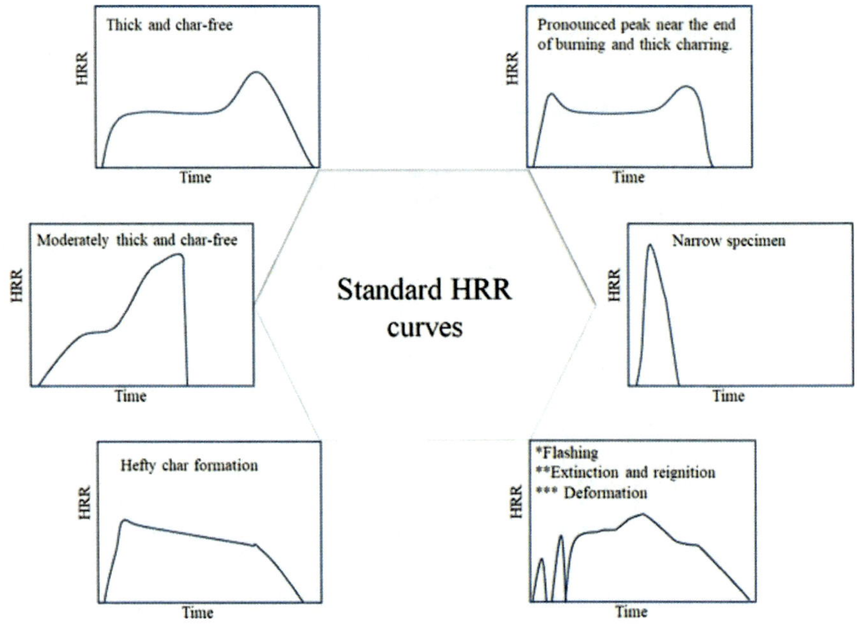

Fig. 5.21 Typical heat release rate (HRR) curves for a variety of burning behaviors

char layer causes a decrease in HRR. The peak value attained at the beginning is equivalent to the average or steady HRR and the PHRR.

(d) Wood is one example of a thermally thick charring material that frequently displays two HRR peaks: one at the start of the measurements, prior to charring, and the other at the end. The second peak may be the result of char that is cracking or an increase in the rate of effective pyrolysis, as seen with the thick, non-charring materials.
(e) Due to the simultaneous pyrolysis of the entire sample, thermally thin samples exhibit a dramatic peak in HRR. The PHRR in this situation is now reliant on the overall firing load.
(f) Some samples have HRR curves with a sort of irregular combustion development. Several causes exist, including flashing (ignition and self-extinction) before a steady flame appears or during the entire measurement or deformation during burning that alters the surface area and/or proximity to the cone heater.

Time to ignition, time to flame out, mass loss rate, total heat released, CO and CO_2 emissions, smoke density, and total smoke released are among the additional metrics that can be made concurrently. As it characterizes the rate of flame propagation and fire spread, the maximum value of the HRR curve is frequently considered when assessing the fire properties [76].

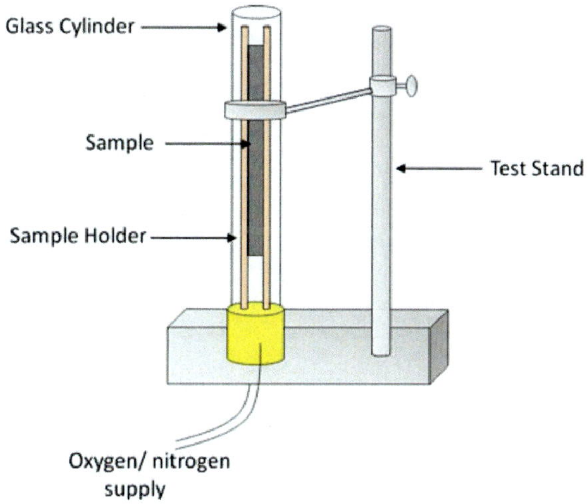

Fig. 5.22 Schematic diagram of limiting oxygen index (LOI) equipment. Reproduced from Maqsood and Seide [78]. Copyright 2020, the Authors

5.2.2 *Limiting Oxygen Index*

LOI is another crucial small-scale fire test, which was first proposed by Fenimore and Martin in 1966. The LOI index value relates to the amount of oxygen that must be present in an O_2/N_2 mixture for a vertical sample to burn steadily for three min or a length of five centimeters in order to meet ISO 4589 requirements. This states that the LOI is determined on $80 \times 10 \times 4$ mm^3 specimens that are positioned vertically in the middle of a glass chimney (Fig. 5.22). At the base of the chimney, the gaseous flow is introduced and homogenized by passing through layers of glass balls. The upper end of the specimen is lit with a burner following a 30-s purge. Materials having a LOI of under 21 are referred to as "combustible," whereas those with a LOI of over 21 are categorized as "self-extinguishing," which means that combustion can only be sustained with an external energy contribution. The value of LOI increases with the flame-retardant quality. The LOI test is characterized by its remarkable repeatability and reproducibility despite being very straightforward to perform. However, the user must be aware that LOI values are determined at room temperature and tend to fall as the temperature rises. Since this test cannot accurately predict how a polymer will behave in a real fire, it will mostly provide information on the flame retardancy of samples that may be compared [77].

5.2.3 *UL-94 Test*

One essential small-scale fire test used to evaluate the flammability of polymer materials is the vertical UL-94 test (IEC 60695-11-10). The UL-94 test (Fig. 5.23)

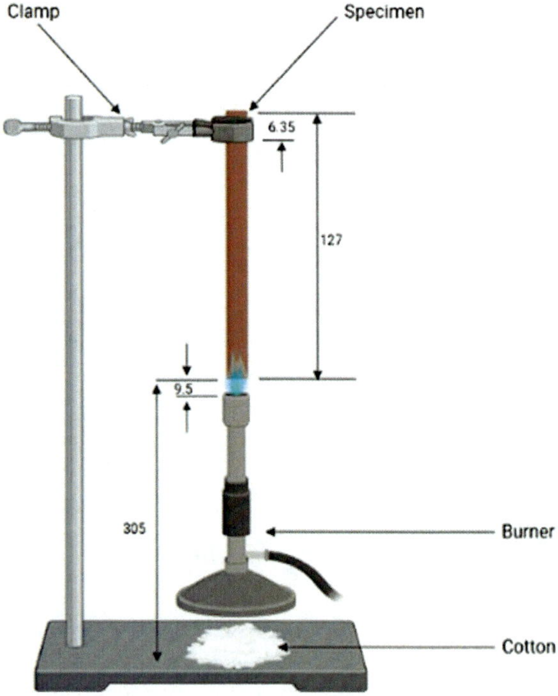

Fig. 5.23 UL-94 V flammability test parameter

considers the effects of melt flow and dripping, which are quite significant in determining the final rating. While extinction by non-flaming dripping is one effective approach to obtain the desired V-1 or V-0 classification, flaming drips drop the UL-94 classification toward HB or V-2. Thus, in the vertical UL-94 scenario, melt flow and dripping are the most significant processes controlling the fire behavior of thermoplastics, in addition to gasification of fuel, charring, effective heat of combustion of the volatiles, and flame inhibition. Achieving a V-0 or V-1 rating, a commonly used term for flame retarded polymeric materials and a prevalent need in electrical engineering and electronic applications is quite challenging in the UL-94 test. However, because the test offers a classification, it is frequently only marginally useful for future development and optimization [79].

5.2.4 Scanning Electron Microscopy

With a wide range of information available through auxiliary analysis modes, SEM is employed for surface morphology study. The quantity and quality of the chars of composite materials can be determined using SEM in flammability experiments, and this can help to explain why the flame retardance efficiency has improved or declined. Morphology is studied by scanning the surface along a line reasonably quickly. Although variable pressure ESEM permits characterization under atmospheric pressure, this technique necessitates a high vacuum for sample analysis. The ability to image at considerably higher magnification is made possible by FESEM (several lakhs). The specimen should have a conducting surface for accurate imaging during this characterization. Even so, in order to increase the generation of SE, which helps to produce a clearer image of surface morphology, materials with insufficient conductivity require coating with a very thin layer (roughly 10 nm) of highly conductive materials like Pt, Ag, Au, or AuPt alloy through sputter deposition on the sample surface.

Furthermore, the presence of the coating substance has no impact on the sample's morphological characteristics. Additionally, this coating layer shields samples sensitive to the beam, such as polymers and biological samples, from the heat produced when high-energy primary electrons impact the sample surface. However, during sample preparation for ESEM characterization, the conductive coating step may be skipped. The filament material in SEM is either tungsten or lanthanum hexaboride (LaB 6). Due to its lower work function, LaB 6 is more effective than tungsten. After passing through the condenser and objective lenses, the electron beam emerging from the filament is focused on the sample surface [80].

In flammability studies, accurate analysis of complex fire residues with respect to their chemical composition and morphology, as well as the examination of their creation during fire, is essential for improving understanding and guiding the development of flame retarded materials. SEM offers great resolution together with a good depth of field, and when an EDX is connected, it can concurrently identify the chemical composition of the microstructure at one specific place. As a result, SEM/EDX is often employed in engineering and material science. Investigating fire residues was comparatively uncommon in the past but is becoming more popular now that viscous smut, char, and slug are not the preferable samples for research in SEM equipment.

The charred remains of a fire are known to act as a barrier to heat and mass movement. The effectiveness of such a residual protection layer is dependent on the features of the char, including its shape, which impacts heat conductivity and gas permeability, in addition to the amount of char present. In actuality, it was discovered that these protective qualities existed largely irrespective of the quantity of thermally stable char [81]. Inorganic additives and synergists can also be added to the residue to modify its characteristics [82]. As a result, the fire residue's intricate, diverse, or progressive morphology is crucial. Creating a fire residue with multicellular structures or closed glassy surface layers is a viable strategy for flame retardancy. Additionally, such physical processes directly impact certain fire properties and serve different purposes in various fire tests [81]. When compared to other fire dangers like total heat evolved, residual protective layers significantly influence characteristics like heat release rate.

The char residues following the LOI test [83], I—Cone [84–86] and UL-94 tests [87, 88] can be examined by SEM in order to gain a better understanding of the microstructure and function of the protective chars.

5.2.5 *Transmission Electron Microscopy*

In contrast to SEM, which detects expelled electrons from sample surfaces, TEM uses an electron gun fired from a filament to pass into the bulk of the sample and collect data on its morphology. Voltage can be supplied up to 200 kV for HRTEM. The distribution and dispersion of graphene-based nanoparticles can be examined from very thin (up to around 10 μm) sections of the sample. The ultra-cryo-microtomy process, which involves cutting thin sections of samples with a diamond knife and storing them in a liquid nitrogen atmosphere to increase their stiffness, is frequently used to prepare samples of graphene-based polymeric materials. Before analysis, the thin pieces are mounted on a copper grid [80].

The dispersion state of graphene-based nanoparticles in the polymer matrix significantly impacts the reinforcement of the integrated performance of polymer nanocomposites. For example, modification of the surface of GO with hyper-branched flame-retardants has been done in order to help improve the interfacial compatibility between the FGO sheets and PS, which is why FGO was well dispersed in the PS matrix. The interlayer spacing of the FGO sheets was noticeably larger than that of the GO sheets following chemical functionalization, which was advantageous for the exfoliation and uniform dispersion of graphene sheets in PS. The PS nanocomposites' improvement in overall performance was greatly aided by FGO's high compatibility and dispersion in the PS matrix [89].

While Hu and colleagues noticed that GO sheets have a folded and wrinkled structure with a few micrometers in diameter when dispersed in XLPE. Compared to GO as per Fig. 5.24a, FGO's surface was rough and covered in addends, which was attributed to the flame-retardant grafted onto GO as depicted in Fig. 5.24b and c [90]. The structural morphology of FGO is presented in Fig. 5.24d.

5.2.6 *XPS*

XPS is a spectroscopic technique that can shed light on the surface of solid materials at the atomic and molecular levels. When the sample surface is subjected to X-ray radiation, surface atoms release electrons after absorbing all of the photon energy. This process is known as XPS. The parameters utilized in an XPS experiment and Eq. 5.49, applied with an additional term, have a very similar relationship [91]:

$$\mathrm{KE} = hv - \mathrm{BE} - w, \tag{5.49}$$

where w is the spectrometer work function.

The KE of the electron is an experimental quantity that the spectrometer can measure, but this is reliant on the photon energy of the X-rays utilized; the electron's BE is the parameter that specifically identifies it, both in terms of its parent element and atomic energy level. Because no two elements have the same binding energy,

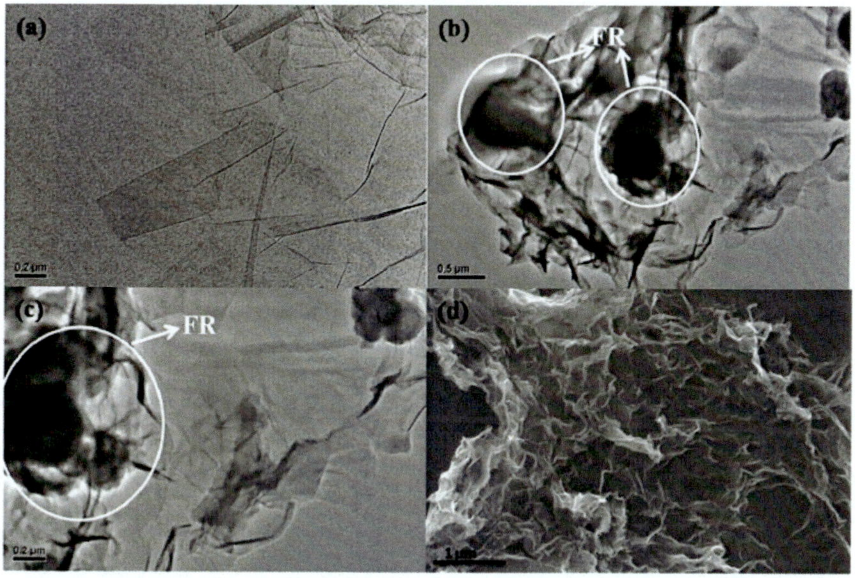

Fig. 5.24 TEM photographs of GO (**a**), FGO (**b**, **c**) and SEM photographs of FGO (**d**). Reproduced with permission from Hu et al. [90]. Copyright 2014, American Chemical Society

the observed kinetic energy allows for the elemental analysis (except for hydrogen and helium). Calculating the electron's BE from Eq. 5.49 is easy because all three other variables are known or at least quantifiable. The user needs to select between a binding energy scale and a kinetic energy scale, depending on which is deemed more appropriate, because the spectrometer's control electronics or data system will conduct this task. Figure 5.25 presents a diagram of the photoemission process.

The diagram (Fig. 5.25) illustrates an incident X-ray striking the material and discharging an electron from an interior shell of the atom. All electrons with a binding energy smaller than the photon energy (hv) will appear in the photoelectron spectrum,

Fig. 5.25 Scheme showing the principle of XPS surface analysis

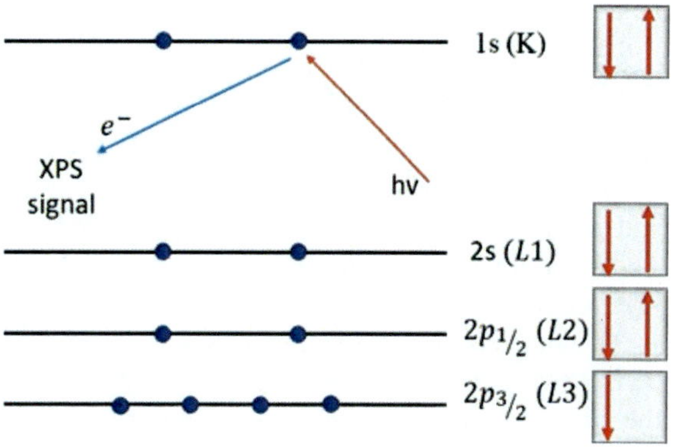

which will recreate the electronic structure of an element rather well. The photoelectrons lines from the core levels are the strongest in an XP spectrum due to the high energy of X-ray radiation. Unresolved lines of low intensity are also present in the region of low binding energy and are caused by photoelectron emission from molecular orbitals (valence band). The characteristic peaks of the spectrum are produced by excited electrons that escape without releasing energy, whereas the background of the spectrum is produced by excited electrons that undergo inelastic scattering. The former is obtained from what is referred to as "sampling depth," which is established by the IMFP of the electrons through the sample and by geometrical parameters. This fluctuates within a few nanometers and determines the surface sensitiveness of XPS.

Various types of carbon, including graphite, graphene, carbon nanotubes, and amorphous and diamond-like thin films, have been analyzed using XPS, a remarkably potent method for characterizing the surface layers. The use of XPS for the analysis of chars has been covered in several articles [92]. The elemental composition of the material surface is identified in these analyses using the binding energy shifts linked to different heteroatoms, such as oxygen, nitrogen, and fluorine. The binding energy shifts associated with particular heteroatoms, such as oxygen, nitrogen, and fluorine, are used in this research to determine the elemental composition of the material's surface. The main carbon–carbon peak, which is the most common heteroatom investigated, is set at 284.4 eV by a rather simple peak fitting for carbons that contain oxygen (here referred to as deconvolution). The shift of the hydroxyl/ether, carbonyl, and carboxylic groups is 1.5, 3, and 4.5 eV greater, respectively. The oxygen concentration determined by the C1s spectra has a fundamental fault in that it rarely agrees with the oxygen concentration determined by detecting each constituent peak separately. The primary carbon–carbon peak is set at 284.4 eV by a relatively straightforward peak fitting for oxygen-containing carbons (referred to here as deconvolution), by far the most common heteroatom studied. Hydroxyl/ether, carbonyl, and carboxylic groups are shifted about 1.5, 3, and 4.5 eV higher, respectively [93]. This methodology has a basic flaw in that the oxygen concentration obtained from the C1s spectra is rarely compatible with the oxygen concentration obtained by separately measuring each constituent peak [92].

5.2.7 *Fourier Transform Infrared (FTIR) Spectroscopy*

As an easy, trustworthy, and affordable method for the chemical study of both organic and inorganic molecules, IR spectroscopy is widely utilized in research and industry. The ability to analyze samples in many states, such as liquids, solids, and gases, is one of the major benefits of IR spectroscopy [94].

The interaction of EM radiation in the IR part of the spectrum (wavelength range, 0.78–1,000 μm; wavenumber range, 10–12,820 cm^{-1}) with a molecular system is the fundamental concept behind IR spectroscopy. This interaction brings about transitions between vibrational levels. The harmonic oscillator is a fundamental model for understanding molecular vibrations. It considers molecules' constituent atoms to be massless spheres [94].

According to Hooke's law, the restoring force of a spring F is proportional to the displacement x of the atoms from their equilibrium position when considering a diatomic molecule (m_1 and m_2 atom masses):

$$F = -kx, \tag{5.50}$$

where k is a measure of the bond strength and is expressed as the spring's force constant (in N/m).

The harmonic oscillator's vibrational frequency ($\sim v$ wavenumber, in cm^{-1}) according to classical mechanics is then determined by

$$v = \frac{1}{2\pi c}\sqrt{\frac{k}{\mu}}, \tag{5.51}$$

The reduced mass (μ) is equal to $m_1 m_2 = (m_1 + m_2)$, k is the force constant, and c is the speed of light.

Only modifications of $\pm$ 1 in the vibrational quantum number are permitted for harmonic oscillators ($n = \pm 1$). An anharmonic function, however, is a more accurate approach to explaining the real change of the potential energy as a function of the displacement of the atoms from their equilibrium location. Because of the anharmonicity, the selection rule is loosened, permitting transitions with $|n| > 1$. Thus, in addition to absorption bands, fundamental transitions can result in additional peaks in the mid-IR spectrum, such as overtone and combination bands ($n = \pm 1$). The most prevalent terms used to describe the motion of atoms during vibration are normal modes of vibration (q_i), in which all of the atoms in the molecule vibrate at the same frequency, and all pass through their equilibrium positions simultaneously. The molecular dipole (μj) connected to the regular mode of vibration must change for a substance to be IR-active (i.e., $\delta\mu/\delta qi \neq 0$). We can distinguish between stretching and deformation vibrations based on the direction of the vibrational movement. The second category can be further divided into bending, twisting and scissoring, wagging, and rocking movements. The symmetry of the vibration is referred to as a further subdivision (i.e., symmetric or antisymmetric, in-plane or out-of-plane) [94].

The usage of flame-retardants in polymer nanocomposite research has involved the use of FTIR. Identifying materials based on graphene and verifying their modification are good examples where FTIR is useful. In the case where GO has been functionalized with a highly branched polymer by Hu et al. [90] utilizing AEPZ and di(acryloyloxyethyl)methylphosphonate. By utilizing FTIR (Fig. 5.26a), they validated the covalent bonding of the hyper-branched flame-retardant grafted onto GO. According to earlier research, the FTIR spectrum of GO's peak at 1730 cm^{-1} was attributed to the stretching vibration of CO, and the characteristic absorption bands at 1620, 1065, and 1220 cm^{-1} were attributed to the stretching vibrations of

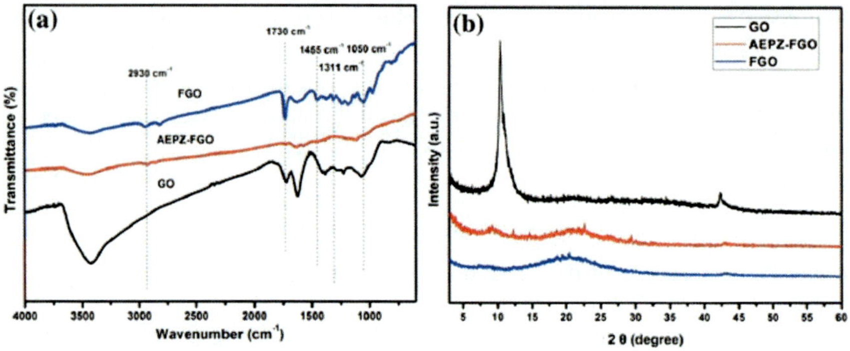

Fig. 5.26 FTIR (**a**) and XRD (**b**) spectrum of GO and functionalized GO. Reproduced with permission from Hu et al. [90], Copyright 2014, American Chemical Society

sp^2 hybridized carbon, C–O, and C–OH, respectively [95]. The new peaks at 2930, 1455, and 1510 cm^{-1} that occur when GO was functionalized with AEPZ were attributed to the stretching vibration of $-CH_2-$, the C–N, and N–H groups, respectively, in AEPZ. They demonstrated that the reduction of GO significantly reduced the C–O peak's strength. The outcomes confirmed that the grafting of AEPZ onto the surface of GO had been successful. The FGO spectrum (Fig. 5.26a) demonstrates the presence of the following specific absorptions: the stretching vibrations of P-O, which have symmetric and asymmetric peaks at 1256, 1050, and 880 cm^{-1}, respectively; peaks at 1640, 1730, and 2930 cm^{-1} are attributed to the stretching of the C=C, C=O, and $-CH_2$-group, respectively, while the peak at 1311 cm^{-1} was ascribed to the deformation vibration of the P–C group [90, 96, 97].

5.2.8 *Raman Spectroscopy*

The technique of RS is employed to investigate the composition, dynamics of change, and functions of various materials. Data with a high spatial resolution is usually collected in conjunction with microscopy [98]. The vibrational method of RS does not use any markers. As a result, it facilitates understanding of the structure of materials and permits an analysis of their molecular composition, playing a crucial role in research [99]. When a photon of input light strikes a particle and creates a scattered photon (Fig. 5.27), RS is intended to quantify the frequency shift of inelastic scattered light [100]. The inelastic scattering from chemically linked structures results in bands in a Raman spectrum [101], while the bands in the acquired Raman spectrum are arranged in accordance with the frequency at which the sample's constituent parts vibrate. Each organic compound and functional group have a distinctive vibration frequency that can be seen as a peak in the Raman spectrum [102]. The distinctive spectral pattern serves as the "fingerprint," allowing us to recognize the molecule,

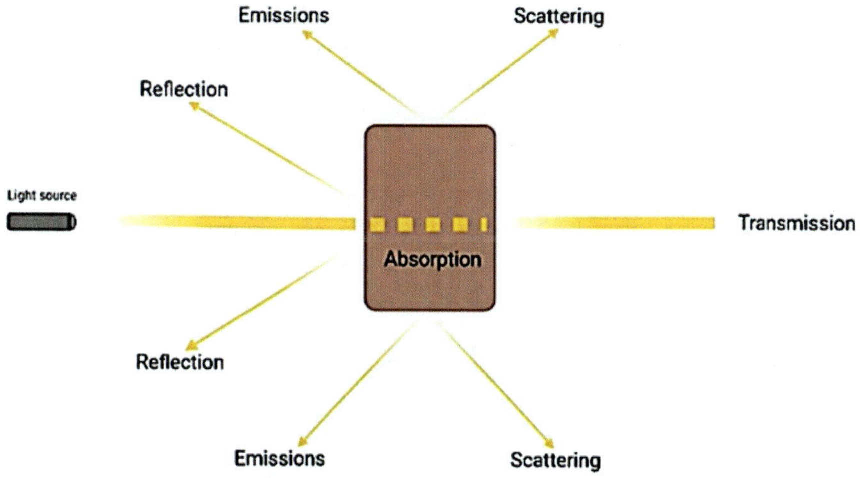

Fig. 5.27 Various optical phenomena brought on by light-medium interaction

and the intensity of the bands allows us to determine its concentration in the material under study [98].

The light scattered by the test material almost always has a component with the same frequency as the input light (Rayleigh scattering, elastic scattering) [103]; however, on occasion, there are variable frequency components connected to the change in photon energy (inelastic scattering, Raman scattering). Stokes Raman scattering is the process of scattering light at a frequency lower than that of the incident photon. In contrast, anti-Stokes Raman scattering is the scattering process at a frequency greater than that of the incident photon. The scattered photon will have a bigger energy by comparing the energy differences between the oscillating energy levels [104, 105]. When the bond of the molecule is originally in an excited vibrational state, anti-Stokes Raman scattering allows the photon to gather energy from the bond [106]. The processes of Raman and Rayleigh scattering are depicted schematically in Fig. 5.28.

In flammability studies, Raman scattering spectrum has been used to assess the quality of the residual char after the cone calorimeter test. The maxima at 1342 and 1596 cm^{-1} are corresponding to the D and G bands, respectively. The D band is associated with defects in amorphous carbon or disordered graphite, whereas the G band is correlated to crystalline graphite or ordered carbon structure. In graphitic materials, the integrated intensity ratio I_D/I_G for the D band and G band is frequently employed to characterize the defect quantity. It is known that a lower I_D/I_G relates to a higher proportion of graphitized carbon atoms in the char. Higher graphitization degree char can be produced with greater thermal stability and better protection of the char layer [107]. In a study by Yuan and colleagues, it has been demonstrated using Raman spectra that the degree of graphitization of the char rises with increasing graphene content in polymer materials. They have indicated that the more graphene residues there were in the char layer, the higher the graphene content. Thus, it has

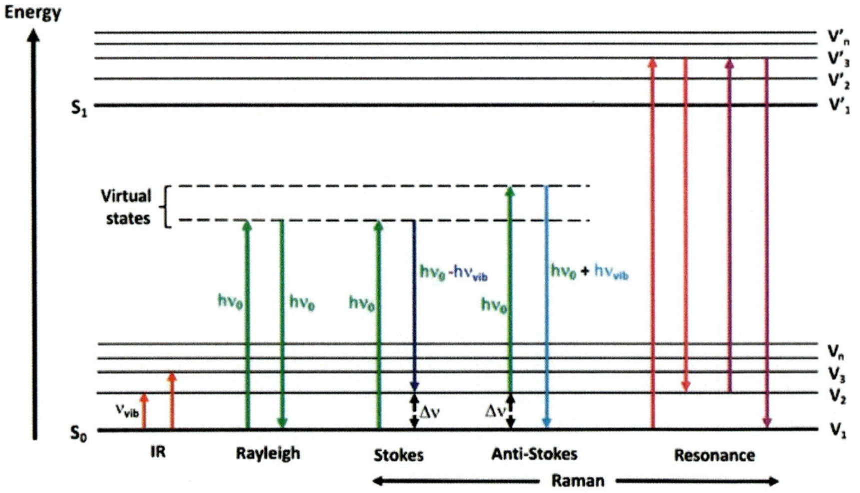

Fig. 5.28 Jablonski energy diagram showing the transitions involved during infrared absorption, Rayleigh, Raman Stokes, anti-Stokes and Resonance Raman scattering. Reproduced from Saletnik et al. [105]. Copyright 2021, the Authors

been established that graphene nanosheets can serve as a template to enhance the graphitic order of the char [85].

5.2.9 X-Ray Diffraction (XRD)

All solids can be roughly categorized into either amorphous or crystalline phases when taking into account the crystalline nature of any material. In crystalline solids, atoms and molecules are distributed in all directions in a regular, periodic fashion, whereas in amorphous materials, this arrangement is random. Each crystalline substance emits a distinct pattern, whereas the same substance emits the same pattern over and over again [108]. X-rays are diffracted and generate a diffraction pattern when they come into contact with a crystalline substance. A pure substance's WXRD pattern is similar to that substance's fingerprint. Today, standards for individual crystalline phase diffraction patterns of over 50,000 inorganic and around 25,000 organic single components have been collected and preserved on magnetic or optical media [109].

Every crystalline substance is composed of an ordered periodic arrangement of atoms. X-rays may have a wide range of effects interacting with crystalline material. Nucleus scattering is kept to a minimum by the elastic or inelastic electron scattering of the material. When the energies of an incoming and an outgoing photon are equal, elastic scattering, also known as Thompson scattering, occurs. Emitted radiation might experience either positive or negative interference. Constructive interference diffraction peaks are gathered and examined.

Contrary to destructive interference, constructive interference abides under Bragg's law. At an angle of q with the tangential surface; the sample is subjected to an X-ray beam with wavelength (λ). The mathematical form of this expression is Bragg's law, which can be demonstrated as follows [110]:

$$2d\text{Sin}\theta = n\lambda, \tag{5.52}$$

where n is an integer, λ represents the radiation's wavelength, d represents the distance between atomic planes, and θ represents the angle between the radiation and the planes it interacts with (Bragg angle). If we rewrite Bragg's law, we obtain

$$\text{Sin}\theta = \frac{n\lambda}{2d}. \tag{5.53}$$

There are two commonly used XRD techniques: the Laue method, which varies λ while maintaining θ, and the powder diffraction method, which maintains λ constant while changing θ. In both situations, the diffraction pattern of the material is determined by measuring the intensity of the diffracted X-ray beam against diffraction angle 2θ. While amorphous solids exhibit broad maxima (humps) instead of sharp maxima (peaks) at the appropriate diffraction angles, crystalline materials exhibit sharp maxima (peaks) at the corresponding diffraction angles.

Miller indices (*hkl*) are used to distinguish between a crystal's different crystallographic planes. The reciprocal values of the intersection of a particular plane with the axes of the crystallographic unit cell are related to these three integral numbers. A cluster of dots represents the diffraction pattern when radiation is applied to a single crystal, and these diffraction spots have a circular form when polycrystalline [110].

Polymers are neither totally crystalline nor fully amorphous; hence, the degree of crystallinity a measurement of how much of the substance is crystalline is used to express the crystallinity. Half of the total intensity is made up of the intensity of the diffracted beam from the crystalline segment (I_c) and the intensity of the diffracted beam from the amorphous region (I_a). The crystallinity (Q) degree is assessed using the following equation:

$$Q = \frac{I_c}{(I_c + I_a)}. \tag{5.54}$$

In flammability studies, XRD can be used to confirm the structural properties of the initial materials such as polymers, graphene-based materials of flame-retardants, effects of modification on the structural properties of materials [90, 111], etc., it can also be used to study exfoliation of nanoparticles in polymer matrices or the structural properties of char materials produced [112].

A great example where the initial properties of graphene-based materials have been studied is shown in Fig. 5.29. In this study, GO is modified with PA. The

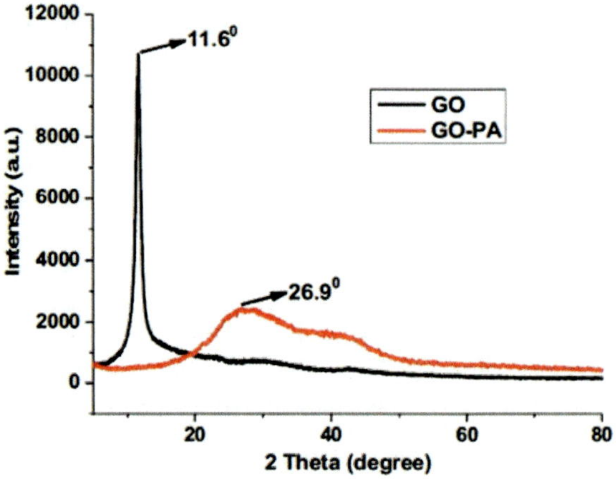

Fig. 5.29 XRD spectra of GO and GO-PA. Reproduced with permission from Kundu et al. [111]. Copyright 2020, Elsevier Science Ltd.

usual diffraction peak at $2\theta = 11.6°$ was identified as belonging to GO in Fig. 5.29 [113]. The spectra of GO-PA demonstrated a broad and diminished peak at 26.9°, indicating a narrower interlayer distance than pure GO, further demonstrating the effective functionalization of PA with GO. Additionally, the diffraction peak intensity of GO-PA, which was lowering, demonstrated that the GO and PA were successfully mixed [114] and reduced and partially destroyed the usual GO stacking [111].

Cheng and colleagues have used XRD to study the char materials from flame-retardant polypropylene (PP)/EG composites with PDPFDE at different temperatures, as shown in Fig. 5.30a and b [112]. The peaks at 26.26 and 55.12 were the properties of the (002) and (004) lattice planes of graphite, and for EG, there was essentially no difference between the two states at room temperature and after heating. The XRD pattern for PDPFDE was noticeably altered after heating. The peaks at $2\theta = 30.18°$, 35.64°, and 62.98° were attributed to (220), (311), and (440) of Fe_3O_4, respectively [115]. The Fe_3C reflections with peaks at 42.89°, 43.82°, 44.70°, 45.05°, and 45.96° had the characteristics of (121), (210), (022), (103), and (211) [116]. Additionally, it contained the graphite phase at reflection at (002) and (004). Although demonstrating lower intensity, the EG/PDPFDE had the same peaks as PDPFDE. Additionally, Fig. 5.30c showed a comparison of XRD analysis of residue char from PP72/PP-g-MA3/EG25 (PP1) and PP72/PP-g-MA3/EG20/PDPFDE5 (PP3) after heating. The graphite peak was present in each of them, but only PP3 revealed weak Fe_3O_4 and Fe_3C phases appearing in the same regions as the EG/PDPFDE system. PP3 produced more denser residue char than PP, while the residual chars of PDPFDE, EG/PDPFDE, and PP3 all had the same phases of Graphite, Fe_3O_4, and Fe_3C. These findings demonstrated that PDPFDE in PP3 significantly enhanced fire retardancy.

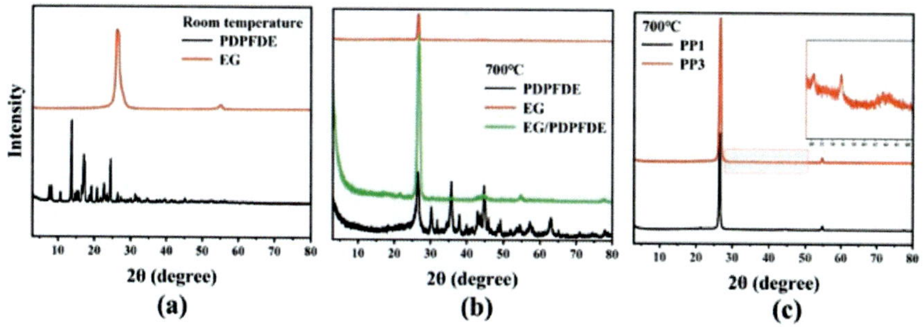

Fig. 5.30 XRD patterns of **a** EG and PDPFDE at room temperature, **b** EG, PDPFDE and EG/PDPFDE after 700 °C heated in a furnace under N_2 atmosphere, **c** PP1 and PP3 after 700 °C heated in furnace under N_2 atmosphere. Reproduced with permission from Cheng et al. [112]. Copyright 2020, Elsevier Science Ltd.

5.2.10 *Mechanical Properties*

Mechanical characterization of the composites is crucial for product design, analysis, and life prediction. The performance of nanocomposites made from various graphene-based materials has improved significantly. Moreover, the mechanical characteristics of the polymeric nanocomposites are mostly controlled by these nanofillers and the improvement in the mechanical properties of the nanocomposites, such as tensile strength, elongation at break, and Young's modulus, can be attributed to the high stiffness of the nanofillers, their high aspect ratio, and their affinity for the polymer matrix. Compared to the standard materials that are typically utilized for structural purposes, nanoparticles have better characteristics. However, there is room for future advancement in this field of materials. Therefore, there is a goal for the manufacturing process, process cost reduction, application, and material properties [80, 117].

During the characterization of tensile properties, a specimen (Fig. 5.31) is positioned in the grip of the tensile testing machine, and the test can be run at a crosshead speed of ~10 mm/min or as necessary. To determine the elastic modulus, load deformation or load strain curves are often plotted or digitally captured throughout the test. The constant strain rate is specified to be influenced by the testing speed. The strain rate should be chosen so that the samples rupture within 1–10 min. The standard strain rate for the strain-controlled tests is 10 min^{-1}, and the standard head displacement rate for the constant head speed tests is 2 mm/min [80].

Before being examined at the same level of preparation, the specimens are prepared in line with the proper moisture profile. Tests conducted under modified settings need to be documented. Additionally, the samples must be stored in a climate-controlled environment while being tested. The specimens are placed in the grips of the tensile testing apparatus with their axes oriented in the direction of the test. The grips are tightened, and the pressure applied to them is recorded [80, 118].

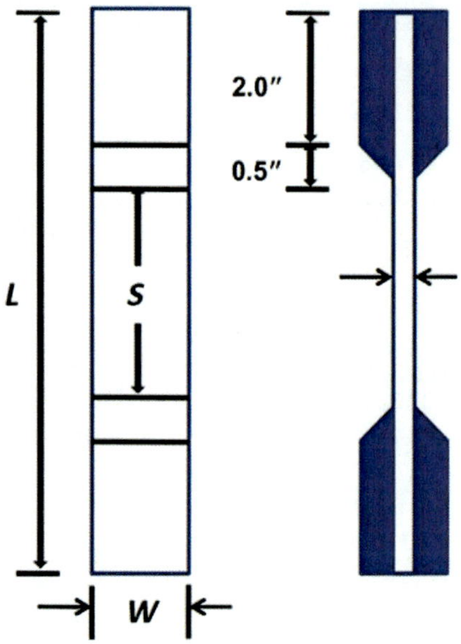

Fig. 5.31 Tensile testing samples as per ASTM D3039. Reproduced with permission from Dey et al. [80]. Copyright 2020, Elsevier Science Ltd.

Agglomerates are frequently formed when additives of the graphite type are used; in certain works, this occurrence is minimized by using ultrasounds to enhance particle dispersion within a solvent before the mixing procedure [119, 120]. In a study by Marset et al. [121], the mechanical behavior of bio-based polyamide 1010 (PA1010) nanocomposites with an increasing proportion of EG was examined. They discovered that adding EG to create composites enhanced the stiffness from the original PA1010's 1701 MPa to the final PA1010/10EG composite's 2164 MPa. The tensile strength for PA1010 was 45.5 MPa; other values varied depending on the amount of EG considered. The PA1010/2.5EGr composite's tensile strength was 50.5 MPa, demonstrating the reinforcing effect. The tensile strength was decreased to 48.4 and 46.5 MPa, respectively, for the 5 and 7.5 wt.% composites due to the difficulty in attaining a good dispersion. Despite the drop relative to PA1010/2.5EG, a better tensile strength was recorded than that for the neat PA1010 polymer. Unlike the plain PA1010, the composite with the maximum amount of EG displayed a tensile strength of 42.9 MPa, demonstrating that the effect of the agglomerates was more significant than the reinforcement obtained with the EG. Although it was possible to improve the material's stiffness and strength, the elongation showed a diminishing tendency as a function of the quantity of EG taken into account. The sample containing 10 wt.% of EGr had an elongation at break of only 5.7% as opposed to 237.4% for PA1010.

Numerous variables could be to blame for the composites' lower elongation after the addition of EG. Since the additive was poorly diffused, the addition of EG reduced the mobility of the polymer chains and prevented the formation of aggregates, which favors the loss of ductile properties. The presence of EG has also been reported as the origin of agglomerate formation due to the dispersion problems that arise, leading to stress concentrators and microcracks that result in the loss of ductile properties [122].

The test outcomes may help identify potential uses or verification of intended applications for flame-retardant polymer nanocomposites.

5.2.11 *Thermal Gravimetric Analysis*

An experimental technique called TGA determines a sample's mass loss as a function of temperature or time. The sample may be heated at a constant heating rate (referred to as dynamic measurement), kept at a constant temperature, or exposed to nonlinear temperature programs like those used in sample-controlled TGA (also known as SCTA) research. A temperature program will be selected according to the type of information about the sample that is required. Furthermore, the environment of the TGA experiment, which may be reactive, oxidizing, or inert, has a considerable impact. It is also possible for the atmosphere to shift during a measurement. A TGA curve that depicts the relationship between mass or mass percentage and temperature and/or time is typically used to represent the results of a TGA measurement. A different and supplementary presentation can be made by using the first derivative of the TGA curve as a function of temperature or time. The differential thermogravimetric curve, also called the DTG curve, shows the rate of mass change. When the sample loses components in various ways or reacts with the environment, mass changes happen, resulting in peaks or steps in the DTG curve or TGA curve [123].

5.2.12 *Tga-Ftir-Ms*

TG-FTIR-MS has been used to examine the gaseous byproducts of the polymer composites during thermal degradation in order to investigate the mechanism of enhanced flame retardancy [89]. The off-gassing materials are transferred via a transfer line to a gas cell, where they are exposed to infrared light in TGA-IR [124]. By measuring the light that is absorbed by various molecular vibrations, infrared spectroscopy can identify these vibrations. Infrared light (Fig. 5.32) from the light sources is divided into two beams by the beam splitter. A fixed mirror and a moving mirror will each receive reflections from a different beam. Next, the two beams are merged again and directed through the sample and toward the detector. A mid-infrared transmission spectrum is produced by Fourier transformation of the resulting interferogram [125]. The functional groups contained in the gases produced when the material deteriorates can be determined from the spectra.

Mass spectrometry (Fig. 5.33) is an effective analytical technique used to measure known materials, identify new chemicals within a sample, and clarify various

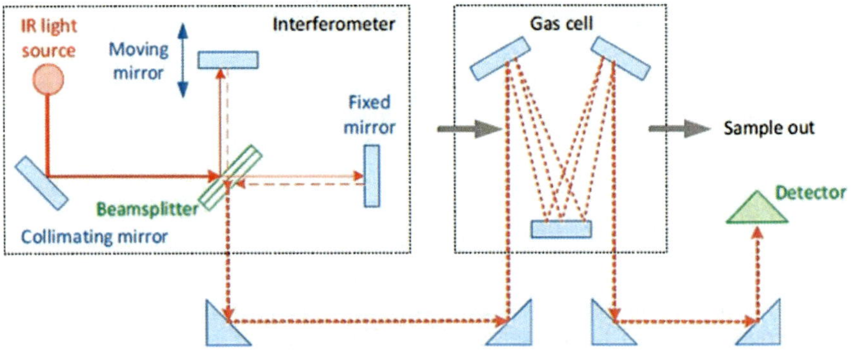

Fig. 5.32 Principle of operation of Fourier transform infrared (FTIR) spectroscopy. IR = infrared. Reproduced from Giechaskiel and Clairotte [126], Copyright 2021, the Authors

molecules' structures and chemical properties. The material is commonly transformed into gaseous ions throughout the entire process, either with or without fragmentation, and these ions are then classified according to their relative abundances and mass-to-charge ratios (*m*/*z*) by a combination of magnetic and electrostatic fields. The field strength is scanned to cause ions with an increasing m/z ratio to enter the detector, which records a mass spectrum. When the TGA is connected to mass spectroscopy, the *m*/*z* ratios are determined, offering a quantitative confirmation of the different gases that are emitted as the polymer material degrades, thus, a more precise composition of the degradation products/gases [125].

A good example of where TGA-IR was utilized to study the evolution of gases in synergistic intumescent flame-retardants (APP) for PP graphene nanocomposites by Yuan et al. [85]. Their research plots the total pyrolysis gas intensity against time curves for PP75/APP25 and PP75/APP19.2/G1.0 using Gram-Schmidt curves (Fig. 5.34a and b). The Gram-Schmidt curves had two main peaks.

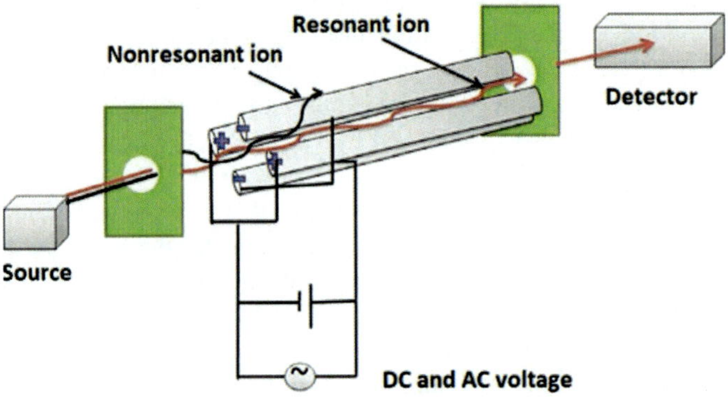

Fig. 5.33 Schematic diagram of MS. Reproduced with permission from Ng et al. [125], Copyright 2002, Wiley

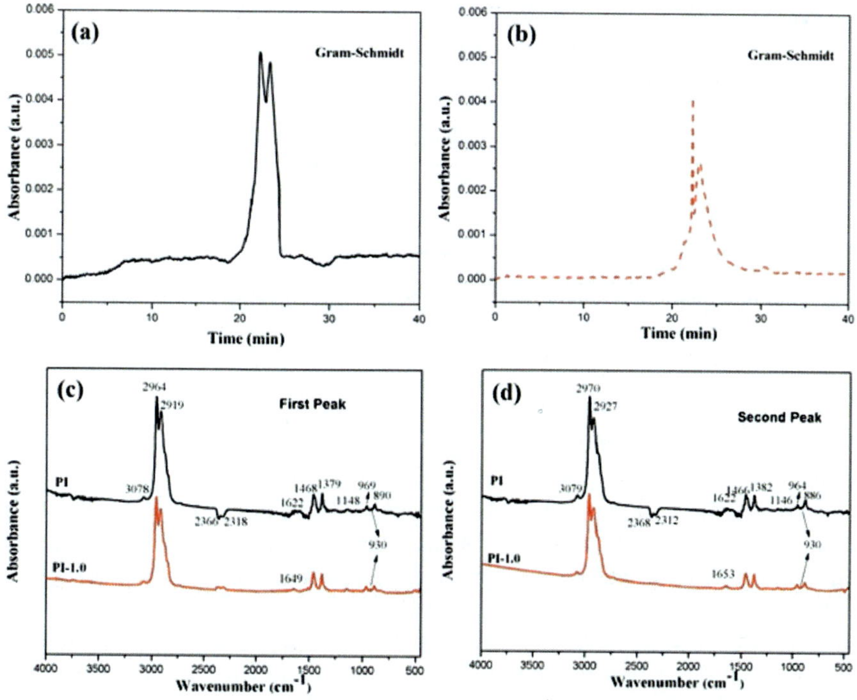

Fig. 5.34 Gram-Schmidt curves of **a** PP75/APP25 and **b** PP75/APP19.2/-1.0; FTIR absorption spectra of the pyrolysis products at the **c** first and **d** second maximum decomposition rate. Reproduced with permission from Yuan et al. [85]. Copyright 2017, Elsevier Science Ltd.

The first and second peak decomposition rates of the pyrolysis products' FTIR spectra are shown in Fig. 5.34c and d, respectively. PP75/APP25 and PP75/APP19.2/G1.0 evolved gas products showed recognizable bands of saturated hydrocarbons (2964, 2919, 1468, 1379, 1148, and 969 cm^{-1}), unsaturated alkane (3078, 1622, and 890 cm^{-1}), CO_2 (2366 and 2318 cm^{-1}), and NH_3 (2366 and 2318 cm^{-1}). At the first maximum decomposition rate was seen from Fig. 5.35c and d that the breakdown products of PP75/APP25 were comparable to those of PP75/APP19.2/G-1.0. This study supported the notion that graphene served as an inert filler and had no impact on the thermal degradation of PP75/APP25. In contrast, the CO_2 signals in the PP75/APP19.2/G-1.0 spectrum entirely vanished at the second maximum rate. The escape of CO_2 and other volatile IFR breakdown products was slowed down by graphene nanosheets' outstanding barrier function and high dispersion.

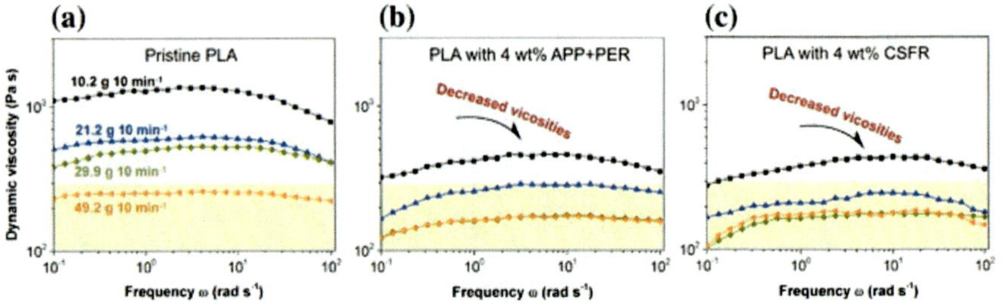

Fig. 5.35 Rheological curves of PLAs. **a** Pristine PLA, **b** the corresponding PLA with 4 wt.% APP + PER, and **c** the corresponding PLA with 4 wt.% CSFR. Reproduced with permission from Yu et al. [131]. Copyright 2022, Elsevier Science Ltd.

5.2.13 Smoke Density

A polymer's flammability and the toxicity of its fire effluents both pose fire risks. It will be more challenging to escape if smoke is released during combustion because it could result in injuries and vision obstruction. Consequently, polymer smoke suppression is quite important. Smoke release curves for neat polymer materials like PP show that the polymer releases smoke soon after burning, and as a consequence, the curve flattens out [85]. Although the smoke data were first gathered in percentage light absorption, they were instantly changed to optical density "OD," which is a more exact concept for evaluating light transmission metrics. Greater smoke suppression ability is generally indicated by lower smoke density. The SDC is usually used to conduct the test, which consists of a sealed test compartment, an electrical furnace, a gas burner, a PM system, and a collimated light beam that travels through the chamber. The exposed sample surface, which is horizontally positioned, receives a steady heat flow from the electrical furnace with a capacity of 25 kW/m^2. The sample holder must be filled with the 76 mm × 76 mm test sample that has been wrapped in aluminum foil. The sample holder's purpose is to reveal a surface (65 mm × 65 mm). Each sample is put through a horizontal test. The PM system monitored the opacity of the smoke produced after it had been allowed to accumulate inside the cage. According to the technique outlined in ISO 5659, the apparatus calibration and testing procedure were carried out in non-flaming conditions [127]. The light transmission was measured every five seconds after burning the sample inside the containment.

A test sample created in a fixed-volume chamber under controlled conditions is used to obtain data on the loss of light transmittance through the smoke and effluent. As shown in Eq. 5.55 [128], the percentage of total transmittance (T) is used to measure the attenuation of a beam of light traveling through smoke.

$$T = \frac{I}{I_o}100, \tag{5.55}$$

where I and I_o stand for the intensity of the transmitted light in the presence and absence of smoke, respectively. The quantitative parameter known as the degree of smoke opacity, or OD (De), which is frequently given in decibels (*db*), is used to calculate the smoke concentration [128]:

$$D_e = 10 \log_{10} \frac{I_o}{I}. \tag{5.56}$$

For a particular path length, the OD is always provided. The OD per meter (DL), *db/m* is as follows:

$$D_L = \frac{10}{L} \log_{10} \frac{I_o}{I}. \tag{5.57}$$

It is advised to use the following relationship between the OD, DL, and visibility in meters (S) through the smoke, as in [129]:

$$S = \frac{1}{D_L}. \tag{5.58}$$

Beer's rules are employed in order to convert the percentage of transmittance acquired on the PM system to a particular OD (D_s):

$$D_s = G \log_{10}\left(\frac{100\%}{T}\right), \tag{5.59}$$

where G is the geometrical chamber factor, which is directly related to the volume of the chamber in which smoke is accumulating (0.5 m^3) and indirectly proportional to the optical path length of light (0.941 m) and the exposed sample surface area creating smoke (65 mm × 65 mm).

5.2.14 Rheology

The study of matter's deformation and flow is known as rheology. Thus, the study of the flow and deformation of polymeric materials is known as polymer rheology. Given the wide range of polymeric materials, we can further categorize polymer rheology into various groups depending on the properties of the polymeric materials. For example, (i) the rheology of homogeneous polymers, (ii) the rheology of miscible polymer blends, (iii) the rheology of immiscible polymer blends, (iv) the rheology of particulate-filled polymers, (v) the rheology of fiberglass-reinforce, (vi) organoclay nanocomposites rheology, (vii) polymeric foams rheology, (viii) thermosets rheology, (ix) block copolymers rheology, and (x) liquid-crystalline polymers rheology. These polymeric materials each exhibit unique rheological characteristics.

In order to explain the experimental results on the rheological behavior of different polymeric materials, multiple hypotheses are needed [130].

In this direction, Yu et al. [131] developed a series of PLA matrices with varying melt flow indexes (MFI), both with and without fire retardants and investigated how the melt viscosity influences a material's fire performance. Pristine PLAs' viscosities were inversely correlated with MFI values (Fig. 5.35a). When exposed to heat or flame, the PLA with a lower viscosity produced more melt drippings and polymer fluid more readily.

5.2.15 Melt Flow Index

The melt viscosity of polymer materials is one of the key factors affecting the development of the char layer. MFI value, which is inversely related to the melt viscosity of polymers, is used to measure melt viscosity indirectly. To correlate the measure of melt viscosity, as depicted in Fig. 5.36, Yuan et al. [85] used the MFI values of PP and its composites at various temperature points. They discovered that as the heating temperature was increased, the MFI values also increased, indicating a rise in melt mobility. The sudden decline in MFI for the neat PP sample was seen at 300 °C and was explained by the oxidation cross-linking reaction [132]. Indicating the plasticization effect of IFR, the MFI values of PP75/APP20/CFA5 (PI) were higher than those of PP. At the same time, a rigid graphene nanofiller can potentially increase the networks of melt viscosity of polymer matrices [133]. According to Fig. 5.36, the MFI values of PI gradually decreased as rGO concentrations were raised. A proposed anti-dripping mechanism in UL-94 and LOI tests was attributed to the increase in melt viscosity brought on by graphene nanosheets [85, 134].

It has been further established that the IFR system's melt viscosity is vital for the successful development of intumescent char [135, 136]. The development of intumescent char is encouraged when the viscosity during polymer degradation is neither excessively high nor low [137]. When the viscosity of the degraded matrix is too low, the intumescence becomes poor because the gases are not confined but disperse and feed the flame. The creation of an enlarged structure results from sluggish diffusion when the viscosity is high enough [138]. When carbonizing materials have an excessively high viscosity, the gases are significantly confined, and the diffusion is massively diminished [139]; as a result, char's intumescence can be prevented. The strongly expanded char in the cone calorimeter tests can be due to the somewhat enhanced viscosity caused by the low loading of graphene. The established anti-dripping mechanism is the graphene-induced increase in melt viscosity [140].

The relationship between the LOI and MFI is seen in Fig. 5.37 [131]. It is evident that pristine PLA exhibited a variety of LOI characteristics as per Fig. 5.37b depending on the MFI value. As an illustration, when the MFI increased from 10.2 to 49.2 g/10 min, the LOI values of various pure PLA increased from 20.0 to 20.9%. While a 0.9% rise in the LOI values of pure PLAs was not particularly noteworthy, Fig. 5.37b showed a more pronounced shift in the melt dripping length from 20 to

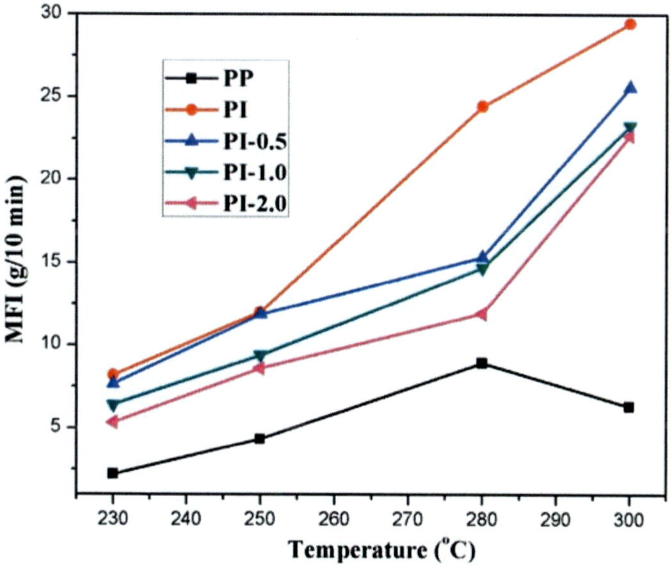

Fig. 5.36 MFI values of PP and its composites. Reproduced with permission from Yuan et al. [85]. Copyright 2017, Elsevier Science Ltd.

34 mm when the MFI value increased from 10.2 to 29.9 g/10 min. This meant that melt drippings could remove more heat from the burning frontline, which helped the PLA attain higher LOI values. Comparable to this, PLA with fire retardants had a more pronounced effect on the LOI value due to greater leaking and MFI values. It was astonishing to discover that the magnitudes of the increment in the LOI values of both fire-retardant systems increasingly grew with the MFI values of parent PLAs in a composite with 4.0 wt.% APP + PER or CSFR as shown in Fig. 5.37c. Intriguingly, the MFI range of 16.6–29.9 g/10 min of pure PLA showed a rapid ascent. A platform developed in the LOI values when the MFI exceeded 29.9 g/10 min, suggesting that the LOI value did not continue to rise as the MFI increased.

Figure 5.38 illustrates the impact of MFI on PLA with and without flame-retardants on UL-94 rating using the association between MFI and UL-94 tests. It has been discovered that the MFI values for pure PLA regulate their final fire-retardant ratings and dripping rates. It took longer for melt droplets to separate from the PLA bulk when the MFI value of pure PLA was <24.4 g/10 min due to its low fluidity, which caused heat retention at the sample's front during testing. The flame ultimately reached the sample clamp with constant heat and enough oxygen sources, leaving no UL-94 ratings in its aftermath. The melt droplets were simpler to form and separate from the bulk polymer when the MFI was higher than 24.4 g/10 min, promoting heat dissipation and cooling the burning frontline. As a result, the combustion duration time was considerably reduced, resulting in a UL-94 V-2 certification. In fire-resistant PLA systems, the controlling impact of MFI on the UL-94 ratings is further demonstrated. Because CSFR can encourage the production of char, which

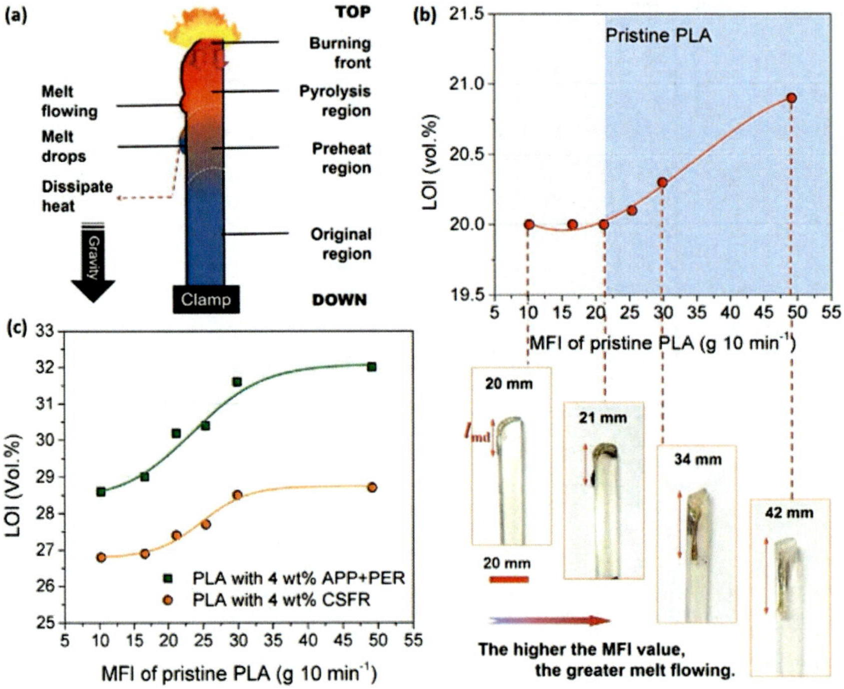

Fig. 5.37 Relationships between LOI and MFI. **a** Combustion process of PLA during LOI tests, **b** LOI of PLA as a function of MFI, **c** LOI of flame-retardant PLA with 4 wt.% APP + PER or CSFR as a function of MFI of the pristine PLA compared with the corresponding PLA matrix. Reproduced with permission from Yu et al. [131]. Copyright 2022, Elsevier Science Ltd.

helped reduce the combustion and melt-down of the PLA matrix, the UL-94 ratings of all fire-retardant PLA systems increased to varying degrees after the addition of 4 wt.% CSFR. Similarly, PLAs with MFI values higher than 24.4 g/10 min received a UL-94 V-0 rating, whereas PLAs with MFI values lower than 24.4 g/10 min only received a V-2 classification. This further underlines the impact of the polymer melt viscosity on the fire performance during UL-94 testing. The APP + PER system demonstrated improved flame retardancy in PLA compared to CSFR (Fig. 5.38).

5.3 Summary

In conclusion, the exploration of techniques for measuring polymer-based electromagnetic interference (EMI) shielding and fire-retarding characteristics is instrumental in ensuring the reliability, safety, and functionality of polymers in diverse

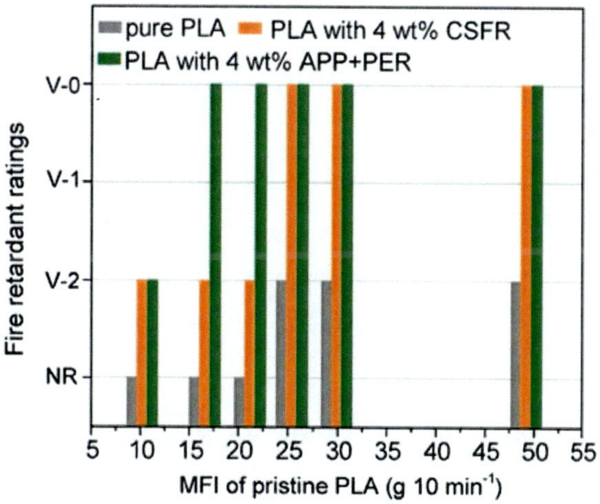

Fig. 5.38 UL-94 tests, correlations between UL-94 rating and MFI of PLA. Reproduced with permission from Yu et al. [131]. Copyright 2022, Elsevier Science Ltd.

applications. The comprehensive evaluation of these properties involves a combination of sophisticated methodologies that provide valuable insights into the performance of polymer materials in challenging environments.

For EMI shielding characteristics, techniques such as shielding effectiveness measurements, surface impedance analysis, and near-field scanning play a crucial role in assessing the ability of polymers to mitigate electromagnetic interference. These methods enable researchers and engineers to quantify the material's efficacy in redirecting or absorbing electromagnetic waves, essential for the optimal functioning of electronic devices.

On the other hand, the evaluation of fire-retarding characteristics involves a range of tests designed to assess a polymer's resistance to ignition, flame spread, and the release of toxic byproducts during combustion. Techniques like CC, LOI measurement, and TGA provide critical data on the fire performance of polymers, contributing to the development of safer materials for various applications.

The integration of both EMI shielding and fire-retarding characteristics measurement techniques reflects a holistic approach to materials design and engineering. By concurrently addressing electromagnetic compatibility and fire safety concerns, researchers and industry professionals can develop polymer materials that meet stringent requirements across multiple dimensions.

As technological advancements continue, the demand for polymers with enhanced EMI shielding and fire-retarding properties grows. Consequently, the refinement and standardization of measurement techniques become imperative to ensure accurate, repeatable, and comparable results across different studies and industries.

In conclusion, the ongoing development and improvement of measurement techniques for polymer-based EMI shielding and fire-retarding characteristics serve as a cornerstone for innovation in materials science. These techniques contribute to the creation of advanced materials that not only support the evolution of modern technologies but also prioritize safety and reliability in the face of diverse and challenging operational conditions.

References

1. A. Dey, S. Mandal, S. Bhandari, C. Pal, J. Tersur Orasugh, D. Chattopadhyay, Chapter 2: characterization methods, in *Fiber-Reinforced Nanocomposites: Fundamentals and Applications*, eds. by B. Han, S. Sharma, T.A. Nguyen, L. Longbiao, K.S. Bhat (Elsevier, Amsterdam, 2020), pp. 7–67
2. S.S. Ray, M. Okamoto, Polymer/layered silicate nanocomposites: a review from preparation to processing. Prog. Polym. Sci. **28**(11), 1539–1641 (2003)
3. S.S. Ray, K. Yamada, M. Okamoto, K. Ueda, New polylactide-layered silicate nanocomposites. 2. Concurrent improvements of material properties, biodegradability and melt rheology. Polymer **44**(3), 857–866 (2003)
4. S. Sinha Ray, K. Yamada, M. Okamoto, K. Ueda, Polylactide-layered silicate nanocomposite: a novel biodegradable material. Nano Lett. **2**(10), 1093–1096 (2002)
5. T. Ramanathan, A.A. Abdala, S. Stankovich, D.A. Dikin, M. Herrera-Alonso, R.D. Piner, D.H. Adamson, H.C. Schniepp, X. Chen, R.S. Ruoff, S.T. Nguyen, I.A. Aksay, R.K. Prud'Homme, L.C. Brinson, Functionalized graphene sheets for polymer nanocomposites. Nat. Nanotechnol. **3**(6), 327–331 (2008)
6. J. Bandyopadhyay, S.S. Ray, The quantitative analysis of nano-clay dispersion in polymer nanocomposites by small angle X-ray scattering combined with electron microscopy. Polymer **51**(6), 1437–1449 (2010)
7. J. Bandyopadhyay, S.S. Ray, V. Ojijo, M. Khoza, Development of a highly nucleated and dimensionally stable isotactic polypropylene/nanoclay composite using reactive blending. Polymer **117**, 37–47 (2017)
8. W. Kai, Y. Hirota, L. Hua, Y. Inoue, Thermal and mechanical properties of a Poly(ε-caprolactone)/graphite oxide composite. J. Appl. Polym. Sci. **107**(3), 1395–1400 (2008)
9. X. Fu, S. Qutubuddin, Polymer-clay nanocomposites: Exfoliation of organophilic montmorillonite nanolayers in polystyrene. Polymer **42**(2), 807–813 (2001)
10. H. Kim, C.W. Macosko, Morphology and Properties of Polyester/Exfoliated Graphite Nanocomposites. Macromolecules **41**(9), 3317–3327 (2008)
11. M. Yoonessi, J.R. Gaier, Highly Conductive Multifunctional Graphene Polycarbonate Nanocomposites. ACS Nano **4**(12), 7211–7220 (2010)
12. Q. Li, Z. Li, M. Chen, Y. Fang, Real-Time Study of Graphene's Phase Transition in Polymer Matrices. Nano Lett. **9**(5), 2129–2132 (2009)
13. M. Hirata, T. Gotou, S. Horiuchi, M. Fujiwara, M. Ohba, Thin-film particles of graphite oxide 1: High-yield synthesis and flexibility of the particles. Carbon **42**(14), 2929–2937 (2004)
14. T.D. Fornes, D.R. Paul, Modeling properties of nylon 6/clay nanocomposites using composite theories. Polymer **44**(17), 4993–5013 (2003)
15. H. Kim, C.W. Macosko, Processing-property relationships of polycarbonate/graphene composites. Polymer **50**(15), 3797–3809 (2009)
16. S.S. Ray, A. Geberekrstos, T.S. Muzata, J.T. Orasugh, *Process-Induced Phase Separation in Polymer Blends: Materials, Characterization, Properties, and Applications.* Carl Hanser Verlag GmbH Co KG (2023)
17. X. Yang, L. Li, S. Shang, X.-M. Tao, Synthesis and characterization of layer-aligned poly(vinyl alcohol)/graphene nanocomposites. Polymer **51**(15), 3431–3435 (2010)
18. D. Cai, M. Song, A simple route to enhance the interface between graphite oxide nanoplatelets and a semi-crystalline polymer for stress transfer. Nanotechnology **20**(31), 315708 (2009)

19. S. Ansari, A. Kelarakis, L. Estevez, E.P. Giannelis, Oriented arrays of graphene in a polymer matrix by in situ reduction of graphite oxide nanosheets. Small **6**(2), 205–209 (2010)
20. N. Du, C.-Y. Zhao, Q. Chen, G. Wu, R. Lu, Preparation and characterization of nylon 6/ graphite composite. Mater. Chem. Phys. **120**(1), 167–171 (2010)
21. J.-Z. Xu, T. Chen, C.-L. Yang, Z.-M. Li, Y.-M. Mao, B.-Q. Zeng, B.S. Hsiao, Isothermal crystallization of poly(l-lactide) induced by graphene nanosheets and carbon nanotubes: a comparative study. Macromolecules **43**(11), 5000–5008 (2010)
22. L. Peponi, A. Tercjak, R. Verdejo, M.A. Lopez-Manchado, I. Mondragon, J.M. Kenny, Confinement of functionalized graphene sheets by triblock copolymers. J. Phys. Chem. C **113**(42), 17973–17978 (2009)
23. P. Das, A.B. Deoghare, S. Ranjan Maity, Synergistically improved thermal stability and electromagnetic interference shielding effectiveness (EMI SE) of in-situ synthesized polyaniline/ sulphur doped reduced graphene oxide (PANI/S-RGO) nanocomposites. Ceram. Int. **48**(8), 11031–11042 (2022)
24. S.H. Ryu, H. Kim, S.-W. Park, S.J. Kwon, S. Kim, H.-R. Lim, B. Park, S.-B. Lee, Y.-H. Choa, Millimeter-scale percolated polyethylene/graphene composites for 5G electromagnetic shielding. ACS Appl. Nano Mater. **5**(6), 8429–8439 (2022)
25. H. Liu, Y. Xu, J.-P. Cao, D. Han, Q. Yang, R. Li, F. Zhao, Skin structured silver/three-dimensional graphene/polydimethylsiloxane composites with exceptional electromagnetic interference shielding effectiveness. Compos. A Appl. Sci. Manuf. **148**, 106476 (2021)
26. X. Sun, X. Liu, X. Shen, Y. Wu, Z. Wang, J.-K. Kim, Graphene foam/carbon nanotube/ poly(dimethyl siloxane) composites for exceptional microwave shielding. Compos. A Appl. Sci. Manuf. **85**, 199–206 (2016)
27. D. Lai, X. Chen, G. Wang, X. Xu, Y. Wang, Highly conductive porous graphene film with excellent folding resilience for exceptional electromagnetic interference shielding. J. Mater. Chem. C **8**(26), 8904–8916 (2020)
28. A. Morales-Rodriguez, P. Reynaud, G. Fantozzi, J. Adrien, E. Maire, Porosity analysis of long-fiber-reinforced ceramic matrix composites using X-ray tomography. Scripta Mater. **60**(6), 388–390 (2009)
29. Q.B. Ho, M. Kontopoulou, Stabilization of the cellular structure of polypropylene foams and secondary nucleation mechanism in the presence of graphene nanoplatelets. Polymer **198**, 122506 (2020)
30. M. Kitano, A. Nishimura, S. Kawai, K. Nishi, in Analysis of Package Cracking During Reflow Soldering Process, 26th Annual Proceedings Reliability Physics Symposium 1988 (IEEE, 1988), pp 90–95
31. K.S. Halake, J. Lee, Superporous thermo-responsive hydrogels by combination of cellulose fibers and aligned micropores. Carbohyd. Polym. **105**, 184–192 (2014)
32. J. Wang, V.M. Ugaz, Using in situ rheology to characterize the microstructure in photopolymerized polyacrylamide gels for DNA electrophoresis. Electrophoresis **27**(17), 3349–3358 (2006)
33. Y. Joliff, L. Belec, J.F. Chailan, Modified water diffusion kinetics in an unidirectional glass/ fibre composite due to the interphase area: Experimental, analytical and numerical approach. Compos. Struct. **97**, 296–303 (2013)
34. H.G. Carter, K.G. Kibler, Langmuir-type model for anomalous moisture diffusion in composite resins. J. Compos. Mater. **12**(2), 118–131 (1978)
35. H. Yang, Z. Li, H. Zou, P. Liu, Preparation of porous polyimide/in-situ reduced graphene oxide composite films for electromagnetic interference shielding. Polym. Adv. Technol. **28**(2), 233–242 (2017)
36. D.U. Shah, P.J. Schubel, Evaluation of cure shrinkage measurement techniques for thermosetting resins. Polym. Test. **29**(6), 629–639 (2010)
37. M. Haider, P. Hubert, L. Lessard, Cure shrinkage characterization and modeling of a polyester resin containing low profile additives. Compos. A Appl. Sci. Manuf. **38**(3), 994–1009 (2007)
38. Y. Li, B. Shen, X. Pei, Y. Zhang, D. Yi, W. Zhai, L. Zhang, X. Wei, W. Zheng, Ultrathin carbon foams for effective electromagnetic interference shielding. Carbon **100**, 375–385 (2016)

39. S.-T. Hsiao, C.-C.M. Ma, H.-W. Tien, W.-H. Liao, Y.-S. Wang, S.-M. Li, Y.-C. Huang, Using a non-covalent modification to prepare a high electromagnetic interference shielding performance graphene nanosheet/water-borne polyurethane composite. Carbon **60**, 57–66 (2013)
40. J.-S. Jang, J. Varischetti, G.W. Lee, J. Suhr, Experimental and analytical investigation of mechanical damping and CTE of both SiO_2 particle and carbon nanofiber reinforced hybrid epoxy composites. Compos. A Appl. Sci. Manuf. **42**(1), 98–103 (2011)
41. H. Zhang, F. Che, T. Lin, W. Zhao, 5: reliability and failure analysis of encapsulated packages, in Modeling, Analysis, Design, and Tests for Electronics Packaging beyond Moore, eds. by H. Zhang, F. Che, T. Lin, W. Zhao (Woodhead Publishing, 2020), pp. 245–292
42. T.D. Dudderar, D.P. Kossives, Y.L. Low, *Integrated Circuit Packages with Improved EMI Characteristics*. Google Patents: 2001
43. H. Ardebili, J. Zhang, M.G. Pecht, *Encapsulation Technologies for Electronic Applications* (William Andrew, 2018)
44. X. Li, Y. Qu, X. Wang, H. Bian, W. Wu, H. Dai, Flexible graphene/silver nanoparticles/aluminum film paper for high-performance electromagnetic interference shielding. Mater. Des. **213**, 110296 (2022)
45. C. Lei, Y. Zhang, D. Liu, K. Wu, Q. Fu, Metal-level robust, folding endurance, and highly temperature-stable MXene-based film with engineered aramid nanofiber for extreme-condition electromagnetic interference shielding applications. ACS Appl. Mater. Interf. **12**(23), 26485–26495 (2020)
46. N.N. Technology, Graphene's use in the aerospace industry. https://nanografi.com/blog/graphenes-use-in-the-aerospace-industry/. Accessed 24 May 2023
47. A.C.D.C. Materials, *Standard Test Method for Tensile Properties of Polymer Matrix Composite Materials* (ASTM international, 2008)
48. J. Joseph, A.K. Koroth, D.A. John, A.M. Sidpara, J. Paul, Highly filled multilayer thermoplastic/graphene conducting composite structures with high strength and thermal stability for electromagnetic interference shielding applications. J. Appl. Polym. Sci. **136**(29), 47792 (2019)
49. A. Standard, D6641/D6641M-14, *Standard Test Method for Compressive Properties of Polymer Matrix Composite Materials Using a Combined Loading Compression (CLC) Test Fixture* (2014)
50. Q. Gao, J. Qin, B. Guo, X. Fan, F. Wang, Y. Zhang, R. Xiao, F. Huang, X. Shi, G. Zhang, High-performance electromagnetic interference shielding epoxy/Ag nanowire/thermal annealed graphene aerogel composite with bicontinuous three-dimensional conductive skeleton. Compos. A Appl. Sci. Manuf. **151**, 106648 (2021)
51. X. Fan, Q. Gao, Y. Gao, G. Zhang, F. Huang, R. Xiao, W. Liu, F. Wang, J. Qin, E. Bilotti, H. Zhang, X. Shi, G. Zhang, Microcellular epoxy/graphene nanocomposites with outstanding electromagnetic interference shielding and mechanical performance by overcoming nanofiller loading/dispersion dichotomy. Compos. Sci. Technol. **215**, 109000 (2021)
52. S. Zhao, Y. Yan, A. Gao, S. Zhao, J. Cui, G. Zhang, Flexible polydimethylsilane nanocomposites enhanced with a three-dimensional graphene/carbon nanotube bicontinuous framework for high-performance electromagnetic interference shielding. ACS Appl. Mater. Interf. **10**(31), 26723–26732 (2018)
53. F. Del Giudice, A.Q. Shen, Shear rheology of graphene oxide dispersions. Curr. Opin. Chem. Eng. **16**, 23–30 (2017)
54. S.E. Zakiyan, H. Azizi, I. Ghasemi, Influence of chain mobility on rheological, dielectric and electromagnetic interference shielding properties of poly methyl-methacrylate composites filled with graphene and carbon nanotube. Compos. Sci. Technol. **142**, 10–19 (2017)
55. P. Song, C. Liang, L. Wang, H. Qiu, H. Gu, J. Kong, J. Gu, Obviously improved electromagnetic interference shielding performances for epoxy composites via constructing honeycomb structural reduced graphene oxide. Compos. Sci. Technol. **181**, 107698 (2019)
56. Y. Huangfu, K. Ruan, H. Qiu, Y. Lu, C. Liang, J. Kong, J. Gu, Fabrication and investigation on the PANI/MWCNT/thermally annealed graphene aerogel/epoxy electromagnetic interference shielding nanocomposites. Compos. A Appl. Sci. Manuf. **121**, 265–272 (2019)

57. C. Liang, H. Qiu, Y. Han, H. Gu, P. Song, L. Wang, J. Kong, D. Cao, J. Gu, Superior electromagnetic interference shielding 3D graphene nanoplatelets/reduced graphene oxide foam/ epoxy nanocomposites with high thermal conductivity. J. Mater. Chem. C **7**(9), 2725–2733 (2019)
58. S. Ganguly, S. Ghosh, P. Das, T.K. Das, S.K. Ghosh, N.C. Das, Poly (N-vinylpyrrolidone)-stabilized colloidal graphene-reinforced poly (ethylene-co-methyl acrylate) to mitigate electromagnetic radiation pollution. Polym. Bull. **77**, 2923–2943 (2020)
59. A.A. Al-Ghamdi, A.A. Al-Ghamdi, Y. Al-Turki, F. Yakuphanoglu, F. El-Tantawy, Electromagnetic shielding properties of graphene/acrylonitrile butadiene rubber nanocomposites for portable and flexible electronic devices. Compos. B Eng. **88**, 212–219 (2016)
60. M. Dai, Y. Zhai, Y. Zhang, A green approach to preparing hydrophobic, electrically conductive textiles based on waterborne polyurethane for electromagnetic interference shielding with low reflectivity. Chem. Eng. J. **421**, 127749 (2021)
61. B. Shen, W. Zhai, M. Tao, J. Ling, W. Zheng, Lightweight, multifunctional polyetherimide/ graphene@Fe_3O_4 composite foams for shielding of electromagnetic pollution. ACS Appl. Mater. Interf. **5**(21), 11383–11391 (2013)
62. P. Bhawal, S. Ganguly, T.K. Das, S. Mondal, S. Choudhury, N.C. Das, Superior electromagnetic interference shielding effectiveness and electro-mechanical properties of EMA-IRGO nanocomposites through the in-situ reduction of GO from melt blended EMA-GO composites. Compos. B Eng. **134**, 46–60 (2018)
63. B.B. Rao, P. Yadav, R. Aepuru, H. Panda, S. Ogale, S. Kale, Single-layer graphene-assembled 3D porous carbon composites with PVA and Fe_3O_4 nano-fillers: an interface-mediated superior dielectric and EMI shielding performance. Phys. Chem. Chem. Phys. **17**(28), 18353–18363 (2015)
64. M.F. Shakir, A.N. Khan, R. Khan, S. Javed, A. Tariq, M. Azeem, A. Riaz, A. Shafqat, H.M. Cheema, M.A. Akram, I. Ahmad, R. Jan, EMI shielding properties of polymer blends with inclusion of graphene nano platelets. Results Phys. **14**, 102365 (2019)
65. A.C.D. Electrical, E.I. Materials, in *Standard Test Method for Measuring the Electromagnetic Shielding Effectiveness of Planar Materials* (American Society for Testing and Materials, 1999)
66. K. Cheng, H. Li, M. Zhu, H. Qiu, J. Yang, In situ polymerization of graphene-polyaniline@ polyimide composite films with high EMI shielding and electrical properties. RSC Adv. **10**(4), 2368–2377 (2020)
67. V. Eswaraiah, V. Sankaranarayanan, S. Ramaprabhu, Functionalized grapheme—PVDF foam composites for EMI shielding. Macromol. Mater. Eng. **296**(10), 894–898 (2011)
68. H. Fang, H. Guo, Y. Hu, Y. Ren, P.-C. Hsu, S.-L. Bai, In-situ grown hollow Fe3O4 onto graphene foam nanocomposites with high EMI shielding effectiveness and thermal conductivity. Compos. Sci. Technol. **188**, 107975 (2020)
69. M. Hamidinejad, B. Zhao, A. Zandieh, N. Moghimian, T. Filleter, C.B. Park, Enhanced electrical and electromagnetic interference shielding properties of polymer-graphene nanoplatelet composites fabricated via supercritical-fluid treatment and physical foaming. ACS Appl. Mater. Interf. **10**(36), 30752–30761 (2018)
70. H. Kim, Y. Miura, C.W. Macosko, Graphene/polyurethane nanocomposites for improved gas barrier and electrical conductivity. Chem. Mater. **22**(11), 3441–3450 (2010)
71. S. Ganguly, S. Ghosh, P. Das, T.K. Das, S.K. Ghosh, N.C. Das, Poly(N-vinylpyrrolidone)-stabilized colloidal graphene-reinforced poly(ethylene-co-methyl acrylate) to mitigate electromagnetic radiation pollution. Polym. Bull. **77**(6), 2923–2943 (2020)
72. X. Gong, H. Yu, X. Chen, P. Xu, H. Wang, Superhydrophobic PDMS/PPy-Ag/graphene/PET films with highly efficient electromagnetic interference shielding, UV shielding, self-cleaning and electrothermal deicing. Materials Today Physics **34**, 101076 (2023)
73. B. Schartel, T.R. Hull, Development of fire-retarded materials—interpretation of cone calorimeter data. Fire Mater. Int. J. **31**(5), 327–354 (2007)
74. T. Fateh, F. Richard, B. Batiot, T. Rogaume, J. Luche, J. Zaida, Characterization of the burning behavior and gaseous emissions of pine needles in a cone calorimeter—FTIR apparatus. Fire Saf. J. **82**, 91–100 (2016)

75. C. Huggett, Estimation of rate of heat release by means of oxygen consumption measurements. Fire Mater. **4**(2), 61–65 (1980)
76. A. Boudenne, L. Ibos, Y. Candau, S. Thomas, *Handbook of Multiphase Polymer Systems* (Wiley, Hoboken, 2011)
77. M. Ferriol, F. Laoutid, J.M.L. Cuesta, Fire retardancy of multiphase polymer systems. Handbook Multiphase Polym. Syst. **12**, 843–864 (2011)
78. M. Maqsood, G. Seide, Biodegradable flame retardants for biodegradable polymer. Biomolecules **10**(7), 1038 (2020)
79. F. Kempel, B. Schartel, J.M. Marti, K.M. Butler, R. Rossi, S.R. Idelsohn, E. Oñate, A. Hofmann, Modelling the vertical UL 94 test: competition and collaboration between melt dripping, gasification and combustion. Fire Mater. **39**(6), 570–584 (2015)
80. A. Dey, S. Mandal, S. Bhandari, C. Pal, J.T. Orasugh, D. Chattopadhyay, Characterization methods, in *Fiber-Reinforced Nanocomposites: Fundamentals and Applications* (Elsevier, Amsterdam, 2020), pp. 7–67
81. H. Sturm, B. Schartel, A. Weiß, U. Braun, SEM/EDX: Advanced investigation of structured fire residues and residue formation. Polym. Test. **31**(5), 606–619 (2012)
82. P. Anna, G. Marosi, S. Bourbigot, M. Le Bras, R. Delobel, Intumescent flame retardant systems of modified rheology. Polym. Degrad. Stab. **77**(2), 243–247 (2002)
83. Z.-M. Zhu, W.-H. Rao, A.-H. Kang, W. Liao, Y.-Z. Wang, Highly effective flame retarded polystyrene by synergistic effects between expandable graphite and aluminum hypophosphite. Polym. Degrad. Stab. **154**, 1–9 (2018)
84. Z. Huang, Z. Wang, Synthesis of a copper hydroxystannate modified graphene oxide nanohybrid and its high performance in flexible polyvinyl chloride with simultaneously improved flame retardancy, smoke suppression and mechanical properties. Polym. Degrad. Stab. **183**, 109425 (2021)
85. B. Yuan, A. Fan, M. Yang, X. Chen, Y. Hu, C. Bao, S. Jiang, Y. Niu, Y. Zhang, S. He, H. Dai, The effects of graphene on the flammability and fire behavior of intumescent flame retardant polypropylene composites at different flame scenarios. Polym. Degrad. Stab. **143**, 42–56 (2017)
86. B. Dittrich, K.-A. Wartig, D. Hofmann, R. Mülhaupt, B. Schartel, Flame retardancy through carbon nanomaterials: carbon black, multiwall nanotubes, expanded graphite, multi-layer graphene and graphene in polypropylene. Polym. Degrad. Stab. **98**(8), 1495–1505 (2013)
87. J. Li, S. Wang, G. Zhang, H. Li, J. Sun, X. Gu, S. Zhang, Burning behavior analysis of polypropylene composite containing poly-siloxane encapsulated expandable graphite. Polym. Degrad. Stab. **202**, 110006 (2022)
88. K.C. Tsai, C.F. Kuan, C.H. Chen, H.C. Kuan, S.W. Hsu, F.M. Lee, C.L. Chiang, Study on thermal degradation and flame retardant property of halogen-free polypropylene composites using XPS and cone calorimeter. J. Appl. Polym. Sci. **127**(2), 1084–1091 (2013)
89. W. Hu, B. Yu, S.-D. Jiang, L. Song, Y. Hu, B. Wang, Hyper-branched polymer grafting graphene oxide as an effective flame retardant and smoke suppressant for polystyrene. J. Hazard. Mater. **300**, 58–66 (2015)
90. W. Hu, J. Zhan, X. Wang, N. Hong, B. Wang, L. Song, A.A. Stec, T.R. Hull, J. Wang, Y. Hu, Effect of functionalized graphene oxide with hyper-branched flame retardant on flammability and thermal stability of cross-linked polyethylene. Ind. Eng. Chem. Res. **53**(8), 3073–3083 (2014)
91. E. De Giglio, N. Ditaranto, L. Sabbatini, 3 Polymer surface chemistry: characterization by XPS. Polym. Surf. Charact. **51**, 73–112 (2014)
92. M. Smith, L. Scudiero, J. Espinal, J.-S. McEwen, M. Garcia-Perez, Improving the deconvolution and interpretation of XPS spectra from chars by ab initio calculations. Carbon **110**, 155–171 (2016)
93. D. Pantea, H. Darmstadt, S. Kaliaguine, C. Roy, Electrical conductivity of conductive carbon blacks: influence of surface chemistry and topology. Appl. Surf. Sci. **217**(1), 181–193 (2003)
94. F. Mangolini, A. Rossi, 4. Attenuated total reflection-Fourier transform infrared spectroscopy: a powerful tool for investigating polymer surfaces and interfaces, in *Polymer Surface Characterization* (De Gruyter, 2014), pp. 113–152

95. Z. Mo, H. Gou, J. He, P. Yang, C. Feng, R. Guo, Controllable synthesis of functional nanocomposites: covalently functionalize graphene sheets with biocompatible L-lysine. Appl. Surf. Sci. **258**(22), 8623–8628 (2012)
96. C. Bao, Y. Guo, L. Song, Y. Kan, X. Qian, Y. Hu, In situ preparation of functionalized graphene oxide/epoxy nanocomposites with effective reinforcements. J. Mater. Chem. **21**(35), 13290–13298 (2011)
97. C. Bao, Y. Guo, B. Yuan, Y. Hu, L. Song, Functionalized graphene oxide for fire safety applications of polymers: a combination of condensed phase flame retardant strategies. J. Mater. Chem. **22**(43), 23057–23063 (2012)
98. N. Gierlinger, M. Schwanninger, The potential of Raman microscopy and Raman imaging in plant research. J. Spectrosc. **21**, 69–89 (2007)
99. O. Maier, V. Oberle, D. Hoekstra, Fluorescent lipid probes: some properties and applications (a review). Chem. Phys. Lipid. **116**(1–2), 3–18 (2002)
100. B. Meng, Y. Zhang, F. Zhang, Material removal mechanism of 6H-SiC studied by nano-scratching with Berkovich indenter. Appl. Phys. A **122**, 1–9 (2016)
101. T. Hänninen, E. Kontturi, T. Vuorinen, Distribution of lignin and its coniferyl alcohol and coniferyl aldehyde groups in *Picea abies* and *Pinus sylvestris* as observed by Raman imaging. Phytochemistry **72**(14–15), 1889–1895 (2011)
102. E. Smith, G. Dent, *Modern Raman Spectroscopy: A Practical Approach* (Wiley, Hoboken, 2019)
103. R.S. Das, Y. Agrawal, Raman spectroscopy: recent advancements, techniques and applications. Vib. Spectrosc. **57**(2), 163–176 (2011)
104. H. Tönshoff, W.V. Schmieden, I. Inasaki, W. König, G. Spur, Abrasive machining of silicon. CIRP Annals **39**(2), 621–635 (1990)
105. A. Saletnik, B. Saletnik, C. Puchalski, Overview of popular techniques of Raman spectroscopy and their potential in the study of plant tissues. Molecules **26**(6), 1537 (2021)
106. A. Kudelski, Raman spectroscopy of surfaces. Surf. Sci. **603**(10–12), 1328–1334 (2009)
107. X. Wang, W. Xing, X. Feng, B. Yu, L. Song, Y. Hu, Functionalization of graphene with grafted polyphosphamide for flame retardant epoxy composites: synthesis, flammability and mechanism. Polym. Chem. **5**(4), 1145–1154 (2014)
108. A. Hull, A new method of chemical analysis. J. Am. Chem. Soc. **41**(8), 1168–1175 (1919)
109. N.J. Joshi, Studies of nano sized high dielectric constant materials, Ph. D. Thesis (2012)
110. A. Bishnoi, S. Kumar, N. Joshi, Wide-angle X-ray diffraction (WXRD): technique for characterization of nanomaterials and polymer nanocomposites, in *Microscopy Methods in Nanomaterials Characterization* (Elsevier, Amsterdam, 2017), pp. 313–337
111. C.K. Kundu, Z. Li, X. Li, Z. Zhang, Y. Hu, Graphene oxide functionalized biomolecules for improved flame retardancy of polyamide 66 fabrics with intact physical properties. Int. J. Biol. Macromol. **156**, 362–371 (2020)
112. Z. Cheng, D. Liao, X. Hu, W. Li, C. Xie, H. Zhang, W. Yang, Synergistic fire retardant effect between expandable graphite and ferrocene-based non-phosphorus polymer on polypropylene. Polym. Degrad. Stab. **178**, 109201 (2020)
113. G. Fadillah, T.A. Saleh, S. Wahyuningsih, E.N.K. Putri, S. Febrianastuti, Electrochemical removal of methylene blue using alginate-modified graphene adsorbents. Chem. Eng. J. **378**, 122140 (2019)
114. A.M. Alansi, M. Al-Qunaibit, I.O. Alade, T.F. Qahtan, T.A. Saleh, Visible-light responsive BiOBr nanoparticles loaded on reduced graphene oxide for photocatalytic degradation of dye. J. Mol. Liq. **253**, 297–304 (2018)
115. X. Wang, F. Pan, Z. Xiang, Q. Zeng, K. Pei, R. Che, W. Lu, Magnetic vortex core-shell Fe_3O_4@C nanorings with enhanced microwave absorption performance. Carbon **157**, 130–139 (2020)
116. J. Sheng, Y. Zhang, L. Liu, B. Quan, N. Zhang, G. Ji, Optimizing electromagnetic wave absorption performance: Design from microscopic bamboo carbon nanotubes to macroscopic patterns. J. Alloy. Compd. **809**, 151866 (2019)
117. K. Schulte, S. Chandrasekaran, C. Viets, B. Fiedler, New functions in polymer composites using a nanoparticle-modified matrix, in *William Andrew is an Imprint* (Elsevier, Amsterdam, 2015)

118. A. Standard, Standard test method for tensile properties of polymer matrix composite materials. *ASTM D3039/DM* **2008,** *3039* (2008)
119. S. Kumar, A. Gupta, Preparation and mechanical properties of nanoclay-MWCNT/Epoxy hybrid nanocomposites. J. Appl. Res. Technol. Eng. **2**(1), 17–21 (2021)
120. Y. Zhan, J. Wu, H. Xia, N. Yan, G. Fei, G. Yuan, Dispersion and exfoliation of graphene in rubber by an ultrasonically-assisted latex mixing and in situ reduction process. Macromol. Mater. Eng. **296**(7), 590–602 (2011)
121. D. Marset, E. Fages, E. Gonga, J. Ivorra-Martinez, L. Sánchez-Nacher, L. Quiles-Carrillo, Development and characterization of high environmentally friendly composites of bio-based polyamide 1010 with enhanced fire retardancy properties by expandable graphite. Polymers **14**(9), 1843 (2022)
122. D. Li, Q. Chen, Y. Yang, Y. Chen, C. Xiao, Effects of flake graphite on property optimisation in thermal conductive composites based on polyamide 66. Plast. Rubber Compos. **46**(6), 266–276 (2017)
123. R. Bottom, Thermogravimetric analysis. Princ. Appl. Therm. Anal. **1**(906), 87–118 (2008)
124. M. Bradley, TGA-IR analysis using the omnic mercury TGA software. Thermo Fisher Sci. (2012)
125. H. Ng, N.M. Saidi, F.S. Omar, K. Ramesh, S. Ramesh, S. Bashir, Thermogravimetric analysis of polymers. Encycl. Polym. Sci. Technol. **53**, 1–29 (2002)
126. B. Giechaskiel, M. Clairotte, Fourier transform infrared (FTIR) spectroscopy for measurements of vehicle exhaust emissions: a review. Appl. Sci. **11**(16), 7416 (2021)
127. I. Standard, 5659-2, Plastics-smoke generation-part 2: determination of optical density by a single-chamber test, in *The International Organization for Standardization: Geneva* (2006)
128. D. Drysdale, *An introduction to fire dynamics* (Wiley, Hoboken, 2011)
129. E.C. Butcher, A.C. Parnell, *Smoke Control in Fire Safety Design* (Spon, 1979)
130. C.D. Han, Relationships between polymer rheology and polymer processing, in *Rheology and Processing of Polymeric Materials: Volume 1: Polymer Rheology* (Oxford University Press)
131. Y. Yu, L. Xi, M. Yao, L. Liu, Y. Zhang, S. Huo, Z. Fang, P. Song, Governing effects of melt viscosity on fire performances of polylactide and its fire-retardant systems. Iscience **25**(3), 103950 (2022)
132. X. Wen, Y. Wang, J. Gong, J. Liu, N. Tian, Y. Wang, Z. Jiang, J. Qiu, T. Tang, Thermal and flammability properties of polypropylene/carbon black nanocomposites. Polym. Degrad. Stab. **97**(5), 793–801 (2012)
133. B. Li, W.-H. Zhong, Review on polymer/graphite nanoplatelet nanocomposites. J. Mater. Sci. **46**, 5595–5614 (2011)
134. X. Wang, L. Song, H. Yang, H. Lu, Y. Hu, Synergistic effect of graphene on antidripping and fire resistance of intumescent flame retardant poly(butylene succinate) composites. Ind. Eng. Chem. Res. **50**(9), 5376–5383 (2011)
135. S. Bourbigot, S. Duquesne, Fire retardant polymers: recent developments and opportunities. J. Mater. Chem. **17**(22), 2283–2300 (2007)
136. S. Duquesne, R. Delobel, M. Le Bras, G. Camino, A comparative study of the mechanism of action of ammonium polyphosphate and expandable graphite in polyurethane. Polym. Degrad. Stab. **77**(2), 333–344 (2002)
137. M. Jimenez, S. Duquesne, S. Bourbigot, Multiscale experimental approach for developing high-performance intumescent coatings. Ind. Eng. Chem. Res. **45**(13), 4500–4508 (2006)
138. S. Duquesne, S. Magnet, C. Jama, R. Delobel, Thermoplastic resins for thin film intumescent coatings–towards a better understanding of their effect on intumescence efficiency. Polym. Degrad. Stab. **88**(1), 63–69 (2005)
139. B. Dittrich, K.-A. Wartig, R. Mülhaupt, B. Schartel, Flame-retardancy properties of intumescent ammonium poly (phosphate) and mineral filler magnesium hydroxide in combination with graphene. Polymers **6**(11), 2875–2895 (2014)
140. X. Wang, L. Song, H. Yang, H. Lu, Y. Hu, Synergistic effect of graphene on antidripping and fire resistance of intumescent flame retardant poly (butylene succinate) composites. Ind. Eng. Chem. Res. **50**(9), 5376–5383 (2011)

Chapter 6
NPs for Polymer-Based EMI Shielding and Fire Retarding Nanocomposites

6.1 Brief Introduction

Welcome to the exploration of a transformative frontier in materials science: The synergistic realm of nanoparticles and polymers aimed at revolutionizing EMI shielding and fire-retardant properties. In this chapter, we embark on a journey through the fascinating interplay of nanotechnology and polymer science, unraveling the potential of these composite materials in addressing critical challenges in industries such as electronics, aerospace, and transportation with regard to these systems ability to provide excellent EMI shielding while maintaining their flame retardancy.

We commence by delving into the world of nanoparticles, miniature entities with extraordinary properties. These nanoscale wonders, ranging from 1 to 100 nm, hold the promise of reshaping material functionalities due to their unique size and surface area characteristics.

This section explores the marriage of nanoparticles with polymer matrices, creating nanocomposites that transcend the limitations of conventional materials. By combining the versatility of polymers with the exceptional properties of nanoparticles, these nanocomposites offer a customizable platform for tailoring materials to meet the specific demands of diverse applications. This covers how the strategic introduction of nanoparticles into polymers amplifies their EMI shielding effectiveness, safeguarding electronic devices from disruptive electromagnetic signals. On the other hand, fire safety is paramount, especially in industries where flammability can have dire consequences. In this chapter, we uncover how the integration of nanoparticles into polymer matrices enhances the fire-retardant properties of materials, offering improved thermal stability and reduced flammability in the face of fire hazards through survey of the current state of research, as scientists endeavor to comprehend the intricate interactions between nanoparticles (with our focus on graphene-based systems) and polymers.

S. S. Ray et al., *Graphene-Bearing Polymer Composites*, Springer Series in Materials Science 340, https://doi.org/10.1007/978-3-031-51924-6_6

As we embark on this chapter, we anticipate a captivating journey into the frontier of graphene-based nanoparticle-infused polymers, where innovation converges with necessity to forge materials that redefine the boundaries of possibility.

6.2 Nano-Reinforcing Fillers

To increase service efficiency and extend product life while lowering costs, fillers have been widely employed in the polymer sector. Subject to the size of the particle, quality of dispersion, volume, particle orientation degree (where necessary), and the interaction between the matrix and reinforcing nanofillers, polymer/filler composites exhibit qualities between pure matrix and nanofiller. There are two fundamental types of polymer nanocomposites fillers: (i) Fillers for reinforcement—these fillers can change a material's stiffness, strength, toughness, abrasion resistance, anti-scratching property, or friction behavior. This class includes carbon black graphene, GP, silica, and others; (2) extending fillers—these fillers extend the favorable qualities and make them endure longer or work more effectively. In the latter, TiO_2, MMT, ZnO, and Al_2O_3 are present. With regard to reinforcing nanofillers for polymeric EMI shields, graphene [1], TiC/MXene [2], MXene [3], G@Ag [3], CNFs/rGO [4], GO@Ni [5], Fe_3O_4 [6], GP [7], etc. have been reportedly used by researchers.

6.2.1 Categories of Nano-Reinforcing Fillers

Comprehension of the structural property correlations of polymeric (nano)composites depends on the ratio of the surface area to the volume of the nano-reinforcing filler (be it blends or non-blends). The surface area/volume ratio of nano-reinforcing fillers is usually three times higher than that of their micron/macro-sized counterparts, and material characteristics can be quite size dependent at the nanoscale. As a result, the previously mentioned polymer nanocomposites' characteristics are dominated by the surface chemistry of the reinforcing nanofillers. The four primary subcategories of nano-reinforcing fillers are nanoparticles, nanofibers, nanotubes, and layered materials, as summarized in Fig. 6.1.

Nanoparticles

In conventional polymer composites, μm-sized particles are frequently adopted as reinforcing fillers to boost elastic modulus and yield strength. However, reducing the size of these particulate reinforcing fillers to the nanometer range can produce amazing, unique material characteristics. For instance, the color of silver NPs varies based on their size and form, whereas gold NPs are red [8]. NPs are mainly called zero-dimensional nanomaterials like graphene quantum dots, carbon dots, Ag NPs,

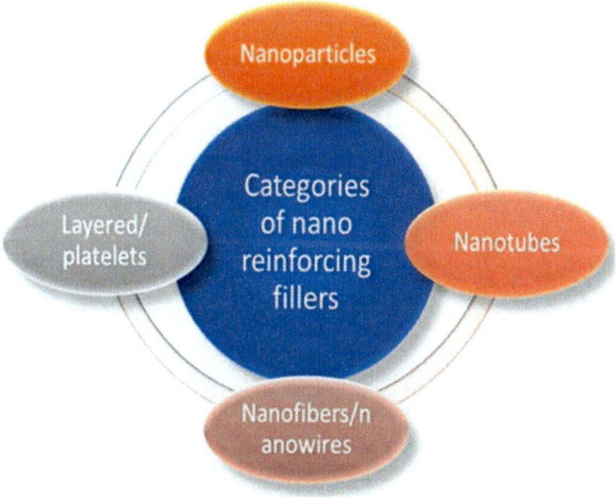

Fig. 6.1 Categories of nano-reinforcing fillers

Au NPs, cellulose nanoballs/nanocrystals, etc. The tiny artificial semiconducting crystals known as quantum dots can carry electrons. When exposed to UV radiation, they can release a range of colored light. Microwave-assisted colloidal synthesis is one of the most common methods of synthesizing these materials. In general, many NPs like graphene or /carbon quantum dots possess quantum effects, which are vital for applications such as energy materials, medical diagnostics, and advanced drug delivery systems.

Layered Platelet Nanomaterials

Within the last ten years, several key types of layered platelet nanomaterials have gained technological attention for polymer nanocomposites, including GP, graphene and its derivatives, Mxene, mica, saponite, and nanoclay. The most economical of these nanomaterials are layered clays made of platelets that are only a few nanometers thick. Commercially accessible nanoclays come in various shapes and sizes that have been altered to be compatible with commodity polymers. MMT makes up most of the mineral NC known as bentonite, which also contains the elements quartz, mica, feldspar, and zeolite. Depending on its origin, bentonite is either mined or quarried to produce [9–11]. When paired with decreased gas permeability and a lower coefficient of thermal expansion, NC nanocomposites offer increased stiffness, strength, toughness, and thermal stability [9–11].

Graphene nanoplatelets have received much interest as promising nanomaterials [12, 13]. Graphene is essentially made of single sheets of exfoliated GP. Similar in thickness to exfoliated NC, but with far better qualities than NC platelets. Even though it is only used in a few commercial materials, graphene is still mostly used in laboratory-scale research. It is particularly interesting due to its incredibly high thermal and electrical conductivity.

Nanotubes

CNTs and nanofibers are two examples of a class of nanomaterial (a structure with one of its dimensions in the nm length scale) with a cylindrical structure but different atomic orientation. These materials differ in size and atomic arrangement: In contrast to carbon nanofibers, which are rod-like structures without a space within their structure, CNTs are typically cylindrical, hollow, and arranged in a tube-like form, as shown in Fig. 6.2. CNTs are categorized into SWCNTs and MWCNTs. SWCNT and MWCNT are compared, as shown in Table 6.1.

Due to their exceptional electrical and mechanical qualities, CNTs are now essential for high-tech sensors, electronic and optical devices, batteries, fuel cells, and catalysts [15]. SWCNTs may provide a combination of qualities unsurpassed by any conventional material when they are purest.

High-quality CNTs cannot yet be produced widely in an industrial setting. All currently used synthesis techniques produce CNTs that are of low purity, and further purification procedures are required to produce CNTs that are >98% pure. Due to their high cost and complexity, it has been recommended that polymer/CNT nanocomposites are more suitable for specialized high-value applications that take advantage of their special qualities, such as electrical and thermal properties, instead of the mechanical reinforcement alone, where reinforcing nanofillers like GNPs, and/or nanoclays are more cost-effective [16].

Consider a piece of GP, which is made of layers of covalently bound carbon atoms arranged in a honeycomb-shaped lattice, to better understand how GP, graphene, and CNTs are related. A sheet of graphene is created by taking a piece of pure GP and

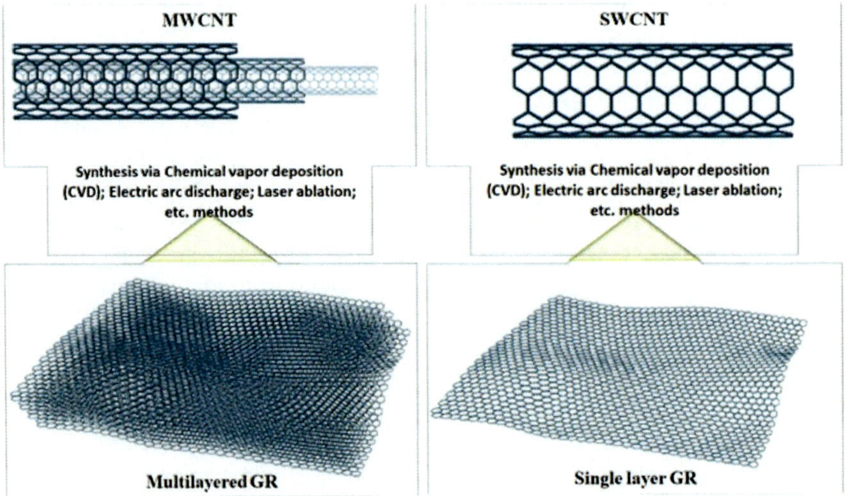

Fig. 6.2 Structural architecture of **a** MWCNTs and **b** SWCNTs. The author created using an amalgamation of BioRender, Microsoft PowerPoint, and Paint software

Table 6.1 Comparison between SWCNT and MWCNT

SWCNT	MWCNT
Single layer of GNP	Multiple layers of GNP
Catalyst is required for synthesis	Can be produced without catalyst
Bulk synthesis is difficult as it requires proper control overgrowth and atmospheric condition	Bulk synthesis is easy
Not fully dispersed, and form bundled structures	Homogeneously dispersed with no apparent bundled formation
Resistivity usually in the range of 10^{-4}–10^{-3} Ω m	Resistivity usually in the range of 1.8×10^{-5}–6.1×10^{-5} Ω m
Purity is poor. Typical SWCNT content in as-prepared samples by chemical vapor deposition (CVD) method is about 30–50 wt.%. However, high purity up to 80% has been reported by using arc discharge synthesis method	Purity is high. Typical MWCNT content in as-prepared samples by CVD method is about 35–90 wt.%
A chance of defect is more during functionalization	A chance of defect is less especially when synthesized by arc-discharged method
Characterization and evaluation are easy	It has very complex structure
It can be easily twisted and is more pliable	It cannot be easily twisted

Reproduced from [14] Copyright 2013, the Authors

separating each layer of carbon atoms one by one. Now, picture rolling this graphene sheet into a tube; this is a CNT. It is referred to as a SWCNT since it only has one layer of graphene. A double-walled CNT (DWCNT) is created by wrapping a second layer of GNP around it, while "multiwalled NTs" (also known as "multiwalled carbon NTs, or MWCNTs) are created when three or more graphene sheets are wrapped around one another. Figure 6.2 summarizes the connection between these materials. Each graphene layer can be "rolled up" into a distinct atomic structure, and these various structures give the CNT various properties, such as metallic or semiconductor. GP, graphene, and CNTs are not manufactured this way. Each of them can be made in various ways, and each way results in a material with somewhat different qualities and variable levels of purity.

Nanofibers/Nanowires

Nanowires are similar to CNTs in that they both have diameters of the order of tens of nanometers, but their length scales range from nanometers to microns and below millimeters. The usual diameter of a CNF is bigger than that of CNT, typically ranging from 50 to 200 nm with various morphologies [17]. As we witnessed with CNTs, different nanofiber morphologies produce various material characteristics. Vapor-grown CNFs have been used to reinforce a variety of polymers, including acrylonitrile–butadiene–styrene, epoxy, PP, polycarbonate, nylon, poly(ether sulfone), poly(ethylene terephthalate) (PET), and poly(phenylene sulfide)

[17]. Other nanofibers that have been of interest are cellulose nanofibers applied in diverse niches and EMI shielding systems [18, 19].

It is important to note that graphene and its derivatives are of great interest to researchers around the globe as a nanoparticulate reinforcing filler of choice because of their attractive properties like excellent conductivity, lightweight, high surface area, outstanding dielectric property performance, good EM waves absorbing ability, flame blanketing property, etc. These properties are vital in aiding the easy formulation of its nanocomposite systems and enhancing the required property performance of the fabricated nanocomposites intended for use as EMI shields or flame-retardants.

6.3 Processing of Graphene-Based Filler-Bearing Polymer Composites

As previously mentioned, polymer composite materials reinforced with NPs can perform well at a low loading of <10 wt.% of nano-reinforcement. In advanced graphene-containing nanocomposite synthesis, issues like lower toughness, poor optical clarity, and increased melt viscosity often associated with high filler content are, therefore, less of a problem. The processes used to create ordinary plastic parts can also be employed to make graphene-bearing nanocomposites. Any proposed method to produce functionalized graphene-based reinforcing filler and create nanocomposites from them must satisfy the following requirements in order to be scaled up: (i) It should be possible to control and reproduce the morphologies of the graphene-based reinforcing filler's achieved properties; (ii) the proposed nanocomposite should be easily prepared using polymer resins and graphene-based reinforcing filler that can be quickly dissolved in ordinary solvents; (iii) the techniques used to create or alter graphene-based reinforcing filler and nanocomposites should be straightforward, repeatable, and provide a high yield of the desired result when carried out in ambient settings with straightforward equipment.

In order to create a nanocomposite (in our example, one based on graphene), there are three important considerations: Choosing the best nano-additive, managing the morphology of the completed nanocomposite, and correctly formulating the chemistry of the composite are all important steps. We must begin by posing the following queries in order to manage each of these factors: What kind of product do we want to produce, and how much of it? Will we be using polar or non-polar polymers? Is the nano-reinforcement compatible with the polymer (or the monomer from which it is formed), or is there a widely accepted solvent in which the polymer dissolves as well as the nano-reinforcing filler can also be dispersed? If the monomer is liquid and miscible with the nano-reinforcing filler, in situ polymerization may be an option. If a common solvent dissolves the polymer and allows for the dispersion of the nano-reinforcement, solution processing may also be an option. Finally, melt processing involves compounding the nano-reinforcement with the polymer while molten.

A homogeneous particle size distribution and minimum aggregation of the nano-reinforcement are the main goals when producing a polymer nanocomposite. As the volume percentage of nano-reinforcements added to the polymer matrix rises, aggregation of the nano-reinforcements typically happens, leading to a loss of mechanical and optical properties. The intended use can frequently limit the choice of potential nano-reinforcements and polymer matrices, and the nano-reinforcement may be expected to improve more than one aspect of the finished material.

Take the application of nanocomposites in EMI shielding and flame retardancy. The selection of matrices is constrained by high dielectric constant polymers dominating the EMI shielding market. There are numerous uses for polymer dielectrics in the electronics sector, with low and high dielectric constants. The material is primarily appropriate for electrical insulation due to its low dielectric constant. This can be used to isolate communication cables and conductors carrying signals from one another for fast signal transmission. It can also be used as an interlayer dielectric to reduce resistance–capacitance (RC) time delays, crosstalk, and power loss. Furthermore, these dielectrics are frequently utilized in dense multilayered integrated circuit boards where they reduce coupling and stray capacitance between nearby metal lines that may otherwise negatively impact the performance of the device. High dielectric constant polymer materials, on the other hand, are used as polarizable media in capacitor devices, as was previously discussed, as well as to aid in the propagation or reflection of electromagnetic waves and other devices like piezoelectric transducers, dielectric amplifiers or rectifiers, and memory components in semiconductor circuits.

The graphene-based reinforcement should have a particle size appropriate for EMI shield application by blocking/absorbing the waves, and in transparent applications, we must ensure that the nanocomposite remains transparent. However, the nano-reinforcement should also improve key properties of the nanocomposites such as the material's mechanical strength, folding endurance, scratch resistance, conductivity, etc. The graphene-based reinforcement for fire-retardant materials must also have functional groups that do not support combustion and the capacity to blanket and extinguish the fire effectively. In actuality, all three major classes of nano-reinforcement NPs, layered materials (nanoclays or graphenes), and CNTs have been used in EMI shields and flame-retardant materials, dispersed in a variety of polymer matrices, demonstrating that more than one formulation solution is typically feasible.

As shown in Fig. 6.3, a maximum value for the attributes achieved will be discovered for every particular polymer formulation, compatibilizer, and nano-reinforcement formulation. It frequently takes talent, expertise, and a lot of trial and error on the side of the formulation scientist to arrive at the best formulation. Statistical design-of-experiments approaches can frequently be used to good effect to more quickly arrive at an ideal formulation if the important factors that affect the formulation can be identified and quantifiable measures to evaluate its success (or failure) can be established [9, 10, 18, 20, 21]. It is also crucial to be mindful that study results for related nanocomposite formulations reported in scholarly articles and research reports occasionally conflict with one another. When comparing such results, it is useful to confirm that the formulations are genuinely equivalent with regard to the preparation, characterization, and starting materials used.

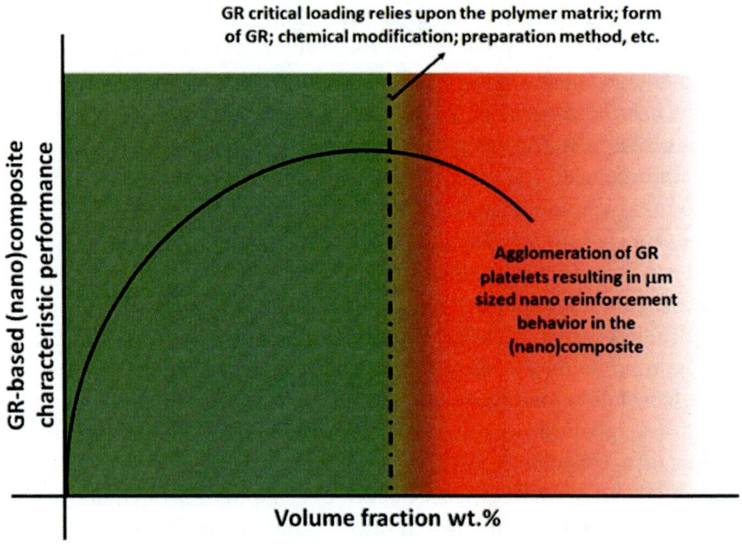

Fig. 6.3 Loading of nano-reinforcing filler affects the material characteristics of GNP nanocomposite. The formulation, as well as the fabrication technique utilized to make the composite, will determine the critical loading to attain maximal characteristics

6.3.1 Processing of Graphene-Bearing Polymer Nanocomposite EMI Shields

A typical processing technique for processing GNP-based composites is melt mixing. The key benefits of this approach are its adaptability, affordability, and environmental friendliness [22, 23]. Although solution blending or in situ polymerization cannot be compared to the dispersion level of GNP in a polymer matrix, it can meet commercial standards [24]. It has a wider range of applications and is more compatible with many recent industrial goods. High temperature and powerful shear pressures are essential for the working parameters of the melt processing process. These techniques enable polymer chains to migrate into the spaces between the NP interlayer and fracture the clumps of NPs [25]. There have been numerous devices employed in the past, including twin-roller mills [26], tri-roll mills [27], internal mixers [28], and extruders [29]. Strong shearing force may produce significant size reduction and structural distortion, which could lower the effectiveness of graphene-based nanofillers for polymers [24]. Additionally, the melt processing approach can only be used to create graphene in a dry powder form, like TrGO, due to the thermal instability of most chemically modified graphene-based fillers.

Melt Processing

Reports on the fabrication of high-performance EMI filters using GNP-bearing polymer nanocomposites have been greatly exploited by researchers and industrialists within the recent decade(s) [30].

High temperatures and shear forces are used in the melt processing operation to disperse the fillers in the polymer network. This process avoids the continuous use of hazardous solvents. A typical melt processing schematic is shown in Fig. 6.4.

Furthermore, melt processing is generally seen as more cost-effective than solution blending. The high temperature for GNP-polymer nanocomposites condenses the polymer stage and enables easy scattering or intercalation of GO platelets. However, due to the enlarged consistency at a high filler stacking, melt processing is less effective in dispersing GNP sheets than dissolvable mixing or in situ polymerization. Both polar and non-polar polymers can be used in the process. This technique can produce various GNP-based nanocomposites, such as peeling GP PMMA, GNP-PP, GO-poly(ethylene-2, 6-naphthalate), and GNP-polycarbonate. Although the poor throughput of artificial rGO in the melt processing process limits the use of graphene nanofiller, GNP formation in mass quantities in heated reduction can still be a suitable option for contemporary scale creation. Despite this, homogeneous scattering in polymeric lattice liquefiers may be hampered by the loss of the practical gathering in warm decreasing, particularly in non-polar polymers.

The most difficult method of combating electromagnetic pollution is the fabrication of high-performance electromagnetic interference shielding effective GNP-polymer nanocomposite. The work by Bhawal and colleagues [32] focused on creating IrGO by combining GO with EMA in situ (Fig. 6.5) to produce composite materials with improved shielding efficiency and regulated electro-mechanical properties. It involved the reduction of GO in polymer matrices, and the efficiency of this

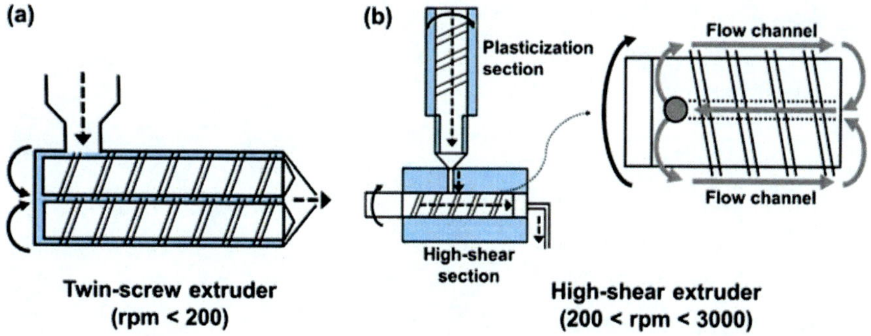

Fig. 6.4 Schematic of the extrusion process using **a** conventional twin-screw extruder and **b** high-shear extruder, which exhibits flow recirculation in a closed barrel. Reproduced with permission from Lee et al. [31]. Copyright 2023, Elsevier Science Ltd.

process was greatly impacted by processing factors, polymer chemistry, and methodology [32]. Making high-performance GNP-polymer nanocomposite electromagnetic interference shields is difficult to combat electromagnetic pollution. The development of EMA-IrGO nanocomposites in a single step with improved shielding effectiveness and modifiable electro-mechanical characteristics could support large-scale production for technologically practical applications.

In a separate investigation, researchers reportedly created melt-compounding-based blends of highly conductive PP and PET with excellent EMI shielding properties. The blends showed co-continuous morphologies in which TRGO nanosheets had been thermodynamically and kinetically pushed into PET polymer, resulting in a twofold percolated structure. This was based on SEM and TEM analyses. With the addition of TRGO, the blend nanocomposites' direct current and alternating current electrical conductivities were enhanced, and a remarkably low electrical percolation threshold of 0.08 vol.% was attained. The samples' complex permittivity and complex permeability were measured in the X-band (8.2–12.4 GHz) in order to look into their electromagnetic properties. Depending on the frequency, a shielding effectiveness of 40–60 dB was achieved, with only 2 vol.% of TRGO being incorporated. The EMI shielding effectiveness rose significantly with increasing TRGO loading

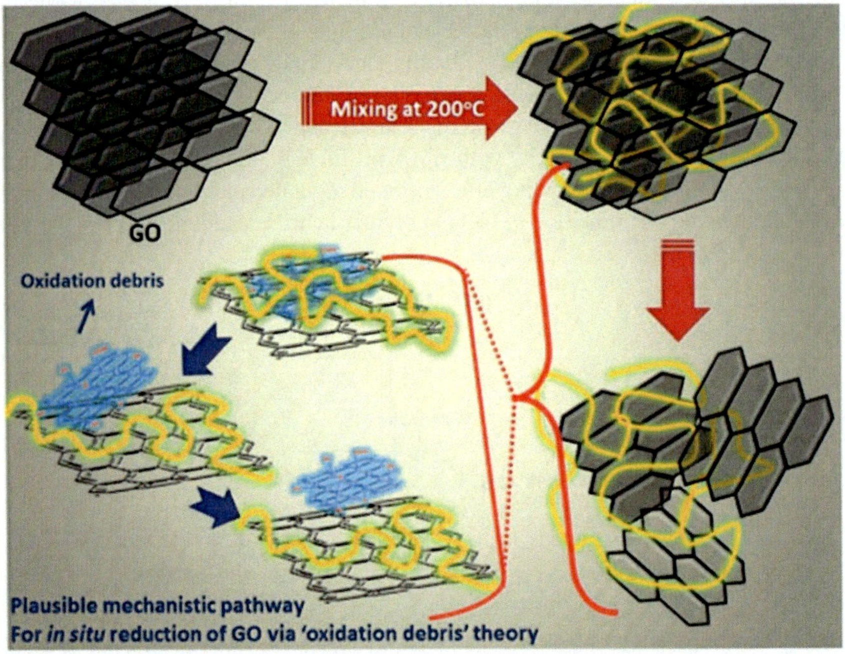

Fig. 6.5 Pictorial representation of in situ reduction mechanism of GO in EMA matrix via single-step melt processing by peeling off the oxidative debris from GO sheets to give segregated rGO architecture for high EMI SE with adjustable electro-mechanical properties. Reproduced with permission from Bhawal et al. [32]. Copyright 2018, Elsevier Science Ltd.

[30]. The decrease of GO through a thermal shock (with a quick heating rate of about ~2000 °C/min) and one-step thermal exfoliation were the methods used by these authors to create GNPs.

Nevertheless, substantial amounts of oxygen were still present in the final reduced sheets. GO sheets were first made using the modified Hummers process from pure GP powder. In order to produce GNPs, the GO was put into a muffle furnace and heated to 1050 °C for 30 s after precipitation and drying in a vacuum oven at 80 °C for 24 h. After that, PET was dried in a vacuum oven for 24 h at 100 °C, and 50/50 weight-composition PP/PET blends were made by melting the materials in a Brabender internal mixer for 12 min at 270 °C. To prevent heat oxidation, 0.1 wt.% of the antioxidant Irganox 1010 was added to the combinations. After 3 min of mixing, different volume fractions of TRGO were added individually to the neat mixture for samples containing GNPs. The weight fraction of GNPs in the composite samples was converted to volume fraction using the theoretical GP density (i.e., 2.28 g.cm^{-3}) (vol.%). Compression molding was used to molded specimens at 270 °C and 15 MPa pressure for microscopy and electrical conductivity analyses [30].

In another interesting study, a group of researchers examined the property performance of PC/SAN/GNPs nanocomposite prepared through a melt processing approach [33]. These researchers processed their blends using a Haake Minilab-II extruder operating at a temperature of 260 °C and an rpm of 60. The reinforcing fillers utilized herewith were G-NiO and G-Ni at different wt.%. The interesting part was the prior utilization of solution intercalation/casting to prepare the blends accordingly. Afterward, the blends dried at 80 °C and then 120 °C were processed in the melt mixing machine [33]. The prior use of solution mixing before melt extrusion enhances the GNPs dispersion/distribution within the matrix blend, leading to improved overall property performance.

In an interesting work, Sadeghi and his colleagues [30] processed a PP/PET/TrGO-polymer blend nanocomposite using melt mixing or compounding to fabricate a high-performance EMI shield(s). They revealed that the EMI SE greatly relied upon the TrGO content inclusion and the frequency as they reported an EMI SE of 40–60 dB with the inclusion of 2 vol.% TrGO.

The interesting thing about melt mixing/blending usage for the processing of GNP-based nanocomposites is that the reduction of GNP-based materials such as GO into highly conductive TrGO, which is beneficial for enhanced EMI SE of the prepared nanocomposites can be achieved in situ during the nanocomposite fabrication [32]. In this vein, Bhawal and his colleagues [32] fabricated an EMA in situ TrGO nanocomposites having excellent EMI SE of 30 dB at just 5 wt.% inclusion of the GNPs. Their fabrication process is depicted in Fig. 6.5.

The use of melt processing/mixing for the preparation of highly efficient GNP-based polymeric EMI filters is a very reliable as well as scalable process that can be easily commercialized, though its demerit(s) is unacceptable dispersal of the reinforcing filler (poor dispersion of Graphene-based reinforcing filler in most instances).

Solution Intercalation/Casting

This method is one of the most popular ways to make polymer-based composites since the polymer dissolves quickly in fluids and natural solvents such as water, ethanol, DCM, toluene, chloroform, and DMF.

This technique involves dissolving the polymer in appropriate solvents before combining it with the ready-made GNP-based dispersion solution. With various graphene-based reinforcements, the polymers, such as PS, polyimides polycarbonate, polycaprolactone, polyacrylamide, polyurethane, and PMMA, have been successfully mixed in solution. Additionally, the simple creation of fluid graphene-based reinforcements by sonication, such as GO platelet suspensions, makes this approach particularly appealing for water-dissolvable polymers. The amount of GNP shedding for solution mixing techniques often controls its scattering in the composite. Accordingly, solution mixing presents a potentially effective approach to cope with GO platelet scattering into specific polymer networks. Surfactants and stage exchange approaches have been used to improve the solution mixing of GNP-polymer nanocomposites, among other techniques. Surfactants may provide distinct composite qualities, even though this is undesirable in some applications. Solution mixing has been largely adopted owing to its facile process [34, 35]: For instance, Esfahani et al. [34] presented excellent findings in their work on the preparation of PU/rGO nanocomposite, where they used DMF solvent for the desolation of the PU at 50 °C. The PU/rGO mixture was added to deionized water dropwise, washed, and dried overnight. The dried samples were immediately followed by compression molding at 200 °C to fabricate membranes having different thicknesses. Their findings were that the composite system “TPU/mTrGO5” attenuated 99.7% (EMI SE of 25 dB) of the incident EM waves, which were due to absorption-based shielding enabled by viscoelastic energy dissipation [34].

An interesting report has also been presented by Zhou et al. [36]: After cleaning with deionized water and drying at 60 °C for 24 h, the authors immersed PP micropowders in isopropyl alcohol for 2 h. A 600 mL aqueous solution containing 12 g Pyrogallol and 1.6 g of ammonium persulfate was then mixed with 100 g of treated PP, and the mixture was maintained, stirring in a three-necked flask at 80 °C for 24 h. The pH of the mixture was then adjusted to 8.2 using NH_3H_2O. After properly drying the powders at 60 °C for 24 h, following being thoroughly cleaned with deionized water, mPP was eventually produced. Once more, GO dispersions were treated with mPP and mechanical stirring for 30 min, after which MWCNTs were added, and the treatment continued for another 30 min. The mixture was then dried at 60 °C for 12 h. Following drying, the composites were placed in a mold with a 3 mm cavity thickness and hot-pressed for 5 min at 190 °C, resulting in the thermal conversion of GO to RGO. Figure 6.6 shows the schematic for creating a composite made of mPP, rGO, and MWCNTs [36].

A facile approach toward the actualization of high-performance EMI shields has been reportedly prepared using an ordered multilayered film casting approach [37]. Figure 6.7 depicts the schematic for creating the ordered multilayer film. The xylene solvent (20 g/L) was used to dissolve the SEBS-g-MAH. Then, 80 g/L of f-BN

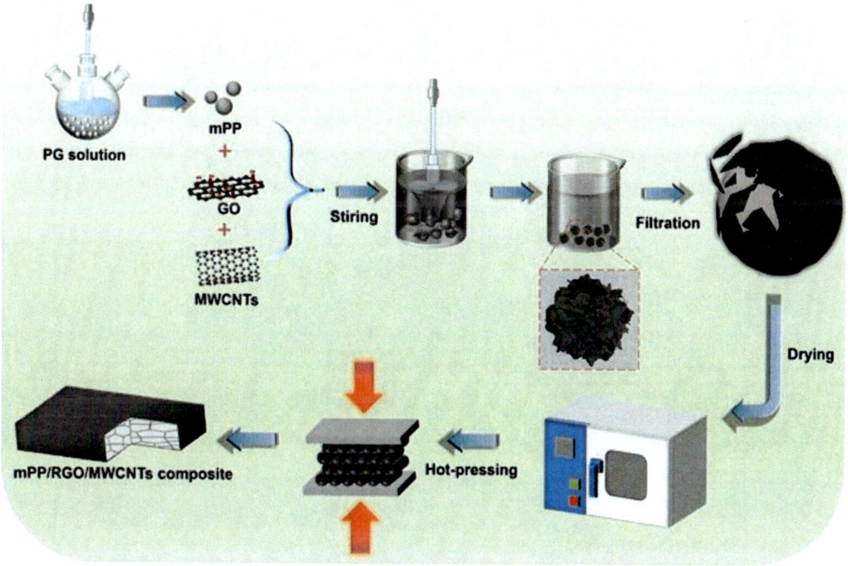

Fig. 6.6 Schematic for the preparation of mPP/rGO/MWCNTs composite. Reproduced with permission from Zhou et al. [36]. Copyright 2023, Elsevier Science Ltd.

particles were added to the solution. In the meantime, the PHDDT was dissolved in a 1:1 volume ratio mixture of phenol and tetrachloromethane (30 g/L). Then, the PHDDT solution was supplemented with the f-GO particles (70 g/L). In step I, as illustrated in Fig. 6.7c, the SEBS-g-MAH/f-BN solution was cast onto a glass plate and left there for 24 h before being dried for 12 h at 80 °C in a vacuum oven. The SEBS-g-MAH/f-BN film was coated with the PHDDT/f-GO solution, and the glass plate cast was then dried for 12 h at 80 °C in a vacuum oven (process II as illustrated in Fig. 6.9c. The multilayer film can be produced by repeating steps I and II above.

In a particular study, Singh et al. [38] prepared an interesting EMI shield thus: To create a homogenous slurry, the necessary wt.% of RGO and NBR was taken and mixed with the solvent xylene (Fig. 6.8). The resulting slurry was utilized to create an aluminum panel covering that was 2–3 mm thick. The coating was given 24 h to cure in the air. As illustrated in Fig. 6.8, this RGO/NBR composite rubber sheet was carefully taken off the aluminum panel and cut into the required size for waveguide tests at various microwave frequency bands. The composite sheets with 2%, 4%, and 10% RGO in the NBR matrix were measured.

Figure 6.9 shows the calculated reflection loss for nanocomposites with different TrGO content and thickness. When demonstrated, the reflection loss rises from 2 to 57 dB at 9.6 GHz as the RGO component in the NBR matrix grows at a thickness of 3 mm. Notably, the reflection loss peak for a sample containing 10% by weight RGO changed from 9.6 to 6.9 GHz when the thickness was increased from 3 to 4 mm. Additionally, it has been noted that a composite of 10 wt.% RGO/NBR exhibits

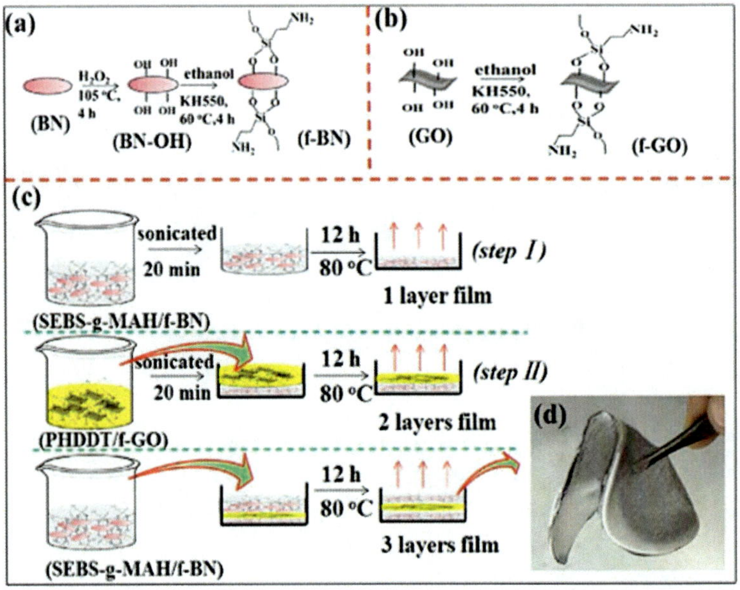

Fig. 6.7 Schematic of the fabrication of the ordered multilayer film: **a** The surface of the BN particles; **b** the surface of the GO particles; **c** the preparation process of the ordered multilayer film; and **d** the optical photograph of the three layers film. Reproduced with permission from Zhang et al. [37]. Copyright 2016, Elsevier Science Ltd.

maximum return loss values of 57 dB (3 mm) and 38 dB (4 mm) for frequencies of 9.6 GHz and 6.9 GHz, correspondingly [38].

The following is an additional explanation of a potential mechanism: In TrGO, dihedral angles are easily produced inside stacks of flakes of TrGO sheets; these angles cause repeated reflections of the microwaves, lengthening their propagation path in the absorber. These multiple microwave reflections, which result in higher losses of electromagnetic energy, may be caused by the fact that molecular motions like ionic conduction and dipolar relaxation are caused when microwaves interact with dielectric materials. The resistive force opposes these molecular motions, so energy is lost as heat [38].

The use of solution intercalation/mixing/casting has been of great interest to researchers owing to its facile nature; though being a batch technique, its lack of continuous production has made it less attractive for most industrial production of advanced polymer materials reinforced with carbon materials like GNPs.

Fig. 6.8 Illustration of preparation of RGO–NBR composite. Reproduced with permission from Singh et al. [38]. Copyright 2012, Elsevier Science Ltd.

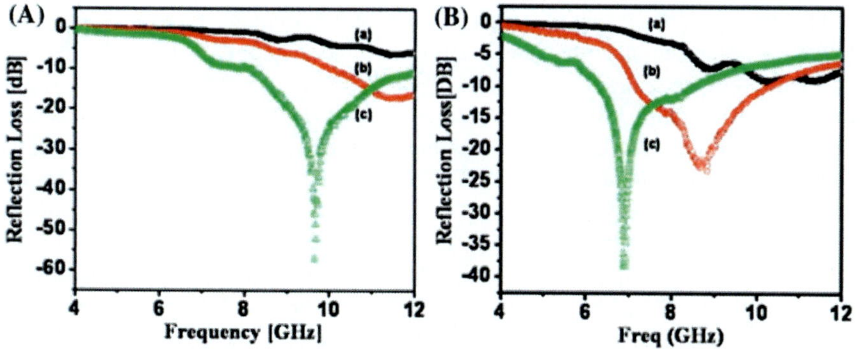

Fig. 6.9 Reflection loss characteristics with thickness: **A** 3 mm, **B** 4 mm for RGO/NBR composites. (a) 2 wt.%, (b) 4 wt.% and (c) 10 wt.%. Reproduced with permission from Singh et al. [38]. Copyright 2012, Elsevier Science Ltd.

In Situ Polymerization

This process involves swelling GNP or a derivative in the liquid monomer before dispersing the appropriate initiator [39]. Either heat or radiation can start the polymerization process. After in situ polymerization, well-dispersed GNP in a polymer matrix is produced by the intercalation of monomers into the layered structure of GP, increasing interlayer spacing and exfoliating GNPs. The covalent attachment between the functionalized sheets and polymer matrix is made feasible by the in situ polymerization technology through several chemical processes. The main disadvantage of this method is the increase in viscosity as the polymerization process progresses, making manipulation difficult and reducing the load fraction [40].

Additionally, the process is sometimes carried out in the presence of solvents, making solvent removal a crucial issue, just like the solvent mixing technique [40]. It has proven possible to produce (nano)composites with covalent cross-connections between the grid and filler using in situ polymerization procedures. Moreover, in situ polymerization has produced non-covalent composites using several other polymers, such as PE, PMMA and poly(pyrrole), PU, nylon 6, PMMA, PBT, PANI, etc. Contrary to solution mixing or melt processing techniques, in situ polymerization systems achieve an anomalous condition of scattering of GNP filler without prior shedding. It has been explained that intercalating monomer, followed by polymerization to isolate the layers, occurs between the layers of GP or GO. This process has mainly been looked at for GP or polymer nanocomposites with GO. For instance, a soluble base metal, a monomer, and GNP sheets with opposite charges can all intercalate GP. The polymerization process is then initiated by the GNPs. GNPs may be lost during polymerization, although single-layer graphene sheets were not seen. In addition to this, the restacking approach is a more modern forced methodology in which nanocomposites are obtained by converting the host material into a colloidal framework and precipitating in the presence of the polymer. This method differs from others in that it does not make use of a previously framed host layered precious stone. These nanocomposites exhibit remarkable modifications in the microstructural stage as well as enhanced material characteristics in relation to both guardian stages.

In situ polymerization approach has been used by several researchers in recent decades for the fabrication of EMI shields [41]. In this regard, Wu and colleagues [41], in their work where they presented the fabrication technique for high-performance EMI shield GNP/PEDOT:PSS blend nanocomposite [41], they started their process by firstly modifying (non-covalently) the pristine GNP with 4-dodecylbenzenesulfonic acid (DBSA: an anionic surfactant compatible with GNP and PEDOT:PSS) aimed at enhancing the wettability as well as adhesion between GNP and the PEDOT:PSS blend (matrix). The mixture was drop-coated onto the functionalized GFs using successive drops of varying total amounts to produce varied PEDOT:PSS-to-GF mass ratios, which were then dried at 80 °C for 0.5 h to evaporate any excess DMSO and water (Fig. 6.10). The dried PEDOT:PSS sheet had 850 S/cm of electrical conductivity. Composites with various PEDOT:PSS contents were created and designated GF/PEDOT:PSS-X, where X denotes the mass ratio of GF to PEDOT:PSS [41]. They found that the GF/PEDOT:PSS composites had a high porosity of 98.8%, an ultralow density of 18.2×10^{-3} g/cm^3, and an increased electrical conductivity of nearly 4 times, from 11.8 to 43.2 S/cm, following the addition of the conductive PEDOT:PSS. With a SE of 91.9 dB and a SSE of 3124 dB cm^3/g, the composites exhibit exceptional EMI shielding performance thanks to their superior electrical conductivity, lightweight porous structure, and efficient charge

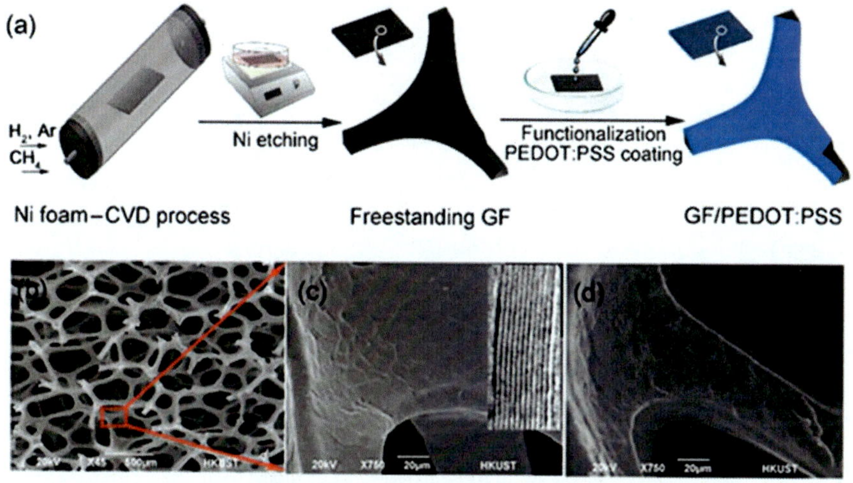

Fig. 6.10 **a** Schematic procedure of the preparation of GF/PEDOT:PSS composites; SEM and TEM images of GFs **b**, **c** before and **d** after PEDOT:PSS coating. Reproduced with permission from Wu et al. [41]. Copyright 2017, ACS Publications

delocalization [41]. Additionally, they proposed that left-handed composites with absolute permittivity and/or permeability larger than one, as well as composites derived from GFs with three-dimensionally interconnected conductive networks, as well as composites with conductive polymer coating, produce significant microwave attenuation by absorption due to their excellent electrical conductivities.

Yet, in another work, Nasir et al. [42] synthesized GO-g-PANI using in situ polymerization thus: A mixture of GO, tween 80, and 100 mL of HCl (1 M) were charged in a glass reactor and sonicated for one hour to obtain a uniform dispersion. The mixture was appropriately infused with monomer "ANI" and initiator "KPS," and the temperature was increased to 70 °C. For 4 h, the solution was mixed. To remove unreacted monomer and homopolymer, the finished product was filtered, washed with ethanol, and dried (Fig. 6.11).

In a separate work, Zhang et al. [43] described a novel method for creating high-performance GNP-based EMI shields. They incorporated GP and modified reduced iron powder into the reaction system created by lignin and isocynate to create lignin-based polyurethane EMI shielding material (FeGLPU). According to their findings, the maximum value of EMI SE reached 21.6 dB, and the highest frequency width larger than 10 dB was 18.3 GHz when the contents of Fe and GNP were fixed at 10% and 5%, respectively. The tensile strength of the material was 11.7 MPa, and conductivity was 4.27 10^{-4} S/m. At a weight loss of 5%, the thermal decomposition temperature was 288 °C. When the lignin concentration was 20%, the maximum EMI SE was 22.5 dB, and the highest frequency at which this occurred was 18.3 GHz. The thermal decomposition temperature at 5% weight loss was 234 °C, and the electrical conductivity was 1.06×10^{-2} S/m. The GNP and the Fe were evenly distributed throughout the lignin-based PU matrix, as shown by the SEM observation. They

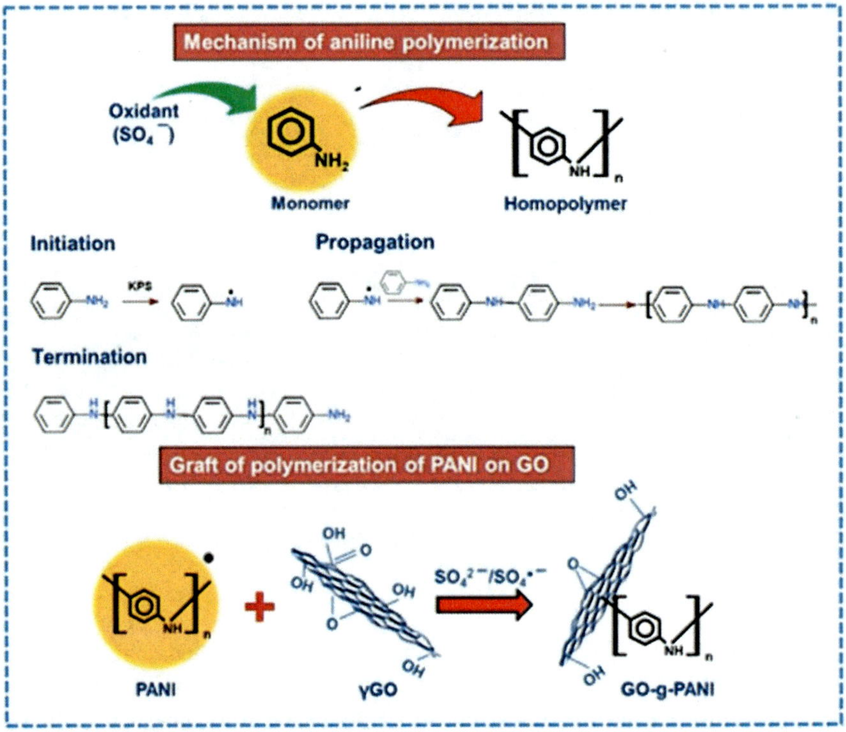

Fig. 6.11 Schematic illustration of the mechanism of graft polymerization of ANI onto GO. Reproduced with permission from Nasir et al. [42]. Copyright 2023, Elsevier Science Ltd.

also showed that the lignin-based polyurethane electromagnetic shielding material has excellent mechanical qualities, great thermal stability, and high electromagnetic shielding properties. This was brought about by the interaction of lignin, GNP, and Fe. Due to its inexpensive cost and exceptional qualities, this electromagnetic shielding material has several potential applications [43]. Table 6.2 reports the comparative presentation of the EMI shielding performance of various graphene-based composite shields prepared by diverse fabrication approaches.

Latex Mixing/Blending

The concept of excluded volume serves as the foundation for the latex mixing technique. The graphene-based nanosheets are forced into the interstitial space between the polymer latex particles, drastically limiting the amount of space that the filler can occupy and yet creating a continuous, segregated network across the whole volume of the sample. To create polymer particles coated with graphene-based nanosheets, a water-born latex suspension and an aqueous filler dispersion are typically combined

Table 6.2 Comparative presentation of the EMI shielding performance of shield or potential prepared by diverse fabrication approaches

Fabrication approach	EMI shield formulation	GNP wt.%	Thickness (mm)	Dominant shielding mechanism	SE (dB)	Reference
In situ polymerization						
	GNP/PEDOT:PSS	5	1.5	Absorption	91.9	[41]
	GO-g-PANI/Paraffin wax	–	30	Absorption	36.2	[42]
	rGO-PANI	1	–	Absorption	−25	[44]
	GNP -$PANI_{10:1}$ @polyimide (PI)	40	0.04	Absorption	21.3	[45]
	PANI@GNP	10	1.5	Absorption	−14.5	[46]
	PEDOT-GNP nanoplatelets (GNPs)	15	–	Absorption	18	[47]
	PANI/S-RGO/Paraffin	10	5	Absorption	−22.5	[48]
	Fe/ GNP /PU/lignin	2	1	Absorption	21.6	[43]
Solution intercalation/casting						
	rGO/PANI-hexaferrite/MWCNT/ABS nanocomposites	5	1	Absorption	21.5	[35]
	mPP(PP micropowders)/RGO/MWCNTs	0.5	3	Absorption	8.3	[36]
	GO/BN/ poly(-hydroxybenzate-co-DOPO-benzenediol dihydrodiphenyl ether terephthalate))-BN/ maleated styrene-ethylene/butylene-styrene)	35	0.02–4	Absorption	29.7	[37]
	GO/nitrile butadiene rubber (NBR)	10	3	Absorption	57	[38]
	GO/PEO	2.6 vol.%	4	Absorption	−38.8 (R_L)	[49]

(continued)

Table 6.2 (continued)

Fabrication approach	EMI shield formulation	GNP wt.%	Thickness (mm)	Dominant shielding mechanism	SE (dB)	Reference
	TrGO/TPU	5 vol.%	1	Absorption	24	[34]
Foaming/coating						
Foaming	PU/GNP	20	2.4 (6.5)	Absorption	17–21 (23–24)	[50]
Dip coating	PUG-5	5	20	Absorption	12.4	[51]
Dip coating	PUG-5	5	40	Absorption	24.6	[51]
Dip coating	PUG-5	5	60	Absorption	34.7	[51]
Dip coating	PUG-10	10	20	Absorption	19	[51]
Dip coating	PUG-10	10	40	Absorption	41.6	[51]
	3D GNP@Fe_3O_4 polydimethylsiloxane (PDMS)	9.72	2	Absorption attenuation & reflection	70.37	[52]
Coating	GNP-PVDF	5	–	Reflection	20	[53]
Foaming	PEI/GNP	0.21	1.5	Absorption attenuation & reflection	44	[54]
Foaming	PI/rGO	16	0.8	Absorption	21	[55]
Foaming	Polyetherimide (PEI)/GNP@Fe_3O_4	10	2.5	Multiple reflections/ absorption	41.5	[56]

(continued)

Table 6.2 (continued)

Fabrication approach	EMI shield formulation	GNP wt.%	Thickness (mm)	Dominant shielding mechanism	SE (dB)	Reference
Coating	GNP/PDMS	0.7	1	Absorption	30	[57]
Foaming	GNP/HDPE	9.8 vol %	3	Multiple reflections/ absorption	31.6	[58]
Melt processing/mixing						
	PP/PET/TRGO	2 vol.%	5	Absorption	60	[30]
	PP/PET/GNP	2 vol.%	5	Absorption	40–60	[30]
	EMA/in situ TrGO	5	5	Absorption	30	[32]
	PP/GNPs	3	400 µm	Absorption	~30	[31]
Others						
Solution intercalation + casting + dip coating + spray coating	Epoxy/CF-30-Fe_3O_4-30GNP	30	1.0 ± 0.1	Absorption	33.1	[59]
Two-roll mixing + vulcanization	FGS/ poly(dimethyl)siloxane (PDMS)	2	100 µm	Absorption	–	[60]
Latex blending	Poly(styrene-butyl acrylate) latex (P(St-BA))-GNS and	5	0.3	Absorption	35	[61]
	Poly(methyl methacrylate-butyl acrylate) latex (P(MMA-BA))-GNS	20	0.05	Absorption	21.5	[61]

(continued)

Table 6.2 (continued)

Fabrication approach	EMI shield formulation	GNP wt.%	Thickness (mm)	Dominant shielding mechanism	SE (dB)	Reference
	AEMA-GN/WPU	5 Vol.%	4	Absorption	38	[62]
	GNP@PVP/WPU	30	0.15	Absorption	30	[63]
	10Ag@FRGO/WPU	5	2	–	35	[64]
	S-GNS/WPU	5 vol.%	2	Absorption	32	[65]

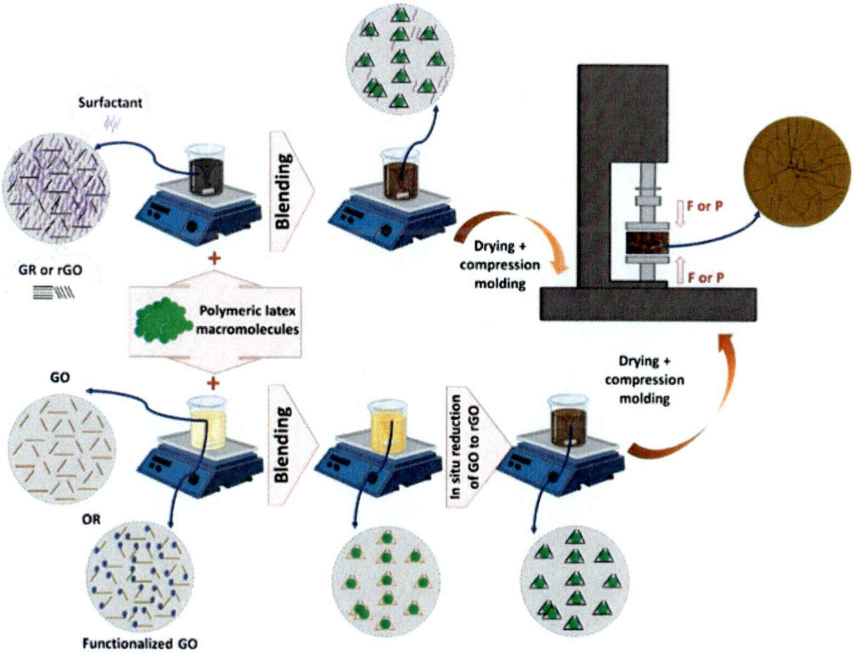

Fig. 6.12 Diagrams outlining the primary phases in setting up the latex-based mixing process for the creation of segregated 3D-GAs. It is suggested that starting materials could be either GNP/rGO or GO nanosheets. The author created with slight modification from BioRender.com

dropwise. These particles are then gathered and hot-pressed as depicted in Fig. 6.12. Even though the fundamental steps of the latex-based approach are fundamentally the same regardless of the kind of polymer and GNPs used, a wide range of synthetic techniques have been proposed in the literature [66].

GNP's strong contact and high surface energy make it challenging to spread evenly within a polymer matrix. The latex mixing method causes GNPs to spread along particular channels and produce a segregated structure, in contrast to conventional dispersion strategies [67]. It has straightforward operation methods. The surface of the polymer particles is first coated with GNP before the final composites are constructed using the appropriate procedures (hot-pressing is the most common). The main disadvantage of this method is that it can only be used with conductive materials [68].

An excellent example of the use of latex compounding as an efficient approach for the fabrication of EMI shields can be obtained from the report by Wei et al. [61]. They suggested using a π-π stacking interface design strategy to create a waterborne polymer/sulfonated GNPs (S-GNS) nanocomposite that is ultrathin, flexible, and easy to self-assemble in latex (Fig. 6.13).

In a different case, a team of researchers created flexible and lightweight GNPs (GNs)/WPU composites that perform very well in terms of electrical conductivity

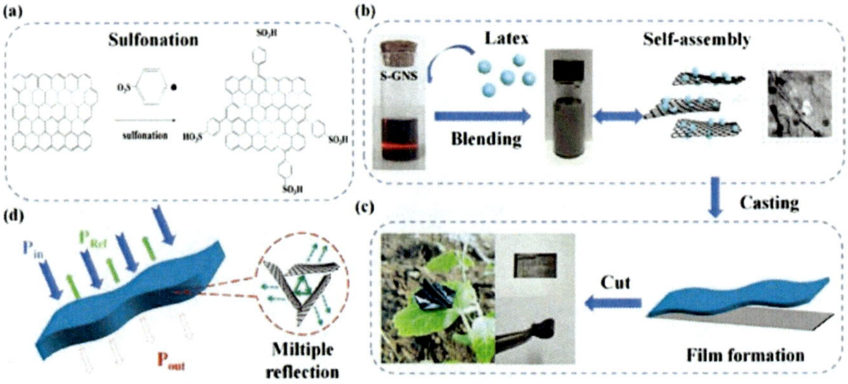

Fig. 6.13 **a** Schematic representation of the preparation of S-GNS, **b** composite latex and **c** composite film; **d** schematic representation of the shielding mechanism of composite films. (A color version of this figure can be viewed online). Reproduced with permission from Wei et al. [61] Copyright 2019, Elsevier Science Ltd.

and electromagnetic shielding. The authors successfully suppressed the restacking and aggregation of the GNs by covalently altering GNs with aminoethyl methacrylate (AEMA; AEMA-GNs) through free radical polymerization. This was made possible by the NH^{3+} functional groups grafted on the AEMA-GNs. The electrostatic attraction that led to the AEMA-GNs' uniform dispersion in the WPU matrix also demonstrated that the AEMA-GNs were compatible with a WPU matrix with grafted sulfonated functional groups. It was claimed that the GNs may create electrically conductive networks because of their homogenous dispersion. Additionally, AEMA-GNs with varying numbers of AEMA segments were added to the WPU matrix, and it was discovered that the surface chemistry of the GNs had an impact on the electrical conductivity and EMI shielding capabilities of composite materials. As shown in Fig. 6.14, their findings showed that AEMA-GN/WPU nanocomposites with GN loading of 5 vol.% had an outstanding electrical conductivity of around 43.64 S/m and EMI SE of 38 dB over the X-band frequency [62].

The initial step in creating the water-dispersible S-GNs was to alter the GNs using aryl diazonium salt. Styrene was added to a waterborne polymer to create an interaction that stacks S-GNs on top of itself. The composite displayed high EMI SE within the X-band and good mechanical characteristics. The property performance of the nanocomposite, such as tensile strength, was improved by 578% with a filler loading of 20 wt. %. A high EMI SE of 21.5 dB at 0.05 mm thickness was produced by adding 25 wt.% of the GNP filler remained constant after 1000 cycles of bending. The composite reportedly presented a specific SE/thickness (SSE/t) of 2663 dB cm^2/g, exceeding previously reported materials with a similar filler loading (25 wt.%). The nanocomposite was suggested as a viable option for coating solid surfaces, demonstrating its enormous potential for widespread use [61].

Yet in another report, Tian and his colleagues [69] reported a facile fabrication of WPU/GO nanocomposites using a latex compounding approach. The flask containing

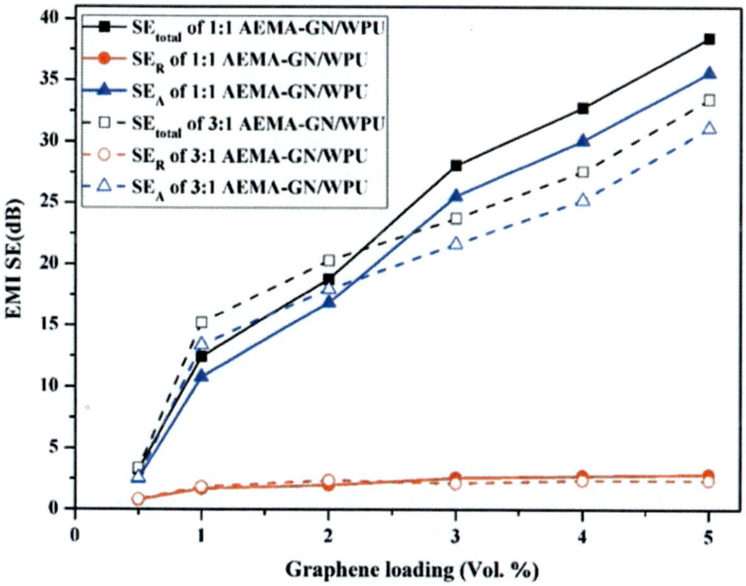

Fig. 6.14 Comparison of the SE_{total}, SE_{A}, and SE_{R} of the 1:1 AEMA-GN/WPU and 3:1 AEMA-GN/WPU composites with different GN loadings at 9.6 GHz. Reproduced with permission from Hsiao et al. [62]. Copyright 2015, American Chemical Society

the WPU emulsion received the GO dispersion while immersed in an ultrasonic bath. The mixture was then supplemented with 10% DETA and stirred at 150 rpm. To obtain NRGO/WPU composites dispersion, the reaction mixture was maintained at 80 °C continuously for 24 h. The nanocomposite sample was created by pouring the NRGO/WPU composites dispersion into Teflon molds and drying them at 50 °C for 24 h in a vacuum oven. The samples were washed twice, once with DI water and once with acetone at 40 °C for 12 h. Three times were spent repeating the washing process outlined above. Finally, vacuum drying the composite samples at 70 °C for 24 h produced the samples [69].

Once more, a team of researchers created GNs/WPU composites that are flexible, light, and have outstanding EMI shielding capability. As the polymer matrix, WPU containing sulfonate functional groups was employed. Restacking and aggregation of the S-GNs were effectively reduced by adsorbing the cationic surfactant (stearyl trimethyl ammonium chloride) on their surface. This attracted sulfonate groups from the WPU matrix. The S-GNs have shown good compatibility with the WPU matrix due to the advantageous interfacial interactions brought on by electrostatic attraction. An electrically conductive network was built as a result of this homogenous dispersion. According to reports, the S-GNs/WPU composite showed a low electrical conductivity percolation threshold as well as an exceptionally improved electrical conductivity of about 5.1 S/m. The WPU composites with 5 vol.% (or around 7.7

wt.% S-GNSs) S-GNs contents had a high EMI shielding efficacy of about 32 dB [65].

This approach has not been widely adopted for the fabrication of high-performance EMI shields yet, though the limited literature has proven its potential as a reliable approach for the preparation of GNP-based EMI shields.

Foaming

A significant type of polymer material known as a porous polymer material is polymer foams created using foaming technology. This type of polymer material has many microscopic foam pores inside its polymer matrix. Polymer foam has many benefits over hefty polymer materials, including low density, effective heat and sound insulation, high specific strength, and corrosion resistance. The polymer business is currently significantly impacted by polymer foam, one of the most often utilized polymer materials. Polymer foam was processed from the polymer using mechanical, physical, and chemical foaming. Foam extrusion molding and foam injection molding are the two main processing methods for polymer foams. When using a supercritical fluid, such as CO_2 or N_2, for molding, polymer foams can be foamed with a lighter weight and greater dimensional stability than their solid equivalents. In numerous industrial applications, polymer foams are frequently employed. For instance, PU soft foam can be utilized in the creation of furniture, composite fabrics, clothes, shoes and hats, car seats, and sports equipment [70], while PU hard foam can be utilized in the creation of refrigerators, freezers, and refrigerated containers [70]. PP foam is frequently utilized in the manufacture of various seat cushions, thermal insulation materials, shockproof materials, packaging materials, and building materials in the military, transportation, and aerospace industries [70]. Phenolic foam is mostly utilized in the construction, automotive, electrical, and electronic industries, iron and steel industries, and aerospace industries [70]. It can satisfy the needs of large-scale sports and entertainment venues, high-rise structures, and fast vehicles for heat insulation and combustion-resistant building materials. With its high thermal insulation, excellent thermal stability, and strong fire resistance, polyimide (PI) foam has several potential uses in the aerospace, aircraft, and marine industries [70]. Polymer foams are constantly evolving, and they are now widely employed in many different industries. Polymer foams, the most significant industrial polymer products, have shown a wide area for expansion and potential in the market for EMI shielding materials.

The preparation of segregated GR-based reinforcing filler can be done in three basic ways. The cell walls and struts of the foam can be viewed as a real polymer nanocomposite once the segregated NP network has developed within the polymeric skeleton, as schematically shown in Fig. 6.17a. The second method shows the other side of the coin: A previously created GNP-based foam can be covered with a homogeneous and seamless polymeric layer later on as shown in Fig. 6.15b. The third method involves using a preformed polymeric foam as a template to direct the creation of segregated GNP-based reinforcing filler as a continuous covering of

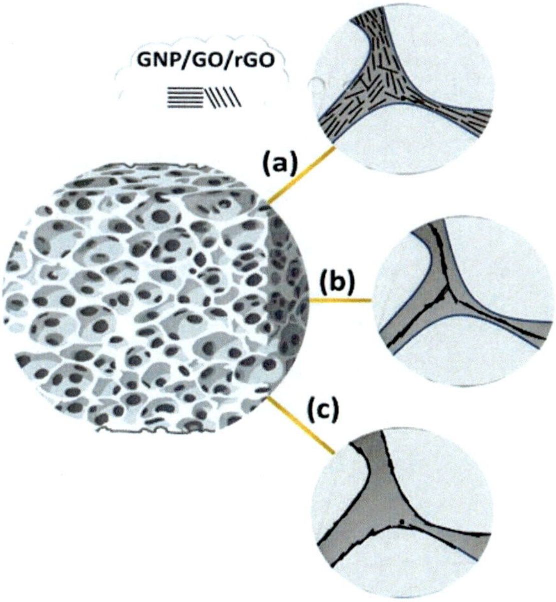

Fig. 6.15 Internal structure of foams made of polymer (nano)materials with segregated GNP-based reinforcing filler is shown schematically: **a** GNP-based reinforcing filler incorporated inside the foam's polymeric structure; **b** foam made of GNP-based reinforcing filler that has already been created and is coated with polymers; **c** nanosheets based on GNP are applied to polymer foams

interconnected nanosheets on the surface of the pores as presented in Fig. 6.15c. The following provides an illustration of the various methods that can be used for each material category's preparation.

For the application of GNP-based nanocomposite foams in EMI shielding, Hamidinejad et al. [58] fabricated high-density polyethylene (HDPE)-GNP nanocomposite foams via injection molding assisting physical foaming using supercritical fluid (SCF) treatment. These authors proved in their study the superior performance of foamed EMI shields over their solid counterparts by preparing side by side both solid samples and foamed samples with similar GNP contents, as depicted in Fig. 6.16 [58]. They also observed that the GNP content required for obtaining or attaining a given EMI SE was less for the foamed nanocomposites in comparison to the solid ones: e.g., GNP vol.% of 19 and 14 was required to attain an EMI SE of 21 dB for the solid and foamed nanocomposite shield.

Also, Li et al. [50] adopted this approach. For example, in Fig. 6.17a, the creation of microcellular PUG composite foams is schematically depicted. In a typical procedure, ample DMF (500 mL) was ultrasonically mixed with GNP (1.304 g) for 10 min. The produced GNP suspension was put into a flask with a round bottom, and 15.0 g of PU pellets were added while the mixture was vigorously stirred at 60 °C for 5 h. PUG composite was created by precipitating the mixture with an aqueous solution. After that, the composite was redissolved in DMF (100 mL) to produce a homogenous PUG solution. The composite was then dried in a vacuum oven at 60 °C for 10 h. In order to create thicker samples, the solution may be poured into an open Teflon mold or cast on a glass substrate after cooling to room temperature. The solution would then be exposed to air with a relative humidity of 80% at room temperature. The

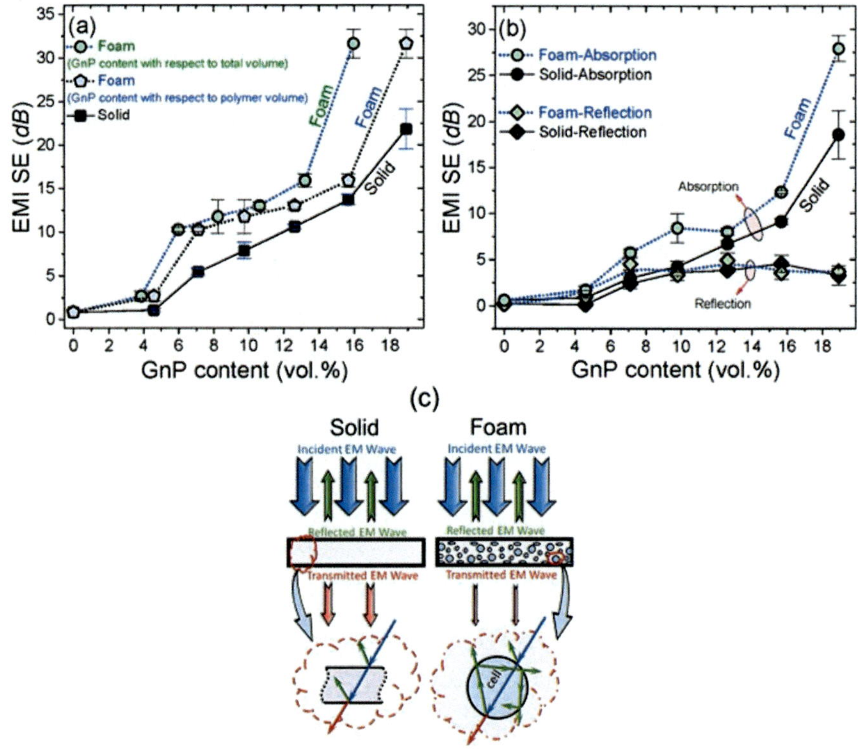

Fig. 6.16 **a** K-band EMI SE of the solid and foamed HDPE-GnP composites as a function of their GnP content; **b** contributions of the reflection and absorption mechanisms to the total K-band EMI SE of the solid and foamed HDPE-GnP composites as a function of their GnP content; and **c** schematic diagrams of the scattering and multiple reflections of the electromagnetic waves. Reproduced with permission from Hamidinejad et al. [58]. Copyright 2018, EACS Publications

finished sample was then submerged in water to wash away any remaining solvent before being dried in a vacuum oven for 24 h at 60 °C. Microcellular PU/8 wt.% GNP foam, known as PUG8, was effectively created as a result. Similarly, PUG foams with successive GNP loadings of 0, 4, 12, 16, and 20 wt.% were created [50].

Intending to fabricate highly efficient porous EMI shields with excellent potential for novel chip packaging, Fang et al. [52] presented a reproduceable approach of growing hollow Fe_3O_4 sphere on a 3D graphene foam surface (via solvothermal technique) followed by filling of the foam with PDMS hybrid system (Fig. 6.18) having outstanding EMI SE of 70.37 dB along with a room temperature thermal conductivity of 28.12 ± 1.212 W/mK. Additionally, the electrical conductivity of the conductive networks inside the hybrid system was found to be 84.02 8.385 S/cm, which is extremely high. Adapting the hybrid system (GF/h-Fe_3O_4/PDMS) as thermal interface materials (TIMs) in chip cooling was proven effective. The full discussion of this facile technique can be accessed in the cited work.

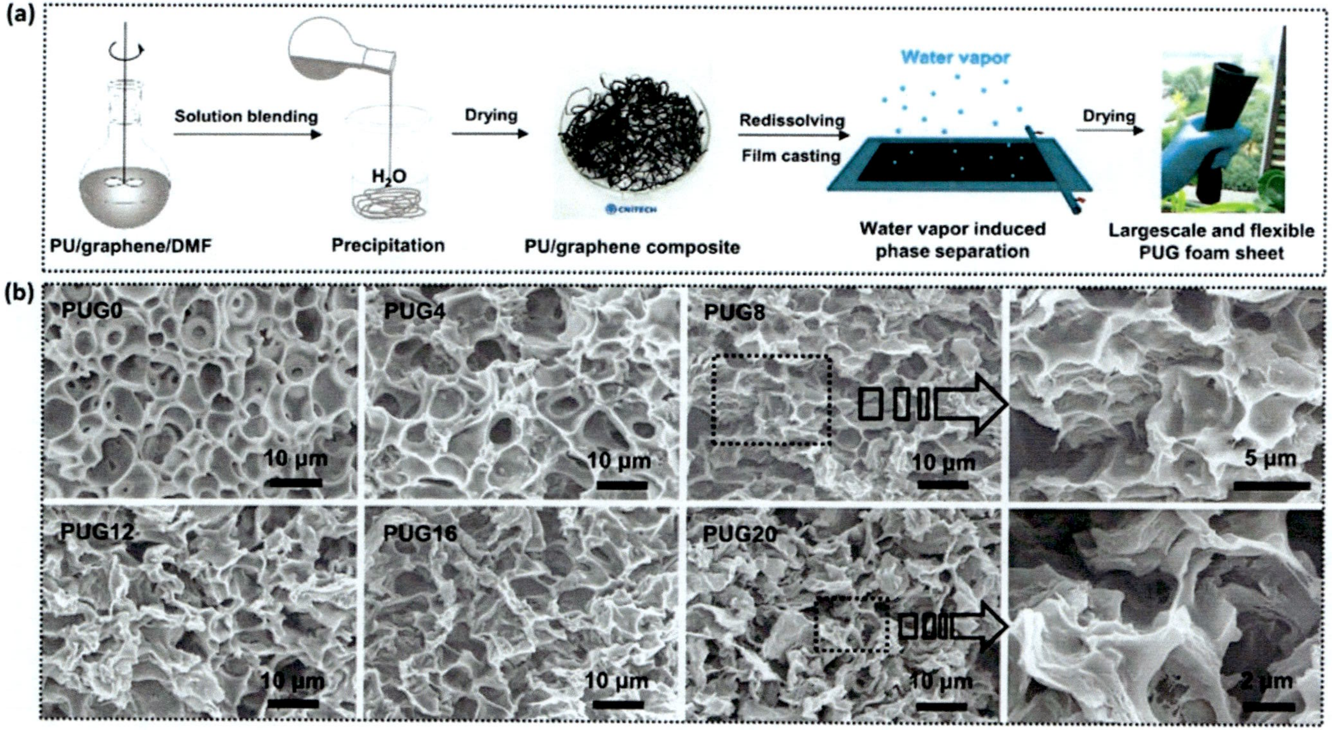

Fig. 6.17 **a** Schematic illustration for the preparation of PUG foams. **b** SEM images of the cell morphology of PUG foams. Reproduced with permission from Li et al. [50] Copyright 2017, Elsevier Science Ltd.

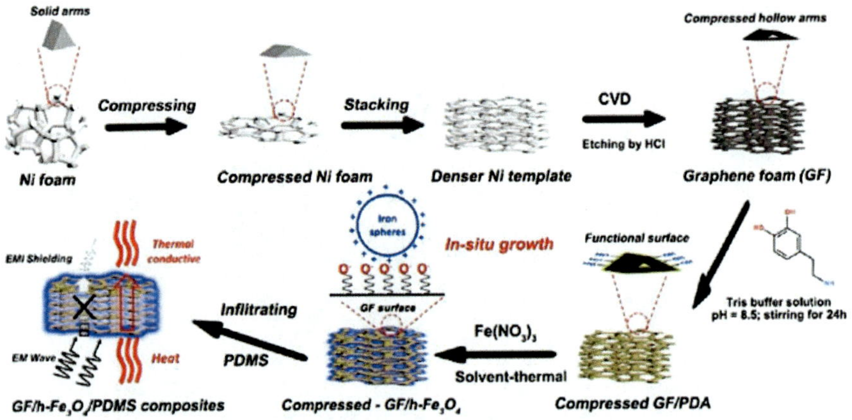

Fig. 6.18 Schematic of the fabrication procedure of GF/h-Fe_3O_4/PDMS composites. Reproduced with permission from Fang et al. [52]. Copyright 2020, Elsevier Science Ltd.

Foam-structured nanocomposites and/or hybrids have been recommended by Banerjee et al. [71] and somewhere else [22, 72] as effective systems for effective shielding of EMI waves. However, there is a need for the design and engineering of foam-structured composites/hybrids, which retain their flexibility, an important parameter in materials aimed at flexible electronics.

Coating

Researchers have utilized diverse coating approaches for the fabrication of GNP-based polymer nanocomposite shields. These approaches include dip coating, vacuum filtration-based coating, electroless deposition, spray deposition, etc. [73]. The preparation and application of low-density PU foams coated with GO have been explicitly reported by Bin et al. [51], as depicted in Fig. 6.19. Their report presented an excellent EMI filter with a high SE of up to 41.6 dB for a 4-cm-thick sample with 10 wt.% GO and 19.9 dB for a 2-cm-thick sample having 5 wt.% GO coating [51].

To create PEI/GNP nanocomposite foams, a WVIPS procedure was applied by Ling et al. [54]. Ultrasonic irradiation and vigorous stirring were used to produce a homogenous PEI/GNP nanocomposite dispersion, as shown in Fig. 6.20. When PEI/GNP dispersion was exposed to air, phase separation occurred due to the diffusion of DMF into the dispersion and the diffusion of water vapor into the air, which was followed by cell nucleation. The entry of solvent and, more crucially, the coalescence of bubbles between one another caused the nuclei to grow larger as the phase separation progressed. The PEI/GNP nanocomposite foams were made after removing the solvent from the cells [54].

These authors described a simple method for creating foams made of lightweight microcellular PEI and GNP nanocomposites with a density of about 0.3 g/cm^3. GNP

Fig. 6.19 **a** Overall fabrication process of the PUG foams, including dip-coating GO sheets onto the PU frameworks and then hydrothermally reducing by hydrazine vapor. **b** A piece of PUG-5 foam with dimensions of 30 × 20 × 15 mm^3 and its fracture surface after being torn from the middle position. **c** A block of PUG-5 foam on the leaf of an asparagus fern. **d**, **e** Optical and SEM images of the original PU sponge and corresponding PUG-5 foam. **f** Compressing and releasing process of the PUG-5 foam, showing excellent compressibility. Reproduced with permission from Bin et al. [51], Copyright 2016, ACS Publications

was shown to be enriched and oriented on cell walls due to the high extensional flow produced during cell development. The electrical conductivity percolation was reduced by this action from 0.21 vol.% for PEI/GNP nanocomposite to 0.18 vol.% for PEI/GNP foam. Additionally, the foaming method greatly improved the EMI shielding performance, going from 17 to 44 dB/(g/cm^3). The PEI/GNP nanocomposite foams also had a high Young's modulus of 180–290 MPa and poor thermal conductivity of 0.065–0.037 W/mK even at 200 °C [54].

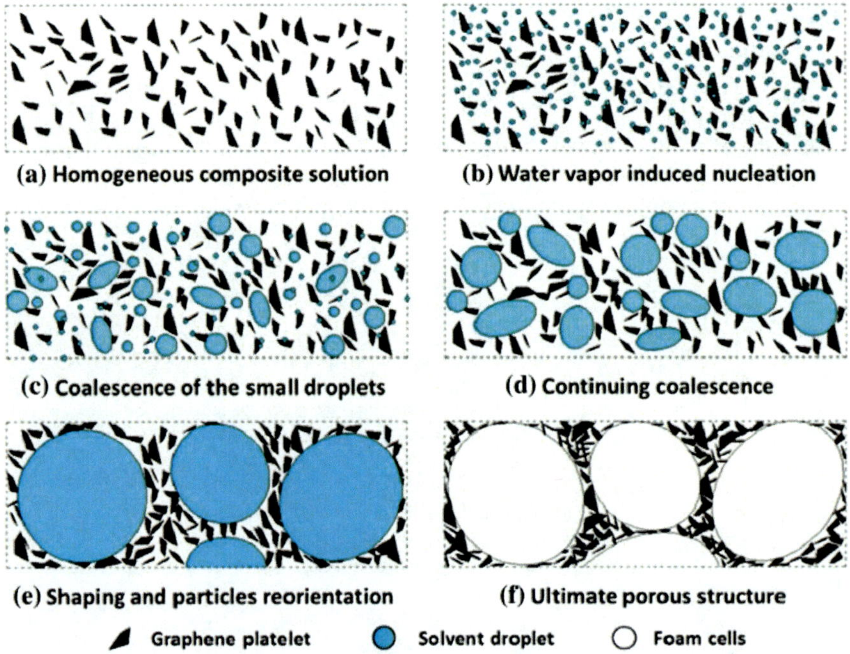

Fig. 6.20 Schematic of WVIPS process to prepare PEI/GNP nanocomposite foam. Reproduced with permission from Ling et al. [54] Copyright 2013, American Chemical Society

f-GNP and PVDF were combined to create novel foam composites. The electrical conductivity and EMI shielding effectiveness of composites with various f-G mass fractions have been studied. With an increase in f-G concentration in an insulating PVDF matrix, the electrical conductivity also increases. A high aspect ratio and high conductivity of the f-G nanofiller, which forms a conductive network in the polymer, are stated to have caused a substantial improvement in conductivity, going from 10^{-16} S/m for insulating PVDF to 10^{-4} S/m for 0.5 wt.% f-G-reinforced PVDF composite. For 5 wt.% of f-G in foam composite, an EMI shielding effectiveness of 20 dB is attained in the X-band (8–12 GHz) region and 18 dB in the broadband (1–8 GHz) zone. The use of lightweight EMI shielding materials made of conductive GNP foam composites for X-band and broadband shielding has been shown, and the mechanism of EMI shielding in f-G/PVDF foam composites has been described [53].

Researchers have proposed foam-based EMI shields as the best-performing filters with regard to their light weightiness and other property performances.

Others

Other processes like lamination have also been adopted to prepare highly functional GNP-based EMI shields [74].

One such approach is the double percolated sandwiched structural GNP-based nanocomposite blends, as exemplified in a recent study where the authors obtained EMI SE up to 21.3 dB [74]. Their process was thus: $Ti_3C_2T_x$ dispersion solution containing 94.97 g was sonicated in an ice bath with a concentration of 21.06 mg/g in a 500 mL three-neck flask. The $Ti_3C_2T_x$ dispersion in the nitrogen atmosphere was exposed to a dropwise addition of 6.00 g of SCS dissolved in 100 mL of anhydrous ethanol. After 4 h of stirring under nitrogen conditions, the solution was centrifuged and washed with deionized water. A final product known as $Ti_3C_2T_x$-SCS was produced after being dried for 24 h at 80 °C. Figure 6.21 depicts the preparation process [74]. GO was then synthesized separately, followed by piperazine-modified ammonium polyphosphate "PA-APP" preparation. The nanocomposite membrane was then fabricated according to the scheme reported in Fig. 6.21.

Other approaches that have been adopted by some researchers for the preparation of high-performance EMI shields are chemical vapor deposition, electrochemical polymerization, etc.

Although not generally utilized by researchers, these approaches possess the required potential for their application toward the fabrication of EMI shields for advanced materials science application.

Other approaches that have been used for the preparation of GNP-based polymeric nanocomposite blends for EMI shielding are mostly the amalgamation of two or more approaches; for example, Xu and his colleagues used solution intercalation + casting + dip coating + spray coating [59]. These authors created a carbon fiber-based electroconductive advanced polymer composite made of epoxy/CF-30-Fe_3O_4-GNPs

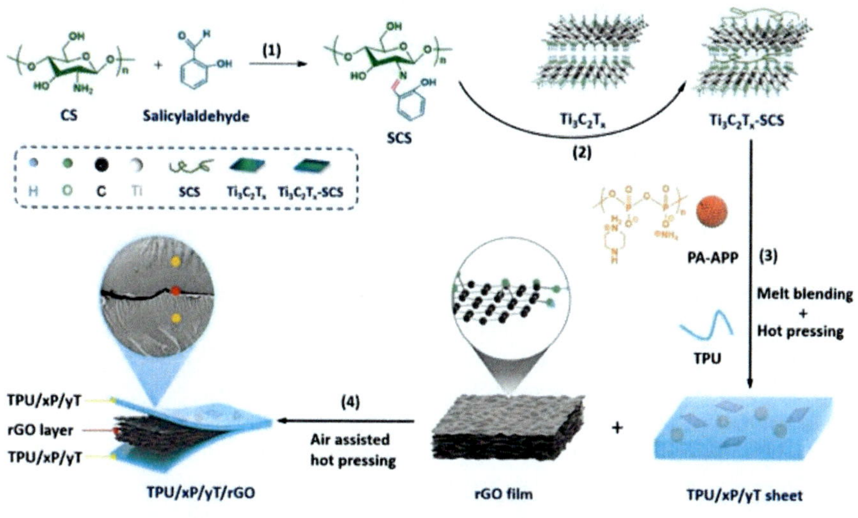

Fig. 6.21 Schematic diagrams for preparation of $Ti_3C_2T_x$-SCS, TPU/xP/yT sheet, and hierarchical TPU/xP/yT/rGO nanocomposite films. Reproduced with permission from Shi et al. [74]. Copyright 2023, Elsevier Science Ltd.

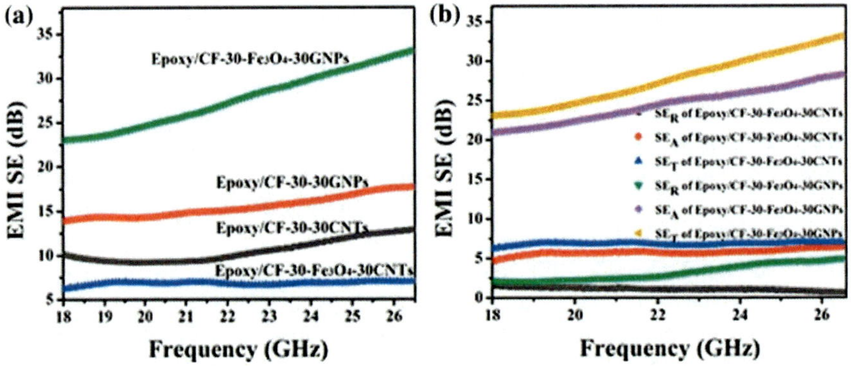

Fig. 6.22 EMI SE performance of sandwich Epoxy/CF-30-30CNT and Epoxy/CF-30-30GNP composites (the average thickness is 2.58 mm) with or without Fe_3O_4 NPTs: **a** EMI SE and **b** SE_T, SE_A, and SE_R (the average thickness of the Fe_3O_4-loaded samples is 2.68 mm). Reproduced with permission from Xu et al. [59]. Copyright 2022, ACS Publications

with an outstanding EMI SE of 33.1 dB, of which over 85.0% is attributed to absorption. It is thought that the combination of GNPs and Fe_3O_4 NPs primarily causes this absorption because of their unique structures and synergetic effects. Additionally, the epoxy infiltration between the many reflections inside the interconnected networks and the randomly distributed loose CFs could potentially enhance the EMI shielding capabilities of GNP-added samples, as shown in Fig. 6.22 [59].

The shielding mechanism, as per these researchers' postulation, was primarily absorption, as shown schematically in Fig. 6.23 [59]. In light of this, they noted that the multifunctional mechanisms of high EMI shielding composites were as follows: Electric conductivity was a crucial component that was attained by incorporating CNTs or GNPs. Additionally, the Fe_3O_4 NPs, which are affordable yet efficient additions that are widely employed in the manufacture of magnetic materials to reduce magnetic losses, offered the magnetic property, another influencing factor that could affect the EMI efficiency. The surface feature is also mentioned as a crucial element that needs to be properly taken into account in the creation of EMI shielding composites. The authors chose two fabrication techniques on the basis of mechanism-oriented design conception. The conductive papers with CNT or GNP additions were simply glued together on one side (way I in Fig. 6.23) to form a single unit. On the opposing side, Fe_3O_4 was sprayed to increase the EMI SE even as epoxy was added to cover the textile pores (route II in Fig. 6.23) [59].

In another instance, Luo and colleagues have reported a novel approach based spraying followed by interface culture repeatedly leading to multiple spraying and interface cultured multilayered, which was finally hot-pressed to obtain high-performance GNP-bacterial cellulose EMI shield as depicted in Fig. 6.24 [75].

These processing methods also have their advantages and disadvantages, as summarized in Table 6.3 [76].

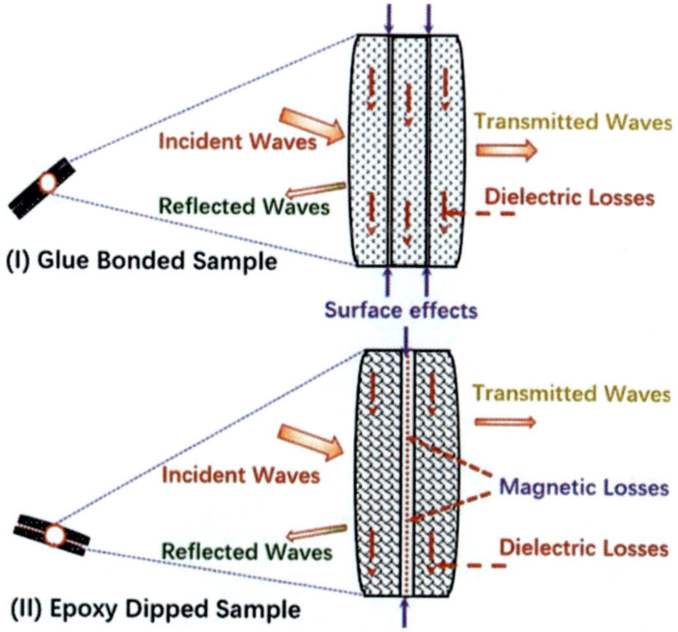

Fig. 6.23 Schematic SEM interference mechanism illustration of the two composites. Reproduced with permission from Xu et al. [59]. Copyright 2022, American Chemical Society

6.3.2 Processing of Graphene-Bearing Flame-Retardant Polymer Nanocomposites

Melt Processing

Dittrich and colleagues [77] have thoroughly examined the characteristics (fire behavior) of the various TrGO nanocomposites concentration dependence. They also contrasted the outcomes for highly exfoliated layered TrGO with those obtained from other commercial carbon nanomaterials, such as carbon black, MWNT, and EG, which represent particles with spherical, tubular, and minimally exfoliated layered morphology. They produced TrGO by first oxidizing the graphite and then reducing it. The required quantity of the carbon nanoparticles CB, MWNT, EG40, and TrGO were dispersed in acetone to produce a nanoparticle concentration of 5 g/L. The dispersion was sonicated in order to remove particle clumping and produce a suitable filler dispersion. The powdered polymer was mixed with the nanoparticle dispersion. In order to prepare the material blend for melt extrusion and injection molding, the solvent was evaporated under decreased pressure, and the mixture was then dried for an entire night at 60 °C. With an average holdup duration of 1 min and a screw speed of 120 rpm, melt extrusion was carried out in a bench-top, co-rotating twin-screw compounder. As the PHRR was decreased, the burning time was prolonged,

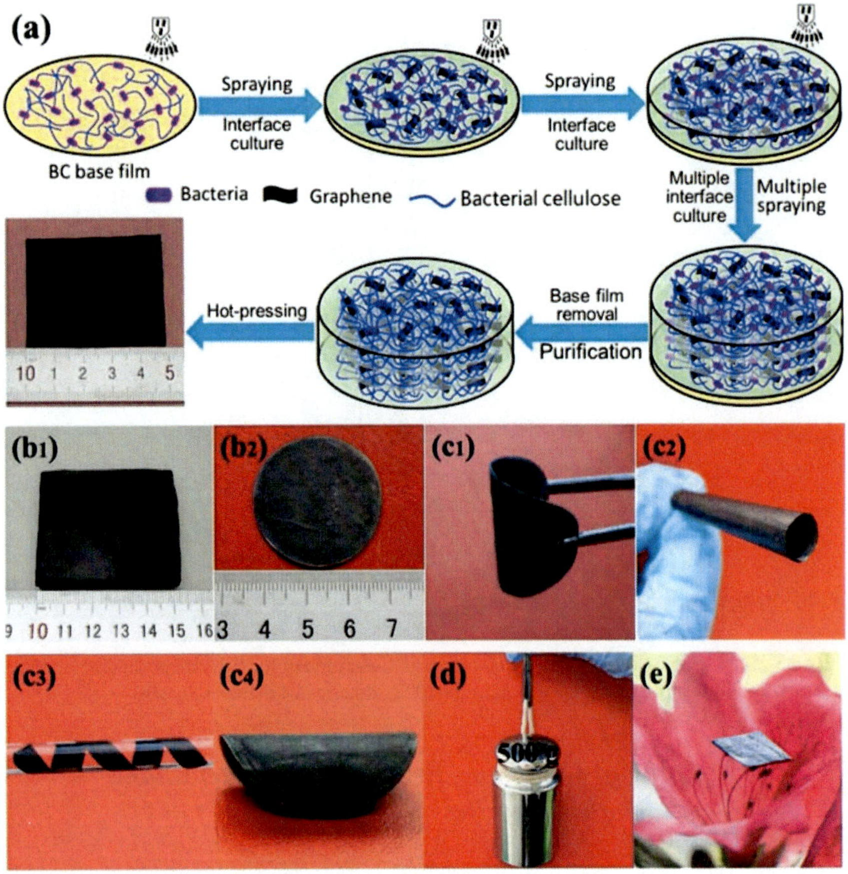

Fig. 6.24 **a** Fabrication processes of GE/BC papers; **b** a photo of large conductive papers; **c** photos showing the papers can be folded and bent without damage; **d** a photo showing the paper has excellent mechanical properties; **e** a photo showing the GE/BC paper is lightweight. Reproduced with permission from Luo et al. [75]. Copyright 2019, Elsevier Science Ltd.

and an efficient residual protective layer was created; they noticed that the samples' burning behavior in the cone calorimeter test also varied significantly with increasing filler content. TRGO's particles clumped together to create a close-packed residue structure, which served as the most effective layer of defense.

While the flammability test and fire behavior under various fire conditions were examined by Yuan et al. [78] to determine the viability of using graphene as an IFR synergist. Before use, they dried the PP, APP, and charring-foaming agent (CFA) in an oven at 80 °C for the entire night. The flame-retardant PP composites were produced using the direct melt processing method. Premixed APP and CFA were in a plastic cup. PP and flame-retardants were melt-mixed for 10 min in a twin-roller mill. After the PP had melted, the combined flame-retardants were put into the mill.

Table 6.3 Merits and demerits of approaches adopted for the fabrication of GNP-based polymer composite EMI shields

Approach	Description	Merits	Demerits	Scale of production
Solution blending	The polymer is combined with GNP-based reinforcing fillers after being properly dissolved in a solvent	Versatile and even distribution of GNP-based reinforcing filler	Heavy use on organic solvents	Bulk
In situ polymerization	The polymer monomers are combined with GNP-based reinforcing fillers, and a temperature, heat, or radiation-activated initiator is used to start the polymerization reaction	High-level dispersion but also extremely efficient	Late stage decrease in polymerization rate	Bulk
Melt compounding	At a temperature above its melting point, polymer is melted as well as combined with fillers made of GNP	Economical and environmentally sound	Unacceptable dispersal of the reinforcing filler	Mass production
Latex blending	The GNP-based reinforcing filler is mixed with the latex before its final preparation into rubber	Does not need GNP to be evenly dispersed throughout the polymer	Use of a lot of surfactant but also a small range of applications	Bulk

The mill's temperature and roller speed were kept at 185 °C and 80 RPM, respectively. They discovered that the presence of graphene reduced the PP75/APP20/CFA5 (PI) reaction to tiny flames, as evidenced by a decline in LOI values and a lower UL-94 rating. The RGO content threshold of 2 wt.% caused a change in the UL-94 classification from V-0 to V-2. While the macroscopic and microscopic structures of the char residues demonstrated how adding graphene could increase the char's compactness for PP composites that were flame-resistant. In studies using cone calorimetry, it was found that low loadings of graphene improved char swelling, which in turn improved insulation and reduced PHRR. PI's fire safety deteriorated as graphene content increased. The relative concentrations of phosphorus and nitrogen in the char layer, as well as the degree of graphitization, might be improved by graphene. Their proposed approach is demonstrated in Fig. 6.25, which shows how adding graphene nanosheets to the char might increase its compactness, degree of graphitization, and thermal stability. The degree of expansion and triazine material content of the char were both increased by graphene. The gases released by the blowing agents caused the carbonizing material's internal pressure to increase, which caused the carbonizing material to expand. The high aspect ratio of graphene was what gave the created intumescent char its enhanced barrier effect and mechanical properties. However, the char's superior mechanical properties enabled it to bear the internal pressure caused by the released gases without cracking. As a result, the residual char

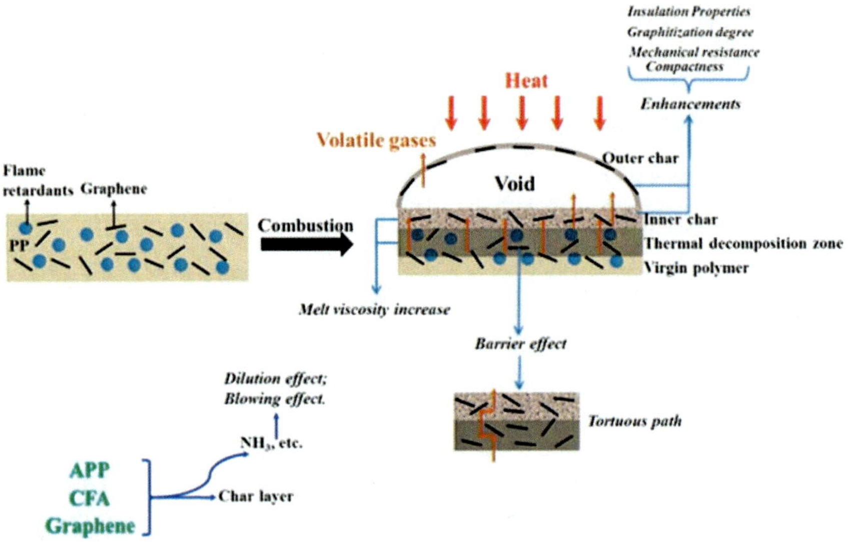

Fig. 6.25 Illustration of the effects of graphene on the formation of intumescent char. Reproduced with permission from Yuan et al. [78]. Copyright 2017, Elsevier Science Ltd.

formed a closed chamber due to the addition of low loadings of graphene to the PI system, improving swelling degree and flame retardancy.

Nevertheless, when more graphene was added, the char's mechanical characteristics were substantially improved, and graphene's barrier performance was more noticeable. The foaming of the carbonizing materials by NH_3 and CO_2 was hampered due to the char's abnormally high mechanical characteristics [79]. It is important to remember that graphene's effective barrier properties can also reduce the amount of fuel and oxygen available and delay the release of flammable byproducts of breakdown. Irradiance used during combustion can have a significant impact on how much char expands via IFR. Thus, the cone calorimeter experiments showed the development of a swelling char layer. The evolved inert and incombustible gases from blowing agents diluted oxygen and combustible gases, and the massive amounts of gases that blowing agents released extinguished flames [80]. The IFR formulation thus performed exceptionally well in terms of flame retardancy. Because the discharge of incombustible gases was impeded when graphene nanosheets were added to PI, the LOI and UL-94 ratings were worsened.

In a different study, Han and colleagues [81] looked into how PS's flame-retardant properties were affected by the TrGOs at various reduction temperatures. To create TrGO, they employed freeze-dried GO powder. The GO powder was heated to different temperatures (200 °C, 500 °C, and 800 °C) in a tube furnace for 10 min. TGO2, TGO5, and TGO8 are the names of these TrGOs, respectively. The heat reduction procedure turned the brownish GO into a fine, puffed black powder. Using an internal mixer, the TrGO and PS were melt-blended for 10 min at 180 °C and 45 rpm to create PS/TrGO composites. TrGO/PS's weight-to-volume ratio was 5:95. The composites were compression molded at 180 °C to prepare sample bars appropriate for cone calorimeter testing. The PS/TGO5 and PS/TGO8 composites demonstrated improved thermal stability and flame-retardant performance. The removal of the functional oxygen groups in the right proportions and the disordered extended layer structure was beneficial for the composites' increase in flame-retardant performance, as shown by the fact that the PHRR for PS/TGO8 nanocomposites shows a 47.5% drop.

Solution Intercalation/Casting

Nanocomposites based on PVA and GNPs have been prepared by solution blending, and their flame-retardant properties have been evaluated by fire tests [82]. The process for creating composites of graphene and PVA was as follows: GO (100 mg/mL) was sonicated into a dispersion in a sodium dodecyl sulfate solution at a concentration of 0.5 wt.%. To prepare composites, this GO solution (1000 mL) was mixed with a certain PVA powder and stirred at 90 °C for 4 h. The obtained uniform mixture was combined with 0.8 mL of hydrazine solution and 3.5 mL of ammonia solution (28%) and agitated for 45 min before being heated in an oil bath for one hour at 95 °C. The mixture was cast onto clean glass plates in order to create composite platelets. The mixture was then heated to 60 °C to promote solvent evaporation. Platelets with thicknesses of up to 4 mm were obtained by doing this operation repeatedly on each disk. PVA-G1, PVA-G2, PVA-G3, PVA-G4 and PVA-G5 were the designations for the composites that contained 1, 2, 3, 4, and 5 wt.% of GNPs, respectively. Comparable samples of PVA filled with 3 wt.% of Na-MMT and MWNTs were designated as PVA-MMT and PVA-MWNTs. Compared to pure PVA, the PHRR of PVA loaded with 3 wt.% GNP was reduced by 49%. With the same concentration, graphene outperformed Na-MMT and MWNTs in terms of flame resistance for the PVA matrix. A solid, dense, and homogenous char produced by combustion could be the cause of such unusual behavior.

To create GO-PPSPB, an intumescent flame-retardant, the surfaces of GO were covalently functionalized with poly(piperazine spirocyclic pentaerythritol bisphosphonate)(PPSPB) [83]. To create GO-PPSPB-based nanocomposites via solvent blending. To achieve this, GO, and GO-PPSPB were chemically reduced using hydrazine to produce CRG and CRG-PPSPB. The common approach for fabricating the sample EVA/CRG-PPSPB composite (1 wt.%) was as follows: With the use of sonication, 800 mL of tetrahydrofuran and 1.0 g of GO-PPSPB powder were mixed for 60 min at room temperature. The aforementioned suspension was then mixed mechanically for 2 h at 60 °C with 99.0 g of EVA resin, and this resin was then added and dissolved. The obtained homogenous dispersion was agitated at 60 °C for 4 h while being combined with 0.8 mL of hydrazine solution and 3.5 mL of ammonia solution (28%). The TFH solvent was removed in the vacuum oven for 12 h at 60 °C. The solid material was then formed into 4.0 ± 0.1-mm-thick sheets for additional measurements by being pressed on a curing machine at 140 °C for

8 min. The PHRR of the EVA/CRG-PPSPB nanocomposites with 1 wt.% CRG-PPSPB was lowered by roughly 56% compared to pure EVA resin. The SEM study demonstrated that the combustion of EVA/CRG-PPSPB nanocomposites resulted in a solid, dense, and consistent intumescent char. It is possible to improve both the dispersion of graphene sheets in the polymer matrix and the flame resistance of the nanocomposites by functionalizing graphene with intumescent flame-retardant PPSPB.

In Situ Polymerization

Bao and colleagues functionalized GO with HCCP and HEA and then used in situ polymerization to introduce it into PS [84]. First, they began stirring while slowly adding drops of the HCCP-THF and TEA-THF mixes to the GO-THF combination. At 15–20 °C for 12 h, the reaction was conducted. The mixture was then sprinkled with HEA (70 g), which had been dissolved in 100 mL of THF. Following a 24 h reaction, THF and deionized water were used to wash the finished FGO. Drying was done on a tiny amount of GO and FGO for 24 h at 80 °C. The FGO was magnetically stirred with styrene (total weight: 25 g) in a three-necked flask to prepare the FGO-PS composites. In an oil bath, the flask was heated to 70 °C. BPO (1 wt.% relative to styrene) was added to the mixture after 30 min of stirring, and the mixture was then agitated again until it became extremely viscous. The mixture was put into a mold with a basic size of 100 cm^2 and heated for 24 h at 80 °C. The mass ratio of the FGO in the composites is indicated by the sample designation PS/*x* percent FGO ($x = 1$, 2, or 3). They outlined how functionalization enhanced Char A formation. Because Char A prevented FGO from being exposed to oxygen, FGO does not burn out and continues to function as Char B. In order to safeguard the underlying polymers, the mixture of Char A and Char B produced a continuous and compact layer of char on the surface of the composites. The fire safety properties of the composites were improved as a result of the improved char, the physical barrier effect of the FGO, and the strong interfacial interactions between the FGO and polymeric matrix, which include peak CO release rate (66% decrease), peak CO_2 release rate (54% decrease), peak heat release rate (53% decrease), thermal degradation rate (30% decrease), and total heat release (38% decrease).

To lessen the flammability and toxicity of PS, GO has also been functionalized with a hyper-branched flame-retardant based on AEPZ and phosphonate derivative [85]. In situ polymerization was used to synthesize PS-FGO nanocomposites. The PS-FGO2.0 preparatory methods were explained as follows: In a three-necked flask with a nitrogen input, mechanical stirrer, and reflux condenser, FGO (0.8 g) was disseminated in styrene (39.2 g) for 30 min using ultrasonication. The mixture was then combined with 0.267 g of benzoyl peroxide initiator, and the flask was heated to 90 °C. Additional benzoyl peroxide (0.133 g) was added and mixed to prepare a homogenous mixture when the liquid became extremely viscous. To eliminate the unreacted monomers, the mixture was put into a mold and heated at 60 °C for 24 h,

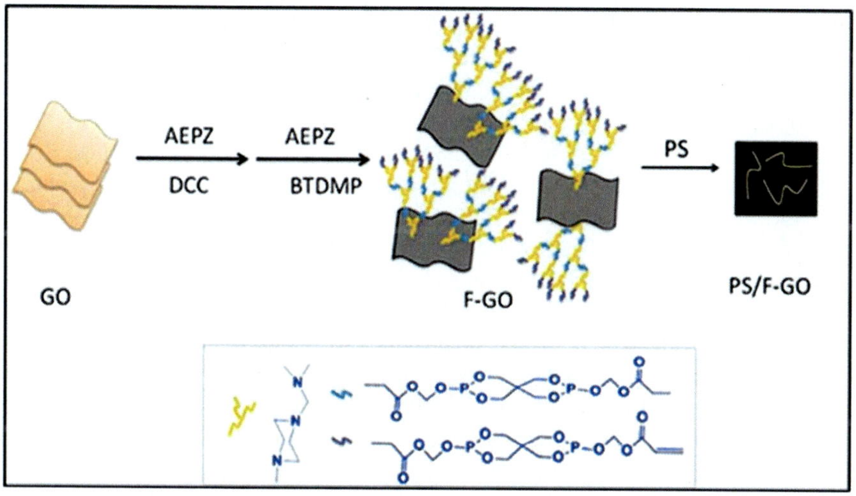

Fig. 6.26 Preparation route of PS-FGO nanocomposites. Reproduced with permission from Hu et al. [85]. Copyright 2015, Elsevier Science Ltd.

80 °C for 24 h, and 120 °C for 2 h. Figure 6.26 depicts the process for making PS-FGO nanocomposites. As a result of the homogeneous dispersion of FGO in the PS matrix, which improved the barrier effect and reduced peak heat release rate, total heat release, and toxic gas release during combustion, the incorporation of FGO into the PS matrix effectively improved the flame retardancy and restrained the toxicity of the volatiles escaped. Additionally, PS-FGO nanocomposites significantly decreased the amount of dangerous and flammable volatiles, such as carbon monoxide, aromatic compounds, and carbonyl compounds, demonstrating that the fire risks associated with PS composites were decreased.

Coating

In a study by Chen and colleagues [86], functionalized GO was made using a flame-retardant chemical that contained phosphorus and nitrogen components. The LBL-assembled flame-retardant multilayer films based on anionic functionalized GO and cationic PEI were applied to the surfaces of PVA films. This was accomplished by soaking PVA films in a PEI aqueous solution (2 mg/mL) for 5 min, washing for 1 min in a purified water bath, and then drying for 10 min at 80 °C while under vacuum to remove the solvent. Following the same rinsing and drying procedures, the PEI-modified PVA membranes were next submerged in the FGO aqueous solution (2 mg/mL) for 5 min. These processes were repeated until the desired number of bilayers (FGO/PEI) were deposited on the PVA substrate. Due to the effective barrier effect of the stacked FGO sheets, the coated PVA's early decomposition temperature increased by more than 30 °C when compared to that of pure PVA, demonstrating significantly better thermal stability. They also assessed the coated samples' flame resistance, and a corresponding method was put forth. The 30-bilayer-coated sample maintained its original form even after extended direct exposure to flame, according to the data, which demonstrated a significant reduction in heat release for coated PVA. They further explained that PVA surfaces developed a composite char that successfully impeded mass and heat transport by combining the physical char from graphene with the chemical char from the catalyzed matrix.

The exploitation of self-assembled coatings comprising GO nanoplates has been recently demonstrated as a promising route to improve the fire safety of flexible PU foams (Fig. 6.27). A study by Maddalena et al. [87] examined the impact of various GO dimensions on the layer-by-layer assembly and characteristics of chitosan/GO coatings applied to polyurethane foams. The neat PU foam (Fig. 6.28) ignited instantly and was quickly devoured by the flames when it came into contact with a little flame. In addition, molten polymer drips that dropped during combustion ignited the cotton beneath the sample. This process, known as melt dripping, poses an extra risk to PU foams since the blazing droplets can quickly spread the fire to adjacent materials. LbL-treated foams displayed a different behavior: After ignition, the flame spread along the edges and the upper surface of the foams, traveling the entire length of the sample, and then extinguished, leaving a coherent residue that maintains the original shape of the foam. The melt-dripping phenomenon was completely suppressed, thus reducing flame propagation risks to other items in a real fire scenario. By comparison with the unmodified foam, 3BL samples showed an increased flame spread rate; the same value is decreased when the BL number is raised to 6, though. They managed to demonstrate that even though the flame spread to the entire surface of the sample, the PU foam was not completely decomposed and volatilized, and the flame extinguished before it had a chance to consume the entire foam by comparing the final residues collected at the end of the test with the coating weight gain. In fact, after-combustion residue studies suggested that the only part of the foam that was truly degraded was the first few millimeters of the layer that was subjected to the flame. This is attributed to the coating's inclusion, which controlled the volatile release while avoiding foam collapse and safeguarding the underlying substance.

In order to increase the flame retardancy and thermal stability of the polymeric composite materials, Uddin and colleagues [88] applied GO thin-film and nanocomposite coatings to aviation composite surfaces. This was accomplished by adhering to the AMS 3095 requirements and the Society of Aerospace Engineers Aerospace Recommended Practice (SAE ARP) 4916. In order to reduce gloss and improve the adhesion of the primer, the surfaces of the laminated panels were lightly sanded using 180- and 240-grit sandpapers. To protect the carbon fibers, provide a dust-free environment, and avoid contamination, sanding was kept to a minimum. In order to reduce gloss and improve the adhesion of the primer, the surfaces of the laminated panels were lightly sanded using 180- and 240-grit sandpapers, whereas to protect the carbon fibers and provide a dust-free environment, avoiding contamination, sanding was kept to a minimum. After a second round of solvent cleaning, epoxy nanocomposites were sprayed onto the surfaces using a HVLP sprayer. A single thin layer

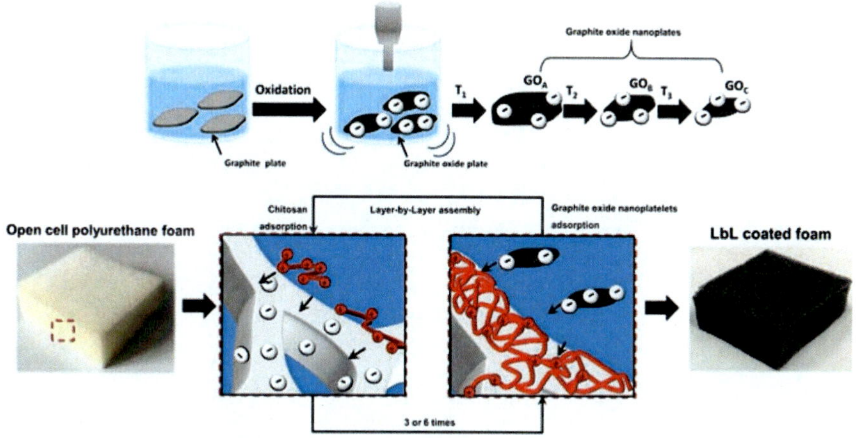

Fig. 6.27 Schematic representations of the preparation process of graphite oxide nanoplates with different sizes and the subsequent layer-by-layer (LbL) assembly on polyurethane (PU) foams. Reproduced from Maddalena et al. [87]. Copyright 2021, the Authors

of primer of the deposit was about 0.0010 inches on the sample surface (25.4 μm). The overall thickness of the primer was about 0.0030 inch after two light applications (76.2 μm). The surface coating was then strengthened by heating the coated composite panels for eight hrs. at 49 °C. Various composite specimens underwent vertical, 12- and 60-s, and 45-degree burn tests. The experimental findings demonstrated that during the 12-s vertical burn tests, the bare polymer composite and the zero % nanoparticle inclusion exhibit the longest average burn lengths and largest average burn areas, respectively. This was because the epoxy resin and combustible epoxy primer components intensified the combustion process' flame and heat. In addition, due to the latent heat that the nanoclay absorbed and slowly released, the nanoclay inclusion samples in the 60 s vertical burn test have longer average burn lengths and bigger average burn areas than those of other samples. As a result, combustion burned more intensely. The 12 s and 60 s vertical tests on the graphene samples yielded encouraging results. Due to the great thermal conductivity of graphene, which aided in quickly dissipating the heat to the air and nearby locations, they had the smallest burn lengths and burn areas. Less heat was produced during combustion as a result. Similarly, the average burn lengths and burn areas of the 0 wt % nano-inclusion samples were larger than anticipated in the 45-degree burn test. Incorporating up to 5 wt.% of graphene into the epoxy primer enhanced the flame retardancy on aerospace composite surfaces.

In a different study, Yang and colleagues [89] described how to easily and affordably create an intelligent fire coating (Fig. 6.29) using a combination of APP and EP. Laser scribing was used to treat the composite, which caused a photothermal action to create a layer of LIG on the APP/EP surface. The flame-resistant APP/EP composite's C–O, C=O, P–O, and N–C bonds were broken during the laser scribing process, and the leftover carbon atoms were reassembled to form the graphene layer.

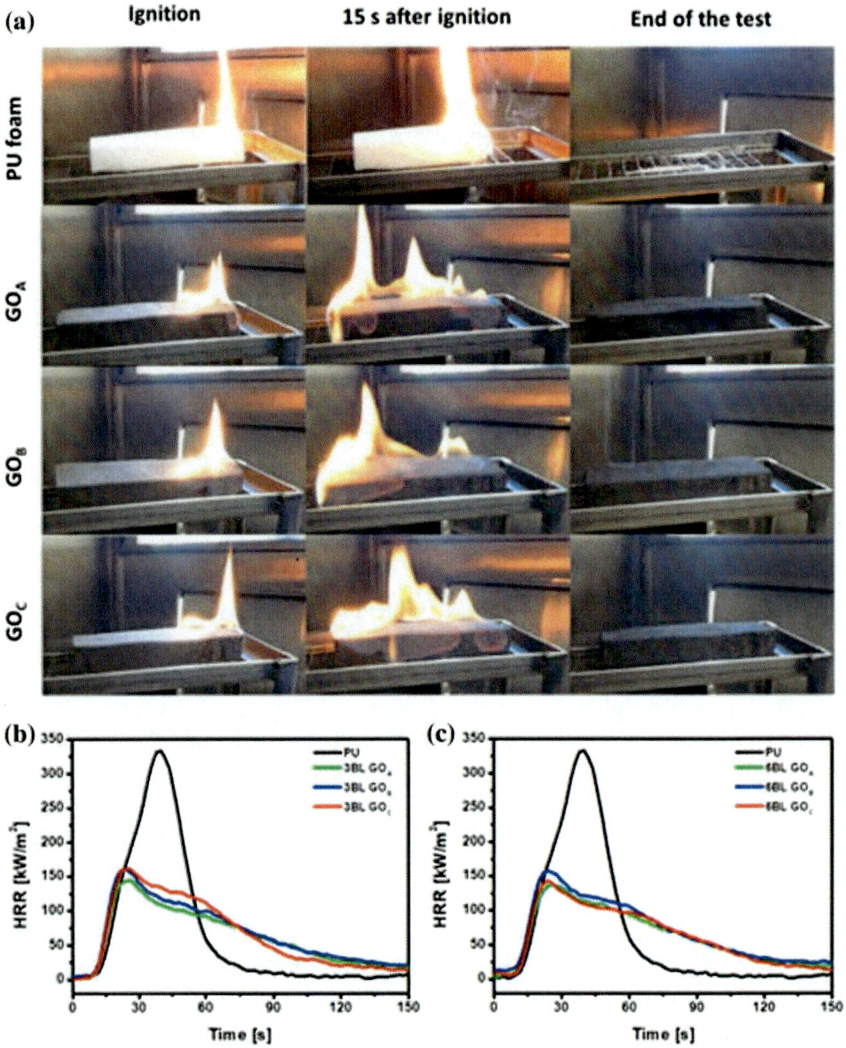

Fig. 6.28 Flame-retardant characterization: **a** Pictures of flammability test in horizontal configuration of untreated PU foam, 3BL CHIT/GOA treated PU foam, 3BL CHIT/GOB PU foam and 3BL CHIT/GOC PU foam. First column: right after ignition; second column: 15 s after ignition; and third column: end of the test. **b** HRR versus time plots of 3BL samples. **c** HRR versus time plots of 6BL samples. Reproduced from Maddalena et al. [87] Copyright 2021, the Authors

The LOI value increased to 37%, and the TTI increased from 21 to 57 s thanks to the intelligent flame protection coating's excellent flame retardancy. The total amount of heat and smoke produced during combustion was effectively reduced by 71.1% and 74.1%, respectively.

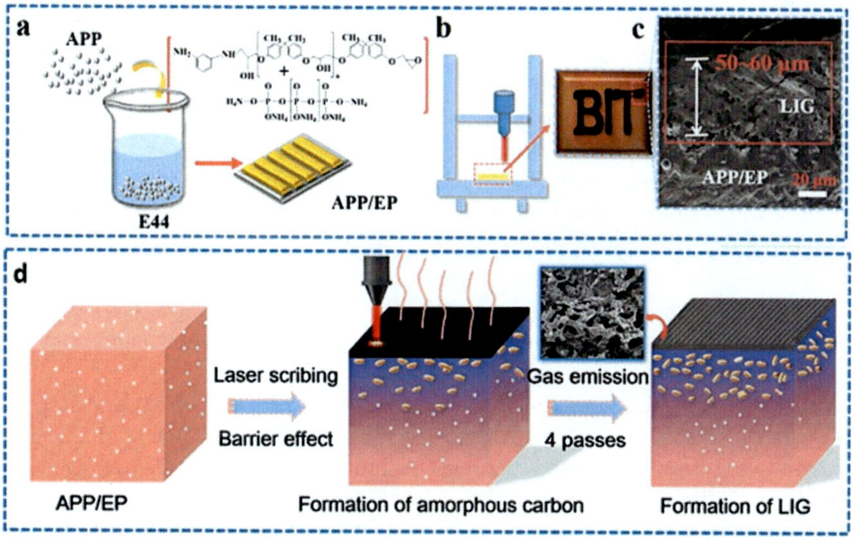

Fig. 6.29 Fabrication of the specimens and the characterization of the laser-induced graphene (LIG). **a** Processing route of the ammonium polyphosphate and epoxy (APP/EP) composites. **b** Processing route for fabricating an LIG layer on the APP/EP surface. The inset shows a photograph of the LIG patterned into "BIT" on the APP/EP. **c** Scanning electron microscopy (SEM) image of the cross-section of the LIG sample. **d** The growth mechanism of the LIG formed on the APP/EP. Reproduced from Yang et al. [89], Copyright 2021, the Authors

6.4 Summary

In conclusion, the utilization of NPs in the development of polymer-based EMI shielding and fire-retarding nanocomposites represents a paradigm shift in materials science, offering a synergistic solution to address the complex challenges associated with electronic device safety and reliability.

The incorporation of NPs, such as graphene, carbon nanotubes, or metal oxides, into polymer matrices introduces a multitude of functionalities that enhance both EMI shielding and fire-retarding characteristics. These nanocomposites exhibit exceptional electrical conductivity, mechanical strength, and thermal stability, creating materials with multifaceted properties crucial for modern technological applications.

For EMI shielding, the unique electrical properties of NPs facilitate the efficient redirection and absorption of electromagnetic waves, ensuring the integrity of electronic devices in environments prone to interference. The synergistic effect of NPs in polymer matrices enhances the overall shielding effectiveness, providing an effective defense against electromagnetic threats.

Simultaneously, the incorporation of NPs contributes to the fire-retarding capabilities of polymers. Whether through the formation of protective char layers, gas-phase flame inhibition, or the creation of intumescent structures, NPs play a pivotal role in slowing down the combustion process and reducing the flammability of the nanocomposite materials.

The combination of EMI shielding and fire-retarding characteristics in a single material underscores the versatility and efficiency of nanocomposites. This dual-functionality approach ensures that materials not only protect electronic devices from electromagnetic interference but also meet stringent safety standards by resisting ignition and retarding flame spread.

As research in this field progresses, the optimization of NP content, dispersion, and compatibility within polymer matrices remains a focal point. Standardization of testing methodologies is essential to facilitate accurate and reproducible assessments of EMI shielding and fire-retarding performance across different nanocomposite formulations.

In conclusion, the integration of NPs into polymer matrices for EMI shielding and fire-retarding nanocomposites opens new avenues for the development of advanced materials that are crucial for the continued evolution of electronic devices in a technologically interconnected world. These nanocomposites represent a promising frontier where innovation in materials science meets the growing demands for safety, reliability, and performance in diverse industrial applications.

References

1. W. Gao, N. Zhao, T. Yu, J. Xi, A. Mao, M. Yuan, H. Bai, C. Gao, High-efficiency electromagnetic interference shielding realized in nacre-mimetic graphene/polymer composite with extremely low graphene loading. Carbon **157**, 570–577 (2020)
2. Y. Li, X. Tian, S.P. Gao, L. Jing, K. Li, H. Yang, F. Fu, J.Y. Lee, Y.X. Guo, J.S. Ho, Reversible crumpling of 2D titanium carbide (MXene) nanocoatings for stretchable electromagnetic shielding and wearable wireless communication. Adv. Func. Mater. **30**(5), 1907451 (2020)
3. X.-Z. Jin, Z.-Y. Yang, C.-H. Huang, J.-H. Yang, Y. Wang, PEDOT:PSS/MXene/PEG composites with remarkable thermal management performance and excellent HF-band & X-band electromagnetic interference shielding efficiency for electronic packaging. Chem. Eng. J. **448**, 137599 (2022)
4. M. Li, F. Han, S. Jiang, M. Zhang, Q. Xu, J. Zhu, A. Ge, L. Liu, Lightweight cellulose nanofibril/reduced graphene oxide aerogels with unidirectional pores for efficient electromagnetic interference shielding. Adv. Mater. Interfaces **8**(24), 2101437 (2021)
5. Z. Chen, Q. Zhang, W. Meng, Z. Wang, X. Han, J. Pu, Nickel-reduced graphene oxide-cellulose nanofiber composite papers for electromagnetic interference shielding. BioResources **15**(1), 814–824 (2020)
6. P. Sambyal, S.K. Dhawan, P. Gairola, S.S. Chauhan, S.P. Gairola, Synergistic effect of polypyrrole/BST/RGO/Fe3O4 composite for enhanced microwave absorption and EMI shielding in X-Band. Curr. Appl. Phys. **18**(5), 611–618 (2018)
7. C. Xing, S. Zhu, Z. Ullah, X. Pan, F. Wu, X. Zuo, J. Liu, M. Chen, W. Li, Q. Li, L. Liu, Ultralight and flexible graphene foam coated with Bacillus subtilis as a highly efficient electromagnetic interference shielding film. Appl. Surf. Sci. **491**, 616–623 (2019)
8. D.M. Ledwith, A.M. Whelan, J.M. Kelly, A rapid, straight-forward method for controlling the morphology of stable silver nanoparticles. J. Mater. Chem. **17**(23), 2459–2464 (2007)
9. S.S. Ray, M. Okamoto, Polymer/layered silicate nanocomposites: a review from preparation to processing. Prog. Polym. Sci. **28**(11), 1539–1641 (2003)

10. S.S. Ray, K. Yamada, M. Okamoto, K. Ueda, New polylactide-layered silicate nanocomposites. 2. Concurrent improvements of material properties, biodegradability and melt rheology. Polymer 44(3), 857–866 (2003)
11. S. Sinha Ray, K. Yamada, M. Okamoto, K. Ueda, Polylactide-layered silicate nanocomposite: a novel biodegradable material. Nano Lett. **2**(10), 1093–1096 (2002)
12. L.T. Temane, J.T. Orasugh, S.S. Ray, Adsorptive removal of pollutants using graphene-based materials for water purification, in *Two-Dimensional Materials for Environmental Applications*. ed. by N. Kumar, R. Gusain, S. Sinha Ray (Springer International Publishing, Cham, 2023), pp.179–244
13. T. Ramanathan, A.A. Abdala, S. Stankovich, D.A. Dikin, M. Herrera-Alonso, R.D. Piner, D.H. Adamson, H.C. Schniepp, X. Chen, R.S. Ruoff, S.T. Nguyen, I.A. Aksay, R.K. Prud'Homme, L.C. Brinson, Functionalized graphene sheets for polymer nanocomposites. Nat. Nanotechnol. **3**(6), 327–331 (2008)
14. N. Saifuddin, A.Z. Raziah, A.R. Junizah, Carbon nanotubes: a review on structure and their interaction with proteins. J. Chem. **2013**, 676815 (2013)
15. J.N. Coleman, U. Khan, W.J. Blau, Y.K. Gun'ko, Small but strong: a review of the mechanical properties of carbon nanotube–polymer composites. *Carbon* 44(9), 1624–1652 (2006)
16. G. Armstrong, M. Ruether, F. Blighe, W. Blau, Functionalised multi-walled carbon nanotubes for epoxy nanocomposites with improved performance. Polym. Int. **58**(9), 1002–1009 (2009)
17. J.T. Orasugh, C. Pal, A.P. Samanta, D. Chattopadhyay, Carbon nanotube and nanofiber reinforced polymer composites, in *Encyclopedia of Materials: Plastics and Polymers*. ed. by M.S.J. Hashmi (Elsevier, Oxford, 2022), pp.837–859
18. S.S. Ray, A. Geberekrstos, T.S. Muzata, J.T. Orasugh, *Process-Induced Phase Separation in Polymer Blends: Materials, Characterization, Properties, and Applications* (Carl Hanser Verlag GmbH Co KG, 2023)
19. A. Gebrekrstos, J.T. Orasugh, T.S. Muzata, S.S. Ray, Cellulose-based sustainable composites: a review of systems for applications in EMI shielding and sensors. Macromol. Mater. Eng. **307**(9), 2200185 (2022)
20. S. Ran, F. Fang, Z. Guo, P. Song, Y. Cai, Z. Fang, H. Wang, Synthesis of decorated graphene with P, N-containing compounds and its flame retardancy and smoke suppression effects on polylactic acid. Compos. B Eng. **170**, 41–50 (2019)
21. K. Malkappa, J. Bandyopadhyay, V. Ojijo, S.S. Ray, Superior flame retardancy, antidripping, and thermomechanical properties of polyamide nanocomposites with graphene-based hybrid flame retardant. J. Appl. Polym. Sci. **139**(37), e52867 (2022)
22. J.T. Orasugh, S.S. Ray, Functional and structural facts of effective electromagnetic interference shielding materials: a review. ACS Omega **8**(9), 8134–8158 (2023)
23. H. Kim, A.A. Abdala, C.W. Macosko, Graphene/polymer nanocomposites. Macromolecules **43**(16), 6515–6530 (2010)
24. H. Kim, Y. Miura, C.W. Macosko, Graphene/polyurethane nanocomposites for improved gas barrier and electrical conductivity. Chem. Mater. **22**(11), 3441–3450 (2010)
25. B. Shen, W. Zhai, C. Chen, D. Lu, J. Wang, W. Zheng, Melt blending in situ enhances the interaction between polystyrene and graphene through π–π stacking. ACS Appl. Mater. Interfaces **3**(8), 3103–3109 (2011)
26. B. Yuan, C. Bao, L. Song, N. Hong, K.M. Liew, Y. Hu, Preparation of functionalized graphene oxide/polypropylene nanocomposite with significantly improved thermal stability and studies on the crystallization behavior and mechanical properties. Chem. Eng. J. **237**, 411–420 (2014)
27. S. Chandrasekaran, C. Seidel, K. Schulte, Preparation and characterization of graphite nanoplatelet (GNP)/epoxy nano-composite: Mechanical, electrical and thermal properties. Eur. Polymer J. **49**(12), 3878–3888 (2013)
28. O.M. Istrate, K.R. Paton, U. Khan, A. O'Neill, A.P. Bell, J.N. Coleman, Reinforcement in melt-processed polymer–graphene composites at extremely low graphene loading level. Carbon **78**, 243–249 (2014)
29. B. Shen, W. Zhai, M. Tao, D. Lu, W. Zheng, Enhanced interfacial interaction between polycarbonate and thermally reduced graphene induced by melt blending. Compos. Sci. Technol. **86**, 109–116 (2013)

30. A. Sadeghi, R. Moeini, J.K. Yeganeh, Highly conductive PP/PET polymer blends with high electromagnetic interference shielding performances in the presence of thermally reduced graphene nanosheets prepared through melt compounding. Polym. Compos. **40**(S2), E1461–E1469 (2019)
31. B. Lee, U. Hwang, J. Kim, S.-H. Kim, K. Choi, I.-K. Park, C. Choi, J. Suhr, J.-D. Nam, Highly dispersed graphene nanoplatelets in polypropylene composites by employing high-shear stress for enhanced dielectric properties and frequency-selective electromagnetic interference shielding capability. Compos. Commun. **37**, 101409 (2023)
32. P. Bhawal, S. Ganguly, T.K. Das, S. Mondal, S. Choudhury, N.C. Das, Superior electromagnetic interference shielding effectiveness and electro-mechanical properties of EMA-IRGO nanocomposites through the in-situ reduction of GO from melt blended EMA-GO composites. Compos. B Eng. **134**, 46–60 (2018)
33. S.P. Pawar, S. Stephen, S. Bose, V. Mittal, Tailored electrical conductivity, electromagnetic shielding and thermal transport in polymeric blends with graphene sheets decorated with nickel nanoparticles. Phys. Chem. Chem. Phys. **17**(22), 14922–14930 (2015)
34. A. Nasr Esfahani, A. Katbab, A. Taeb, L. Simon, M.A. Pope, Correlation between mechanical dissipation and improved X-band electromagnetic shielding capabilities of amine functionalized graphene/thermoplastic polyurethane composites. Eur. Polymer J. **95**, 520–538 (2017)
35. G.P. Abhilash, D. Sharma, S. Bose, C. Shivakumara, PANI-wrapped $BaFe_{12}O_{19}$ and $SrFe_{12}O_{19}$ with rGO composite materials for electromagnetic interference shielding applications. Heliyon **9**(3), e13648 (2023)
36. J. Zhou, C. Liu, L. Xia, L. Wang, C. Qi, G. Zhang, Z. Tan, B. Ren, B. Yuan, Bridge-graphene connecting polymer composite with a distinctive segregated structure for simultaneously improving electromagnetic interference shielding and flame-retardant properties. Colloids Surf. A **661**, 130853 (2023)
37. X. Zhang, X. Zhang, M. Yang, S. Yang, H. Wu, S. Guo, Y. Wang, Ordered multilayer film of (graphene oxide/polymer and boron nitride/polymer) nanocomposites: an ideal EMI shielding material with excellent electrical insulation and high thermal conductivity. Compos. Sci. Technol. **136**, 104–110 (2016)
38. V.K. Singh, A. Shukla, M.K. Patra, L. Saini, R.K. Jani, S.R. Vadera, N. Kumar, Microwave absorbing properties of a thermally reduced graphene oxide/nitrile butadiene rubber composite. Carbon **50**(6), 2202–2208 (2012)
39. J.T. Orasugh, C. Pal, M.S. Ali, D. Chattopadhyay, Electromagnetic interference shielding property of polymer-graphene composites, in *Polymer Nanocomposites Containing Graphene*, ed. by M. Rahaman, L. Nayak, I.A. Hussein, N.C. Das (Woodhead Publishing, 2022), pp 211–243
40. D.G. Galpayage Dona, M. Wang, M. Liu, N. Motta, E. Waclawik, C. Yan, Recent advances in fabrication and characterization of graphene-polymer nanocomposites. Graphene **1**(2), 30–49 (2012)
41. Y. Wu, Z. Wang, X. Liu, X. Shen, Q. Zheng, Q. Xue, J.-K. Kim, Ultralight graphene foam/conductive polymer composites for exceptional electromagnetic interference shielding. ACS Appl. Mater. Interfaces **9**(10), 9059–9069 (2017)
42. A. Nasir, A. Raza, M. Tahir, T. Yasin, M. Nadeem, B. Ahmad, Synthesis and study of polyaniline grafted graphene oxide nanohybrids. Mater. Res. Bull. **157**, 112006 (2023)
43. J. Zhang, Y. Qi, Y. Zhang, J. Duan, B. Liu, B. Liu, Z. Sun, Y. Xu, W. Hu, N. Zhang, Lignin based flexible electromagnetic shielding PU synergized with graphite. Fibers Polym. **22**(1), 1–8 (2021)
44. H. Wadhwa, S. Mahendia, S. Kumar, Microwave synthesized graphene-PAni nanocomposites for EMI shielding, in *AIP Conference Proceedings* (AIP Publishing LLC, 2019), p. 030123
45. K. Cheng, H. Li, M. Zhu, H. Qiu, J. Yang, In situ polymerization of graphene-polyaniline@polyimide composite films with high EMI shielding and electrical properties. RSC Adv. **10**(4), 2368–2377 (2020)
46. S. Khasim, Polyaniline-Graphene nanoplatelet composite films with improved conductivity for high performance X-band microwave shielding applications. Results Phys. **12**, 1073–1081 (2019)

47. M. Taj, S.R. Manohara, B. Siddlingeshwar, N. Raghavendra, M. Faisal, U.V. Khadke, Anticorrosion and electromagnetic interference shielding performance of bifunctional PEDOT-graphene nanocomposites. Diamond Relat. Mater. **132**, 109690 (2023)
48. P. Das, A.B. Deoghare, S. Ranjan Maity, Synergistically improved thermal stability and electromagnetic interference shielding effectiveness (EMI SE) of in-situ synthesized polyaniline/sulphur doped reduced graphene oxide (PANI/S-RGO) nanocomposites. Ceram. Int. **48**(8), 11031–11042 (2022)
49. X. Bai, Y. Zhai, Y. Zhang, Green approach to prepare graphene-based composites with high microwave absorption capacity. J. Phys. Chem. C **115**(23), 11673–11677 (2011)
50. Y. Li, B. Shen, D. Yi, L. Zhang, W. Zhai, X. Wei, W. Zheng, The influence of gradient and sandwich configurations on the electromagnetic interference shielding performance of multilayered thermoplastic polyurethane/graphene composite foams. Compos. Sci. Technol. **138**, 209–216 (2017)
51. B. Shen, Y. Li, W. Zhai, W. Zheng, Compressible graphene-coated polymer foams with ultralow density for adjustable electromagnetic interference (EMI) shielding. ACS Appl. Mater. Interfaces **8**(12), 8050–8057 (2016)
52. H. Fang, H. Guo, Y. Hu, Y. Ren, P.-C. Hsu, S.-L. Bai, In-situ grown hollow Fe3O4 onto graphene foam nanocomposites with high EMI shielding effectiveness and thermal conductivity. Compos. Sci. Technol. **188**, 107975 (2020)
53. V. Eswaraiah, V. Sankaranarayanan, S. Ramaprabhu, Functionalized graphene–PVDF foam composites for EMI shielding. Macromol. Mater. Eng. **296**(10), 894–898 (2011)
54. J. Ling, W. Zhai, W. Feng, B. Shen, J. Zhang, W.G. Zheng, Facile preparation of lightweight microcellular polyetherimide/graphene composite foams for electromagnetic interference shielding. ACS Appl. Mater. Interfaces 5(7), 2677–2684 (2013)
55. Y. Li, X. Pei, B. Shen, W. Zhai, L. Zhang, W. Zheng, Polyimide/graphene composite foam sheets with ultrahigh thermostability for electromagnetic interference shielding. RSC Adv. **5**(31), 24342–24351 (2015)
56. B. Shen, W. Zhai, M. Tao, J. Ling, W. Zheng, Lightweight, multifunctional polyetherimide/graphene@Fe_3O_4 composite foams for shielding of electromagnetic pollution. ACS Appl. Mater. Interfaces **5**(21), 11383–11391 (2013)
57. Z. Chen, C. Xu, C. Ma, W. Ren, H.-M. Cheng, Lightweight and flexible graphene foam composites for high-performance electromagnetic interference shielding. Adv. Mater. **25**(9), 1296–1300 (2013)
58. M. Hamidinejad, B. Zhao, A. Zandieh, N. Moghimian, T. Filleter, C.B. Park, Enhanced electrical and electromagnetic interference shielding properties of polymer-graphene nanoplatelet composites fabricated via supercritical-fluid treatment and physical foaming. ACS Appl. Mater. Interfaces **10**(36), 30752–30761 (2018)
59. J. Xu, R. Chen, Z. Yun, Z. Bai, K. Li, S. Shi, J. Hou, X. Guo, X. Zhang, J. Chen, Lightweight epoxy/cotton fiber-based nanocomposites with carbon and Fe_3O_4 for electromagnetic interference shielding. ACS Omega **7**(17), 15215–15222 (2022)
60. L.J. Romasanta, M. Hernández, M.A. López-Manchado, R. Verdejo, Functionalised graphene sheets as effective high dielectric constant fillers. Nanoscale Res. Lett. **6**(1), 508 (2011)
61. L. Wei, W. Zhang, J. Ma, S.-L. Bai, Y. Ren, C. Liu, D. Simion, J. Qin, π-π stacking interface design for improving the strength and electromagnetic interference shielding of ultrathin and flexible water-borne polymer/sulfonated graphene composites. Carbon **149**, 679–692 (2019)
62. S.-T. Hsiao, C.-C.M. Ma, H.-W. Tien, W.-H. Liao, Y.-S. Wang, S.-M. Li, C.-Y. Yang, S.-C. Lin, R.-B. Yang, Effect of covalent modification of graphene nanosheets on the electrical property and electromagnetic interference shielding performance of a water-borne polyurethane composite. ACS Appl. Mater. Interfaces **7**(4), 2817–2826 (2015)
63. W. Yang, H. Bai, B. Jiang, C. Wang, W. Ye, Z. Li, C. Xu, X. Wang, Y. Li, Flexible and densified graphene/waterborne polyurethane composite film with thermal conducting property for high performance electromagnetic interference shielding. Nano Res. **15**(11), 9926–9935 (2022)

64. S.-C. Lin, C.-C.M. Ma, S.-T. Hsiao, Y.-S. Wang, C.-Y. Yang, W.-H. Liao, S.-M. Li, J.-A. Wang, T.-Y. Cheng, C.-W. Lin, R.-B. Yang, Electromagnetic interference shielding performance of waterborne polyurethane composites filled with silver nanoparticles deposited on functionalized graphene. Appl. Surf. Sci. **385**, 436–444 (2016)
65. S.-T. Hsiao, C.-C.M. Ma, H.-W. Tien, W.-H. Liao, Y.-S. Wang, S.-M. Li, Y.-C. Huang, Using a non-covalent modification to prepare a high electromagnetic interference shielding performance graphene nanosheet/water-borne polyurethane composite. Carbon **60**, 57–66 (2013)
66. M. Salzano de Luna, Y. Wang, T. Zhai, L. Verdolotti, G.G. Buonocore, M. Lavorgna, H. Xia, Nanocomposite polymeric materials with 3D graphene-based architectures: from design strategies to tailored properties and potential applications. Progr. Polym. Sci.89, 213–249 (2019)
67. J. Du, L. Zhao, Y. Zeng, L. Zhang, F. Li, P. Liu, C. Liu, Comparison of electrical properties between multi-walled carbon nanotube and graphene nanosheet/high density polyethylene composites with a segregated network structure. Carbon **49**(4), 1094–1100 (2011)
68. M. Li, C. Gao, H. Hu, Z. Zhao, Electrical conductivity of thermally reduced graphene oxide/polymer composites with a segregated structure. Carbon **65**, 371–373 (2013)
69. K. Tian, Z. Su, H. Wang, X. Tian, W. Huang, C. Xiao, N-doped reduced graphene oxide/waterborne polyurethane composites prepared by in situ chemical reduction of graphene oxide. Compos. A Appl. Sci. Manuf.A Appl. Sci. Manuf. **94**, 41–49 (2017)
70. A. Ghosh, J.T. Orasugh, D. Chattopadhyay, S.S. Ray, Polymers for foams and their emerging applications, in *Specialty Polymers* (CRC Press), pp. 295–309)
71. R. Banerjee, A. Gebrekrstos, J.T. Orasugh, S.S. Ray, Nanocarbon-containing polymer composite foams: a review of systems for applications in electromagnetic interference shielding, energy storage, and piezoresistive sensors. Ind. Eng. Chem. Res. **62**(18), 6807–6842 (2023)
72. J.T. Orasugh, S.S. Ray, Nanocellulose-graphene oxide-based nanocomposite for adsorptive water treatment, in *Functional Polymer Nanocomposites for Wastewater Treatment*. ed. by M.J. Hato, S. Sinha Ray (Springer International Publishing, Cham, 2022), pp.1–53
73. J.T. Orasugh, S.S. Ray, Graphene-based electrospun fibrous materials with enhanced EMI shielding: recent developments and future perspectives. ACS Omega **7**(38), 33699–33718 (2022)
74. Y. Shi, A. Yao, J. Han, H. Wang, Y. Feng, L. Fu, F. Yang, P. Song, Architecting fire safe hierarchical polymer nanocomposite films with excellent electromagnetic interference shielding via interface engineering. J. Colloid Interface Sci. **640**, 179–191 (2023)
75. H. Luo, J. Xie, L. Xiong, Y. Zhu, Z. Yang, Y. Wan, Fabrication of flexible, ultra-strong, and highly conductive bacterial cellulose-based paper by engineering dispersion of graphene nanosheets. Compos. B Eng. **162**, 484–490 (2019)
76. M. Zhang, Y. Li, Z. Su, G. Wei, Recent advances in the synthesis and applications of graphene–polymer nanocomposites. Polym. Chem. **6**(34), 6107–6124 (2015)
77. B. Dittrich, K.A. Wartig, D. Hofmann, R. Mülhaupt, B. Schartel, The influence of layered, spherical, and tubular carbon nanomaterials' concentration on the flame retardancy of polypropylene. Polym. Compos. **36**(7), 1230–1241 (2015)
78. B. Yuan, A. Fan, M. Yang, X. Chen, Y. Hu, C. Bao, S. Jiang, Y. Niu, Y. Zhang, S. He, H. Dai, The effects of graphene on the flammability and fire behavior of intumescent flame retardant polypropylene composites at different flame scenarios. Polym. Degrad. Stab. **143**, 42–56 (2017)
79. S. Pack, T. Kashiwagi, C. Cao, C.S. Korach, M. Lewin, M.H. Rafailovich, Role of surface interactions in the synergizing polymer/clay flame retardant properties. Macromolecules **43**(12), 5338–5351 (2010)
80. D. Chattopadhyay, D.C. Webster, Thermal stability and flame retardancy of polyurethanes. Prog. Polym. Sci. **34**(10), 1068–1133 (2009)
81. Y. Han, T. Wang, X. Gao, T. Li, Q. Zhang, Preparation of thermally reduced graphene oxide and the influence of its reduction temperature on the thermal, mechanical, flame retardant performances of PS nanocomposites. Compos. A Appl. Sci. Manuf. **84**, 336–343 (2016)

82. G. Huang, J. Gao, X. Wang, H. Liang, C. Ge, How can graphene reduce the flammability of polymer nanocomposites? Mater. Lett. **66**(1), 187–189 (2012)
83. G. Huang, S. Chen, S. Tang, J. Gao, A novel intumescent flame retardant-functionalized graphene: nanocomposite synthesis, characterization, and flammability properties. Mater. Chem. Phys. **135**(2), 938–947 (2012)
84. C. Bao, Y. Guo, B. Yuan, Y. Hu, L. Song, Functionalized graphene oxide for fire safety applications of polymers: a combination of condensed phase flame retardant strategies. J. Mater. Chem. **22**(43), 23057–23063 (2012)
85. W. Hu, B. Yu, S.-D. Jiang, L. Song, Y. Hu, B. Wang, Hyper-branched polymer grafting graphene oxide as an effective flame retardant and smoke suppressant for polystyrene. J. Hazard. Mater. **300**, 58–66 (2015)
86. W. Chen, P. Liu, L. Min, Y. Zhou, Y. Liu, Q. Wang, W. Duan, Non-covalently functionalized graphene oxide-based coating to enhance thermal stability and flame retardancy of PVA film. Nano-micro Lett. **10**, 1–13 (2018)
87. L. Maddalena, J. Gomez, A. Fina, F. Carosio, Effects of graphite oxide nanoparticle size on the functional properties of layer-by-layer coated flexible foams. Nanomaterials **11**(2), 266 (2021)
88. M.N. Uddin, L. Le, R. Nair, R. Asmatulu, Effects of graphene oxide thin films and nanocomposite coatings on flame retardancy and thermal stability of aircraft composites: a comparative study. J. Eng. Mater. Technol. **141**(3), 031004 (2019)
89. W. Yang, Y. Liu, J. Wei, X. Li, N. Li, J. Liu, An intelligent fire-protection coating based on ammonium polyphosphate/epoxy composites and laser-induced graphene. Polymers **13**(6), 984 (2021)

Chapter 7
Applications

7.1 EMI Shielding Nanocomposites

All electronic equipment produces electromagnetic radiation and can be transferred from one electronic device to another through the air. The movement of electromagnetic waves and interference can harm crucial electronics in some fields of application and possibly be dangerous. EMI shields are generally employed for shielding and/or preventing EM radiation for the shielded object from disturbing the outside environment or shielding the object from outside/external EM radiation. Automotive, the military (for control panels, signal intelligence, signal blocking, consumer electronics, radio military packs), information technology and telecommunications, the medical field, and export manufacturing are among the industry's most prone to EMI.

GNP-polymer blends, or non-blend EMI shields, have demonstrated satisfactory with results presented in the literature, as discussed earlier. EMI shields made from GNP-polymer nanocomposites can serve well as board-level shields; conductive elastomer shields; fabric over form and conductive foam gaskets; enclosure doors finger strips, cabinets, as well as boxes; wire meshes; microwave absorbers in sheet/membrane form; foam in place gaskets; etc., which are of great importance in automobile, aerospace, electronics, etc., industries.

Practically proposed application niches of GNP-based polymer nanocomposite EMI shields include protective high-strength aircraft parts [1, 2], flexible circuit boards [3], smart textiles, laptops, as well as cellphones [2], electronic packaging, etc. These practical instances include systems such as Silicon-GNP (presenting an astonishing EMI SE of 54 dB) [4], Zero Gauss Chamber (MuChamber), etc.

With regard to literature, diverse GNP-polymer composite EMI shielding systems have been applied in addition to that; however, these reports have shown variation in EMI shielding performance, which is largely due to the intrinsic properties of the polymer but also largely influenced by the nature of graphene-based nanofiller used as the functional reinforcing nanofiller [5]. Table 7.1 presents the EMI SE property

S. S. Ray et al., *Graphene-Bearing Polymer Composites*, Springer Series in Materials Science 340, https://doi.org/10.1007/978-3-031-51924-6_7

performance of diverse polymer matrix systems functionalized with graphene-based nanofillers.

With reference to Table 7.1, it can be deduced that literature has pointed out that using single polymeric matrices to fabricate graphene-based nanocomposite systems for EMI shielding applications produces results different from their blended counterparts. Also, it can be deduced from Table 7.1 that diverse polymeric matrices produce different results for EMI SE, largely due to the interfacial interaction between the graphene-based filler and the chosen processing approach (which was discussed earlier).

Again, Table 7.1 shows that PEDOT/GNP (GNP represents graphene-based nanofiller) systems: 18–56.5 dB; PDMS/GNP systems: 30–80 dB; PPy/GNP systems: 11–48 dB; PANI/GNP systems: 14–40 dB; PE and/or PEG/GNP systems: 21.8–60.2 dB; PS/GNP systems: 20–48.2 dB; PP/GNP systems: 8.3–120.6 dB; rubber/GNP systems: 23.46–77 dB; PU/GNP systems: 17–41.6 dB; PVDF/GNP systems: 20–69.7 dB; PEI/PI/GNP systems: 6.37–44 dB; cellulose/GNP systems: 22.6–58.4 dB; PEN/GNP systems: 25.8–77 dB; epoxy/GNP systems: 22.6–58 dB; PMMA/GNP systems: 17–63 dB; PLA/GNP systems: 11.6–38 dB; and other polymer/GNP systems: 19.5–51 dB. The order is thus: PP/GNP systems > PDMS/GNP systems > PEN/GNP systems ≥ Rubber/GNP systems > PVDF/GNP systems > PMMA/GNP systems > PE and/or PEG/GNP systems > cellulose/GNP systems > epoxy/GNP systems > PEDOT/GNP systems > polymer/GNP systems > PS/GNP systems > PPy/GNP systems > PEI/PI/GNP systems > PU/GNP systems > PANI/GNP systems > PLA/GNP systems. On the other hand, diverse blends present EMI SE for their systems, including GNP nanofillers within the range of 12–99 dB.

With consideration to the majority of formulated GNP-based polymer nanocomposite shielding mechanisms being dominated by absorption, it is apparent that military-targeted EMI shields can be successfully fabricated using GNP and its derivatives as reinforcing nanofillers, especially stealth vehicles, war tanks, warships, stealth cruise missiles, etc. Another interesting area where GNP-based polymer nanocomposite shields will find convenient application is civil shields like human body protection, high precision instruments, protection of high-tech buildings, and so on.

We believe that with the proper formulation of structurally and chemically compatible polymeric matrices in a blend or single matrix systems, researchers will be able to fabricate superior EMI shields/filters that will be able to serve the rising demand for these systems, especially with the present challenge been posed by the introduction of 5G network.

Table 7.1 Comparative presentation of the EMI shielding performance of shield or potential prepared by diverse fabrication approaches

Polymer matrix used	Formulation	GNP-based filler wt.%	Thickness (mm)	Dominant shielding mechanism	SE (dB)	References
Single matrix						
PEDOT	PEDOT/GNPs	15	–	Absorption	18	[6]
	GN/PEDOT/Fe_3O_4	1 mg/mL	2.9	Absorption	−56.5	[7]
	PEDOT-GN-$NiFe_2O_4$	–	2	Absorption	45.4	[8]
	GN/PEDOT/CoFe2O4	–	2.4	Absorption	−43.2	[9]
	PEDOT/rGO	20	2.5	Absorption	34.7	[10]
PDMS	Fe_3O_4@$Ti_3C_2T_X$/GNP/polydimethylsiloxane (PDMS)	–	1	Absorption	80	[11]
	BS-GF-PDMS	–	1	Absorption	56	[12]
	3D GNP@Fe_3O_4/PDMS	9.72	2	Absorption attenuation and reflection	70.37	[13]
	PDMS/rGO-AF	21.7	1.547	Absorption	53	[14]
	PDMS/Ag/rGO-AF/rGO-F	–	–	Reflection and absorption	67.3	[15]
	PDMS/GF/h-Fe_3O_4	12	2	Reflection and absorption	70.37	[13]
	PDMS/Fe_3O_4/GNP	–	1	Absorption	32.4	[16]
	PDMS/GNP	0.42	5	Absorption	65	[17]
	PDMS/GNP	0.7	1	Absorption	30	[18]
	FGS/poly(dimethyl)siloxane (PDMS)	2	100 μm	Absorption	–	[19]
PPy	PPy/GNP	0.2	2	Absorption	33	[20]
	PNT/GO-EDA-Fc-EDA-GO-7:1	–	3	Reflection and absorption	28.73	[21]
	PPy/GNP	5	–	Absorption	26	[22]
	PPyFc-rGO	17.5	3	Absorption	−11	[23]

(continued)

Table 7.1 (continued)

Polymer matrix used	Formulation	GNP-based filler wt.%	Thickness (mm)	Dominant shielding mechanism	SE (dB)	References
	PPy/rGO/Fe_3O_4/BST	–	2.5	Multiple scattering, absorption, and reflection	48	[24]
PANI						
	rGO-PANI	1	–	Absorption	−25	[25]
	GNP-$PANI_{10:1}$@polyimide (PI)	40	0.04	Absorption	21.3	[26]
	PANNI@GNP	10	1.5	Absorption	−14.5	[27]
	PANI/GNP	5	–	Absorption	40	[22]
	PEDOT-GNPs	15	–	Absorption	18	[6]
PE and/or PEG	GNP/HDPE	9.8 vol %	3	Multiple reflections/absorption	31.6	[28]
	HDPE/bagasse fibers (BF)/xGO	–	3	Absorption	56.2–60.2	[29]
	PE/rGO	0.66 vol.%	2.5	Absorption	48	[30]
	Fused deposition modeling (FDM) LLDPE/GNP	10	2	Absorption	32.4	[31]
	UHMWPE/FeCo (ferrocobalt)@rGO	8.97 vol.%	2	Reflections/absorption	21.8	[32]
	GNP-PEG	7.5	2.35	Absorption	−43	[33]
Polystyrene (PS)	PS/rGO	3.47 vol.%	2.5	Absorption	45	[34]
	PS/rGO/Fe_3O_4	2.24	–	Absorption	30	[35]
	PS/rGO	10	2	Absorption	29.7	[36]
	Polystyrene-*block*-poly(ethylene-*ran*-butylene)-*block*-polystyrene (SEBS)/rGO@Ni	50	2.3	Absorption and reflection	48.2	[37]
	PS/GP nanoplatelets (GNPs)	35	3	Multireflection and absorption	33	[38]
	PS/rGO	7.5 vol.%	2	Absorption	24.5	[39]

(continued)

Table 7.1 (continued)

Polymer matrix used	Formulation	GNP-based filler wt.%	Thickness (mm)	Dominant shielding mechanism	SE (dB)	References
	PS/GNP	10	–	Absorption	−31	[40]
	PS/GNP/MWCNTs	1.5	3	Absorption	20	[41]
	Poly(styrene-butyl acrylate) latex (P(St-BA))-GNS	5	0.3	Absorption	35	[42]
PP	mPP(PP micro-powders)/RGO/MWCNTs	0.5	3	Absorption	8.3	[43]
	PP/GNPs	3	400 μm	Absorption	~ 30	[44]
	GNP/PEDOT:PSS	5	1.5	Absorption	91.9	[45]
	PP/rGO	20	2	Absorption	50	[46]
	PP/rGO	5	2	Absorption	−10.2	[47]
	PP/rGO/Fe_2O_4	10	0.5	Absorption	71.3	[48]
	PP/rGO/$NiFe_2O_4$ (NFD@600-rGO-PP)	5	2	Absorption	45.56	[49]
	PP/rGO/$NiFe_2O_4$ (NFD@800-rGO-PP)	5	2	Absorption	36.43	[49]
	PP/rGO/$NiFe_2O_4$ (NFD@1000-rGO-PP)	5	2	Absorption	35.71	[49]
	PP/GNP	13	2.56	Absorption	120.6	[50]
Rubber	GO/nitrile butadiene rubber (NBR)	10	3	Absorption	57	[51]
	NR/Fe_3O_4@rGO	–	1.8	Reflection and absorption	43	[52]
	NBR/GNP	1	2	Absorption	48	[53]
	NBR/GNP	4	2	Absorption	77	[53]
	Fe_3O_4@rGO/NR	–	1	Reflection and absorption	26.4	[54]
	Silicone rubber (SR)/GNP	3	1.7	Reflection and absorption	23.46–36.01	[55]
	SR/GNP/MWCNTs	2.77	–	Absorption	42	[56]
	NR/CNFs/rGO	0.8	0.1	Absorption	25.81	[57]

(continued)

Table 7.1 (continued)

Polymer matrix used	Formulation	GNP-based filler wt.%	Thickness (mm)	Dominant shielding mechanism	SE (dB)	References
PU	TrGO/TPU	5 vol.%	1	Absorption	24	[58]
	S-GNS/WPU	5 vol.%	2	Absorption	32	[59]
	PU/GNP	20	2.4 (6.5)	Absorption	17–21 (23–24)	[60]
	PUG-5	5	20	Absorption	12.4	[61]
	PUG-5	5	40	Absorption	24.6	[61]
	PUG-5	5	60	Absorption	34.7	[61]
	PUG-10	10	20	Absorption	19	[61]
	PUG-10	10	40	Absorption	41.6	[61]
	10Ag@FRGO/WPU	5	2	–	35	[62]
	PU/rGO	6.5	1.8	Absorption	21.8	[63]
PVDF	GNP-PVDF	5	–	Reflection	20	[64]
	PVDF/RGO/$BaCo_2Fe_{16}O_{27}$	10	0.2	Absorption	35.94	[65]
	PVDF/GNP/CNTs	6	1	Absorption	46.6	[66]
	PVDF/GNP-Ni-CNT	2.75	0.6	Reflection	46.4	[67]
	PVDF/GNP	5	–	Reflection	20	[64]
	PVDF-CNT-GNP	–	0.3	Reflection	63.3	[68]
	PVDF/10wt%-GNP	10	3	Absorption	37.4	[69]
	PVDF/GNP	20	0.3	Reflection	69.7	[70]
Polyetherimide (PEI) and/ or polyimide (PI)	Polyetherimide (PEI)/GNP	0.21	1.5	Absorption attenuation and reflection	44	[71]
	PEI/GNP@Fe_3O_4	10	2.5	Multiple reflections/absorption	41.5	[72]
	PI/rGO	8	500 μm	Absorption	15.1	[73]

(continued)

Table 7.1 (continued)

Polymer matrix used	Formulation	GNP-based filler wt.%	Thickness (mm)	Dominant shielding mechanism	SE (dB)	References
	PI/rGO/MWCNTs	8	500 μm	Absorption	18.2	[73]
	PEI@diglycidyl ether of the bisphenol A/polyetherimide (DGEBA)/rGO	3	2	Absorption	25.8	[74]
	PEI/GO	1.5	–	Absorption	23	[75]
	PI/rGO	16	0.8	Absorption	21	[76]
	PEI/rGO	0.66 vol.%	2 μm	Absorption	6.37	[77]
	PEI/rGO	2.5	0.086	Absorption and reflection	26	[78]
Cellulose	BCs/GNP	18.4	–	–	–	[79]
	GO/cellulose "GN/CNA-5"	–	2	Absorption	58.4	[80]
	CNF/GNP	25	35 μm	Reflection	27.4	[81]
	GNP/CNF	118 μm	13 μm	Reflection	43	[82]
	CNF/rGO@Ni	50	~ 20 μm	Absorption	32.2	[83]
	GNs/BCs-Janus	–	24 μm	Reflection and absorption	48	[84]
	CNFs/rGO	10 mg/ml	3–5	Absorption	33	[85]
	CNFs/GO@Ni	–	105 μm	Absorption and reflection	22.6	[86]
	CNFs/rGO	50	23 μm	Absorption	26.3	[87]
Poly(arylene ether nitrile)(PEN)	PEN/GNP/CNT/Fe_3O_4	–	2	Absorption	38	[88]
	PEN/GNP (PEN/5G3C)	5	1.6	Absorption	25.8	[89]
	PEN/GNP	–	48 μm	Absorption	77	[90]
Epoxy	Epoxy/GNP	0.5	3	Absorption	−23.45	[91]
	EP/f-rGO@Ni	5 vol.%	–	Absorption-reflection-reabsorption	40.82	[92]

(continued)

Table 7.1 (continued)

Polymer matrix used	Formulation	GNP-based filler wt.%	Thickness (mm)	Dominant shielding mechanism	SE (dB)	References
	Epoxy/rGO/MWCNTs	5	1.5	Absorption	22.6	[93]
	Epoxy/rGO	0.5	0.3	Absorption	−22.6	[94]
	Epoxy/Ag/rGO	0.44 vol.%	3	Multiple internal reflections and absorption	58	[95]
	Epoxy/rGO	5	6	Absorption	24.78	[96]
	GO/PEO	2.6 vol.%	4	Absorption	−38.8 (R_L)	[97]
	Epoxy/CF-30-Fe_3O_4-30GNP	30	1.0 ± 0.1	Absorption	33.1	[98]
PMMA	Poly(methyl methacrylate-butyl acrylate) latex (P(MMA-BA))-GNS	20	0.05	Absorption	21.5	[42]
	PMMA/GNP@Ni (N4)	40	1.75	Absorption and reflection	38	[99]
	PMMA/rGO@ strontium hexaferrite	9	3.2	Absorption	17	[100]
	PMMA/GNP	4.2 vol.%	2	Absorption	30	[101]
	PMMA/rGO@magnetite NPTs	2.6 vol.%	2.9	Absorption	63	[102]
	PMMA/MX@rGO	2 vol.%	2	Absorption	61	[103]
	PMMA/rGO/CNTs	10	2	Absorption	30	[104]
PLA	PLA/rGO/OPEFB (oil palm empty fruit bunch)	8	6	Absorption	31.2	[105]
	PLA/GNP	9.08 vol.%	2	Absorption	34.9	[106]
	PLA/GNP	15	1.5	Absorption	11.6	[107]
	PLA/GNP	6	2	Absorption	14.7	[108]
	PLA/GNPs/CNTs (4 wt.%)	2	4	Absorption	38	[109]
	PLA/GNP	10	3	Absorption	34.7	[110]
Others	Ethylene methyl acrylate (EMA)/in situ TrGO	5	5	Absorption	30	[111]

(continued)

Table 7.1 (continued)

Polymer matrix used	Formulation	GNP-based filler wt.%	Thickness (mm)	Dominant shielding mechanism	SE (dB)	References
	EVA (poly(ethylene-co-vinyl acetate))/rGO/PdNi	5	–	Absorption	30	[112]
	Polycarbonate (PC)/EMA/GNP	15 phr	–	Absorption	−51	[113]
	Polycarbonate (PC)/EMA/GNP	7 phr		Absorption	−34	[114]
	Poly(N-vinylpyrrolidone) (PVP)/GNP/PE-co-MA	3	1–2	Absorption	30	[115]
	PC/EMA/rGO	15 phr	3	Absorption	−30	[116]
Poly(vinyl alcohol) PVA	PVA/GNP	8	2.43	Absorption	32	[117]
	PVA/GO/Fe_3O_4	0.08	200 μm	Absorption	40.7	[118]
	PVA/rGO	–	0.3 mm	Reflection	25	[119]
	PVA/Mxene/GNP@Fe_3O_4	–	1	Absorption	36	[120]
	PVA/GNP	0.5 vol.%	1	Absorption	19.5	[121]
Blends						
PEDOT:PSS:PEG	PEDOT:PSS/MXene/PEG	1.22	2	Reflection and absorption	29.8	[122]
PEDOT:PSS	PEDOT:PSS/GNP@Ag	–	~ 30 μm	Reflection and absorption	26	[123]
BCs:ARAMID	Aramid NFs/BCs/GNP/Ni (ANF/BC-3/GNs/Ni-300)	–	0.052 μm	Reflection	51.2	[124]
PANI:PS	PANI:PS/GO	0.09	0.85	Reflection	40	[125]
PANI:ABS	rGO/PANI-hexaferrite/MWCNT/ABS	5	1	Absorption	21.5	[126]
PANI:paraffin	GO-g-PANI/Paraffin wax	–	30	Absorption	36.2	[127]
PEDOT:SEBS-g-maleated (MAH)	GO/BN/poly(-hydroxybenzate-co-DOPO-benzenediol dihydrodiphenyl ether terephthalate))-BN/maleated styrene-ethylene/butylene-styrene)	35	0.02–4	Absorption	29.7	[128]
s	PVDF/PEDOT-block-PEG/GNP/CuO	2	> 1	Absorption	17	[129]

(continued)

Table 7.1 (continued)

Polymer matrix used	Formulation	GNP-based filler wt.%	Thickness (mm)	Dominant shielding mechanism	SE (dB)	References
	Epoxy (EP)/PVDF/NiCo-GNP	~ 7.5	–	Reflection	34.62	[130]
PVA:PPy	PVA/IL(1-butyl-3-methylimidazolium tetrafluoroborate)-PPy-rGO	10	1	Absorption	32.9	[131]
CNCs:EPDM	*rGO@CNCs/EPDM*	–	2	Absorption	24.1	[132]
Chitosan (CTS):PVA	CTS/GO/PVA	30	60–80 μm	Absorption	12	[133]
PP:PET	PP/PET/TRGO	2 vol.%	5	Absorption	60	[134]
PP:PET	PP/PET/GNP	2 vol.%	5	Absorption	40–60	[134]
PANI:Paraffin	PANI/S-RGO/Paraffin	10	5	Absorption	−22.5	[135]
PU:lignin	Fe/GNP/PU/lignin	2	1	Absorption	21.6	[136]
Cellulose:NR	NR:Cellulose/rGO	0.8	0.13	Absorption	25.81	[57]
PLA:PEG	PLA:PEG/rGO	0.8	6	Absorption and multiple internal reflection	22.5	[137]
PU:AEMA	Aminoethyl methacrylate (AEMA)-GN/WPU	5 Vol.%	4	Absorption	38	[138]
PU:PVP	GNP@PVP/WPU	30	0.15	Absorption	30	[139]
PPy:PP	PPy/GNFs/IONPs	0.5	15 μm	Absorption	24	[140]
PVC:PANI	PVC:PANI/GNP	10	0.1	Absorption	99	[141]

7.2 Application of Polymer-Based Fire Retarding Nanocomposites

7.2.1 PVA

Huang and colleagues [142] have worked on nanocomposites based on PVA and GNPs using polymer solution blending. In this work, they compared the performance of graphene nanosheets to Na-MMT and MWNTs. The flame retardancy capabilities of PVA were only minimally enhanced by Na-MMT and MWNTs despite their outstanding nanodispersion in PVA. Even though the THR, AMLR, and TTI showed minimal change, the PHRR of PVA-MMT (Fig. 7.1) decreased by 30% when compared to pure PVA. For the PVA matrix, MWNTs, a carbon-based nanofiller, had superior flame retardancy than Na-MMT. The PHRR and THR of PVA-MWNTs were reduced by 35% and 10% relative to pure PVA. The TTI of PVA-MWNTs was 6 s longer than that of pure PVA. In comparison with PVA-MMT and PVA-MWNTs, the PVA composite based on 3 wt.% graphene nanosheets (PVA-G3) showed a different fire behavior. PVA-G3 has substantially lower PHRR, THR, ASEA, and AMLR values than PVA-MMT and PVA-MWNTs. TTI was also considerably longer for PVA-G3 than PVA-MMT and PVA-MWNTs. The PHRR of PVA-G3 was 49% lower than that of pure PVA.

In another study, Huang and colleagues [143] prepared a flame-retardant PVA/MPP-GNP nanocomposite based on the solvent blending method. They showed fine dispersion (Fig. 7.2) of graphene sheets in the PVA/G1 and PVA/G1/MPP10

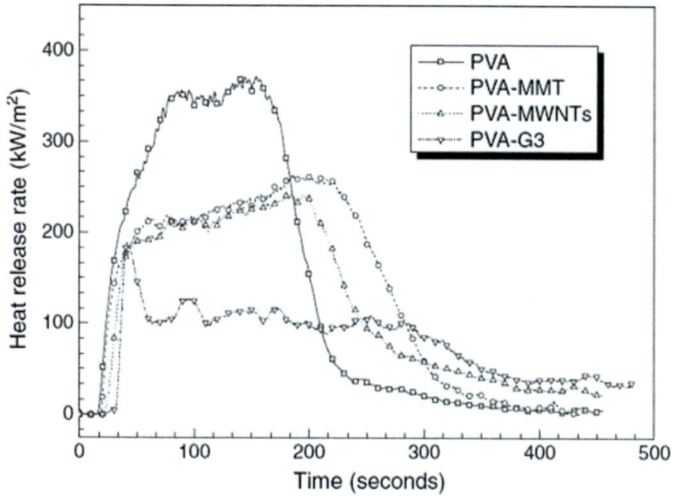

Fig. 7.1 Cone calorimeter experiments, heat release rate versus time curves: comparison of the effect of the different nanofillers (Na-MMT, MWNTs, and graphene). Reproduced with permission from Huang et al. [142]. Copyright 2012, Elsevier Science Ltd.

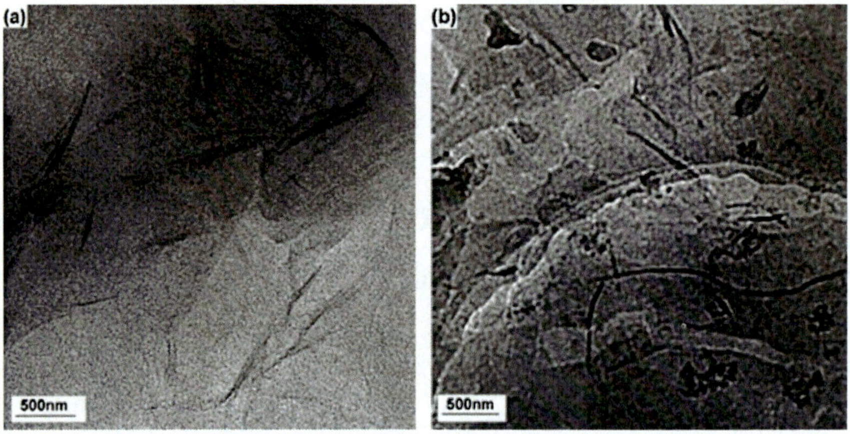

Fig. 7.2 TEM images of **a** PVA/G1 and **b** PVA/G1/MPP10. Reproduced with permission from Huang et al. [143], Copyright 2012, Elsevier Science Ltd.

nanocomposites. A stack of two to twenty-five partially intercalated graphene sheets, which is aligned multilayer graphene, was observed as discrete black lines in PVA/G1. In contrast, the SEM image of PVA/G1/MPP10 showed that most graphene sheets and flame-retardant MPP particles were uniformly distributed across the polymeric matrix.

In comparison with pure PVA (Table 7.3), PHRR and AHRR of PVA/G1 were reduced by 43% and 46% even though THR and ASEA remain almost the same. The TTI of PVA/G1 was 7 s longer than that of pure PVA. For PVA/MPP composite, the PHRR, AHRR, and AMLR are reduced significantly with the addition of MPP. The PHRR was reduced by 25% and 64% for PVA/MPP10 and PVA/MPP20 relative to pure PVA. The THR was reduced by 34% and 67% for PVA/MPP10 and PVA/MPP20; the TTI of PVA/MPP composites was longer than that of pure PVA. The reduction of THR indicates that some PVA chains participated in the carbonization process due to the flame retardancy of MPP. Interestingly, the AHRR of PVA/MPP10 is almost equal to that of PVA/G1, further showing the flame retardancy of GNPs with a low loading content [143].

7.2.2 PP

The effects of carbon additions with various particle sizes and shapes on the flame retardancy and mechanical attributes of isotactic PP have been studied by Dittrich and colleagues [144]. CB, MWNT, and EG were examples of spherical, tubular, and platelet-like carbon fillers. TRGO and MLG, which had a few graphene layers, were also contrasted. They subjected the melt-extruded PP and its composites to an external flux of 50 kW/m^2, and they noticed that adding different carbon nanomaterials caused ignition to occur sooner after 23–25 s as opposed to 33 s for the pure PP. The HRR curve's shape (Fig. 7.3) was also changed when these carbon-based nanoparticles were added to PP at a concentration of 5.0 wt.%, depending on the type of nanofiller used. In contrast to PP combustion, which produced an HRR curve pattern characteristic of non-charring materials, the HRR curve of the carbon nanocomposites changed in the direction of a pattern typical of charring/residue-forming materials. After combustion, the PHRR decreased significantly, the HRR plateaued, and the burning time got longer. The list of materials below is arranged according to the order in which the HRR curve for residue-forming substances changed appearance: According to PP/5.0EG40, PP/5.0MWNT, PP/5.0CB, and PP/5.0TRGO, there has been an increase in the creation of protective residue layers that are efficient during combustion. Carbon nanocomposites with different morphologies decreased the PHRR of the nanocomposites in the following order: for EG40, MWNT, CB, and TRGO, 855 Kw/m^2 (a 57% reduction), 765 Kw/m^2 (a 62% reduction), and 532 Kw/m^2 (a 74% reduction) from 2011 Kw/m^2. The addition of carbon nanomaterials EG40, EG60, MLG250, and TRGO to PP decreased the PHRR of the nanocomposites. EG40 and EG60 had poor dispersion and lowered PHRR by 44% (to 1129 kW/m^2) and 45% (to 1104 kW/m^2), respectively, when compared to PP alone. Due to much-enhanced dispersion, the PHRR of the PP/5.0MLG250 and PP/5.0TRGO systems, respectively, reduced by 72% (MLG250; to 570 kW/m^2) and 74% (TRGO; to 532 kW/m^2) [144].

Depending on the type of nanofiller utilized, a significant variation in melt flow behavior for the PP nanocomposites was seen during the vertical UL-94 set-up. The vertical UL-94 residues displayed the same behavior for PP as shown in Fig. 7.4a, poorly dispersed PP/5.0EG40 as per Fig. 7.4d. Due to the steady flow of the molten pyrolyzing polymer, the viscosity of both materials was equivalent over the examined range of angular frequency as in Fig. 7.4g. Melt flow was not observed in the more

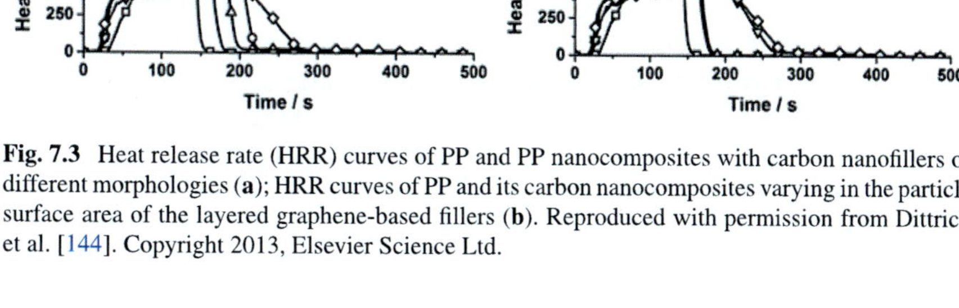

Fig. 7.3 Heat release rate (HRR) curves of PP and PP nanocomposites with carbon nanofillers of different morphologies (**a**); HRR curves of PP and its carbon nanocomposites varying in the particle surface area of the layered graphene-based fillers (**b**). Reproduced with permission from Dittrich et al. [144]. Copyright 2013, Elsevier Science Ltd.

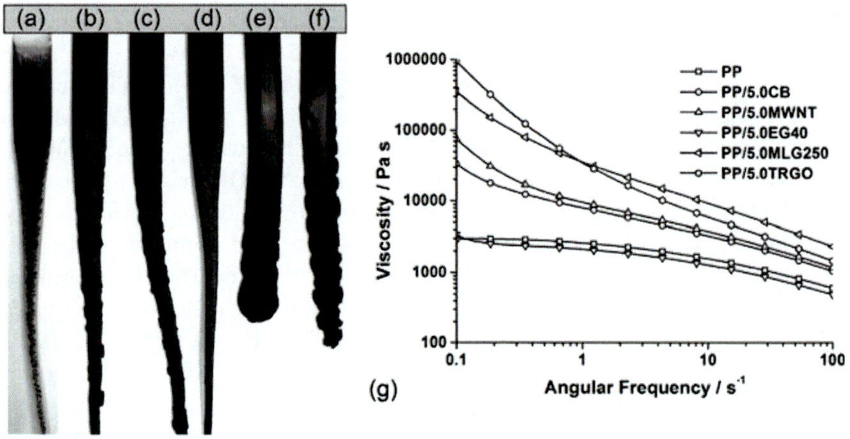

Fig. 7.4 Residues of **a** PP, **b** PP/5.0CB, **c** PP/5.0MWNT, **d** PP/5.0EG40, **e** PP/5.0MLG250, and **f** PP/5.0TRGO after vertical UL-94 testing, the result of PP/5.0EG60 is similar to (**d**); melt viscosity depending on angular frequency (**g**), showing rheological behavior of materials **a–f**. Reproduced with permission from Dittrich et al. [144]. Copyright 2013, Elsevier Science Ltd.

evenly dispersed nanocomposite systems containing CB as per Fig. 7.4b, MWNT as depicted Fig. 7.4c, MLG250 as seen in Fig. 7.4e, and TRGO as per Fig. 7.4f. The largest impact was demonstrated by the well-exfoliated layered particles in PP/5.0MLG250 and PP/5.0TRGO. The order of CB < MWNT < MLG250 < TRGO considerably increased each nanocomposites melt viscosity at low shear rates by up to two decades. The carbon nanoparticles are disseminated at nanoscales and serve as anti-dripping agents without significantly affecting processing at higher shear rates (Fig. 7.4g).

When looking at the flame-retardant properties of PP and its composites with 5 wt.% of carbon-based materials, it was noticed that pure PP reached a LOI of 19%, while the addition of all carbon-based materials led to a slight increase in the LOI; resulting in 20.7, 19.8, 19.4, 19.0, 20.9 and 20.4% for composites with CB, MWNT, EG40, EG60, MLG250, and TRGO, respectively [144].

The flame retardancy of impact-modified isotactic PP (PP-FR), which also contains antimony trioxide and a typical brominated FR, was studied by Hofmann and coworkers using functionalized graphene nanosheets TRGO and MLG 250 produced from thermally reduced graphite oxide [145]. Using a cone calorimeter and an external heat flow of 50 kW/m^2, the burning behavior of PP-FR and its nanocomposites, including 5.0 wt.% CB, CNT, EG 40, MLG250, and TRGO, was examined. They also noticed that the nanocomposites ignited faster than PP-FR due to the carbon nanoparticles' greater capacity to absorb heat from the cone heater's radiation. However, they did not report any significant difference in the cone calorimeter results as compared to those of Dittrich et al. [144].

In another work by Huang et al. [146], IFR, CNTs, and RGO were added to the PP matrix to create a unique PP nanocomposite. IFR was made up of 80% melamine polyphosphate and 20% pentaerythritol. They noticed that neat PP burned more quickly after ignition, achieving a PHRR (Fig. 7.5) of around 1242 kW/m^2. While the TTI and AMLR of the nanocomposites barely differed from the pure PP at about 42 ± 4 s, the PHRR values in the PP/CNTs and PP/RGO nanocomposites were 538 and 486 kW/m^2, respectively, significantly lower than that of neat PP. While the TTI of the PP/CNTs/RGO nanocomposites was 6 s slower than for pure PP, and the PHRR indicated a 63% (465 kW/m^2) reduction. The combination of CNTs and graphene greatly reduced the flammability of the PP matrix, as shown by the fact that PP/CNTs/RGO had lower PHRR (465 kW/m^2), AMLR (0.046 g/s), and average specific extinction area (ASEA—439 m^2/kg) values than PP/CNTs and PP/RGO. In comparison with pure PP, the PHRR of the PP/IFR/CNTs/RGO composites demonstrates an 83% (212 kW/m^2) reduction, while TTI is 40 s delayed.

The LOI analysis demonstrated that neat PP was combustible because it only had an 18.8% LOI value and could pass the UL-94 test. Despite LOI values of 20.6, 20.1 and 21.0% for PP/CNTs, PP/RGO, and PP/CNTs/RGO, respectively, they failed the UL-94 test. By adding 20 wt.% IFR, the LOI value of the PP/IFR sample rises to

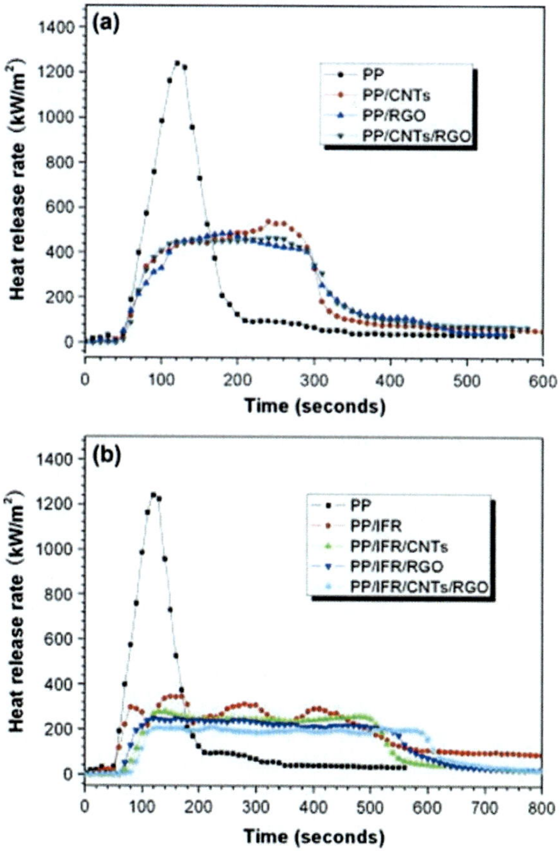

Fig. 7.5 Heat release rate curves for PP composites (**a**) and intumescent flame-retardant PP composites (**b**) at 35 kW/m^2. (For interpretation of the references to color in this figure legend, the reader is referred to the web version of this article). Reproduced with permission from Huang et al. [146]. Copyright 2014, Elsevier Science Ltd.

29.2% and achieves a V-1 rating. The LOI values of PP/IFR/CNTs and PP/IFR/RGO were higher than those of PP/IFR when CNTs or GNPs were added. While the findings of the LOI test and the cone measurements were in agreement, proving that the inclusion of carbon-based nanofillers had enhanced the PP/IFR system's flame retardancy. The LOI value of the PP/IFR/CNTs/RGO sample was indeed significantly higher than that of neat PP (31.4%) and could achieve a V-0 rating, showing that the combination of CNTs and graphene played a significant role in improving the flame retardancy of the PP/IFR system and that there was a clear synergistic effect between CNTs, graphene, and IFR in the PP matrix [146].

Concurrently, Yu et al. [147] developed a new functionalized GO (FGO)/phosphoramide oligomer flame-retardant that was made using in situ polymerization. To enhance the fire safety of PP. A nanocomposite flame-retardant (FRs-FGO) comprising exfoliated graphene was synthesized by modifying the GO with DDM and then incorporating it in situ into a phosphoramide oligomer. The FRs-FGO was then mixed with PP, and at the same time, maleic anhydride was grafted onto PP. Pure PP was found to burn extremely quickly after ignition with a PHRR as high as 1199 kW/m^2. The 20 wt.% FRs considerably reduced the PHRR of PP1, dropping it to 620 kW/m^2. Once the FRs were replaced in part by 2 wt.% GO, the PHRR of PP2 was further reduced to 473 kW/m^2. In comparison with PP1 and PP2, PP3 exhibits the lowest PHRR (397 kW/m^2) when 20 wt.% of FRs-FGO nanocomposite flame-retardant is added. Compared to pure PP, the PHRR values for PP1, PP2, and PP3 were decreased by 48.3%, 60.6%, and 66.9%, respectively.

In a study by Yuan et al. [148], melt mixing was used to integrate melamine FGO, which was synthesized via a non-covalent technique, into the PP matrix. Comparisons were also made between the GO/PP composite and FGO's dispersion state in PP. Due to improved heat absorption in the surface layer of these composites containing carbon nanomaterials, the graphene-based PP composites showed an earlier igniting behavior than plain PP (50 s). As the loading of FGO was increased, the PHRR values of the FGO/PP composites dropped (892–739 kW/m^2). The decrease in PHRR was considered a sign that the flame retardancy had improved. The 2 wt.% FGO/PP nanocomposite (739 kW/m^2) was found to have a 29% lower PHRR than neat PP (1044 kW/m^2), however, with 2 wt.% GO/PP, there is no discernible decrease in PHRR (979 kW/m^2). It was clear that the PP composite made from FGO had better flame-retardant qualities than the GO/PP composite, while the combination of GO and FGO did not result in a noticeably lower THR than neat PP.

The effectiveness of using graphene as an effective synergist for IFR was later explored by Yuan et al. [149]. APP and CFA, which serve as carbonization and blowing agents, were the two main components of the IFR. The flammability and heat release during combustion of the flame-retardant PP composites were examined after the addition of RGO nanosheets to the APP/CFA/PP flame-retardant system. With a clear HRR peak of 1025 kW/m^2 (Fig. 7.6), neat PP burned fiercely after a fire. IFR/PP composite ignition started earlier than with ordinary PP. The PHRR and combustion duration of IFR/PP composites were much less than those of pure PP. It is significant to note that there are two peaks on the HRR curve for the IFR/PP composite. There was a correlation between the formation of the intumescent protective shield, which was the cause of the initial peak at about 70 s, and the breakdown of the intumescent layer, which led to a reduction in protection and an increase in HRR—the addition of 25 wt.% IFR resulted in a decrease in PHRR and THR of 79.2% and 18.3%, respectively, compared to the neat PP. When IFR was partially substituted with 0.5 wt.% RGO, PHRR was remarkably further lowered; it decreased from 213 kW/m^2 for PI to 140 kW/m^2, exhibiting the improved barrier efficacy of the intumescent char by graphene nanosheets. It is clear that as RGO loading was increased in PP/APP/CFA/RGO nanocomposites, the combustion period between the two peaks gradually shrank. This was attributed to the high thermal conductivity filler's contribution to the improved thermal conductivity of PP/APP/CFA/RGO nanocomposites. The PHRR values of the related nanocomposites were 156 and 262 kW/m^2 when the additional concentrations of RGO were 1.0 and 2.0 wt.%, respectively. According to this, higher loadings of graphene had an antagonistic effect on the IFR and could not be used to create a flame-retardant synergism. Notably, the THR value of PP/APP/CFA/RGO nanocomposites with 1% RGO was further reduced to 86.0 MJ/m^2, which was less than that of the PP/APP/CFA nanocomposite.

The attachment of graphene nanosheets to the surface of modified APP (Fig. 7.7) through hydrogen bonding interactions was further investigated by Yuan and coworkers [150] with the aim of enhancing the dispersion of RGO in the PP matrix. This straightforward approach avoided the modification or solvent blending of poisonous or damaging solvents. To determine the roles of well-dispersed and poorly dispersed graphene in the burning behavior of IFR composites under various flame situations, the effects of graphene's dispersion state on the response to tiny

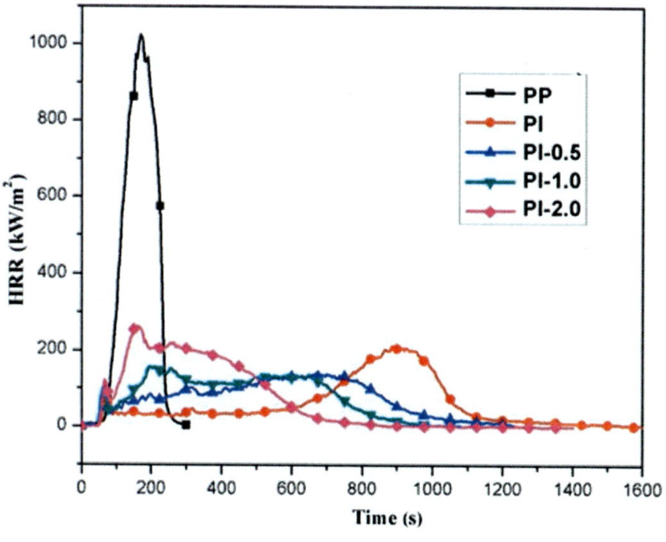

Fig. 7.6 HRR curves as a function of time. Reproduced with permission from Yuan et al. [149]. Copyright 2017, Elsevier Science Ltd.

flame and fire behavior of IFR/PP composites are examined. Large graphene agglomerates were visible in TEM images of PP/IFR nanocomposites that contained 2 w.% of unmodified RGO. Although some smaller agglomerates were still present, it was possible to see distributed graphene nanosheets in the TEM images of PP/IFR with 2 wt.% of modified RGO. The wrapping process helps improve graphene's dispersion in the nonpolar PP matrix. In contrast to the sample of graphene that had been organically changed and blended with a solvent, the graphene dispersion condition is undesirable. The low LOI value of 17.0% and lack of a UL-94 rating for PP indicate that it is a highly flammable polymer. The significantly improved LOI value and UL-94 rating show that the PA-APP and CFA formulation is an effective flame-retardant for PP. As pristine RGO loading increases, LOI gradually decreases. However, still obtaining a UL-94 V0 grade. The addition of graphene also had a negative impact on the LOI and UL-94 values for PA-APP with RGO. For instance, the LOI value decreases from 35.0% to 33.0%, 30.5%, and 26.5%, respectively, when IFR is replaced by 0.5, 1.0, and 2.0 wt.% RGO. When 2.0 wt.% of RGO was mixed with PA-APP@G, significant dripping was also observed during flammability testing, and only V-2 classification was obtained. Graphene inclusion has a negative impact on the flammability of the PA-APP/CFA flame-retardant formulation findings. It was inferred from the comparison of the flammability results of PP/IFR-G and PP/IFR-MG that enhanced dispersed graphene results in noticeably weakened LOI values and UL-94 ratings.

The masterbatch compounding-melt processing method has been used by Trusiano and colleagues [151] to prepare PP/GNP nanocomposites. With and without PP grafted maleic anhydride compatibilization (PP-g-MA), they observed combustion of pristine PP and its GNP nanocomposites with 3wt.% compatibilizer using a cone calorimeter at an external heat flux of 25, 35, and 50 kW/m^2, respectively. In comparison with pure PP, the results demonstrated that adding just 1 wt.% of GNPs

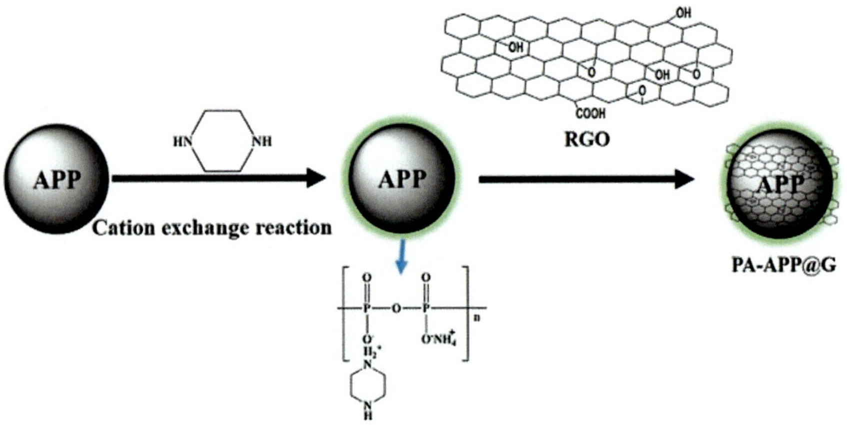

Fig. 7.7 Illustration of the synthetic route of PA-APP@G. Reproduced with permission from Yuan et al. [150]. Copyright 2018, Elsevier Science Ltd.

to the PP matrix was adequate to reduce the PHRR drastically. It was also noted that the PHRR reduced as the heat flux increased, reaching 21% less than the PP matrix at a heat flux of 50 Kw/m^2. The PHRR gradually declined until it was roughly 40% less than it had been in the neat PP as the proportion of GNPs rose, the compatibilizer was added, and the heat flux was increased [151]. They discovered that the higher-loading samples produced char throughout the combustion process, which formed an elastic shell around the sample and inflated as it burned. They explained that chars with higher flexibility could extend further and were more effective at trapping the hot gases that caused burning [151, 152]. They further explained that GNPs were recognized to absorb and re-emit heat from an oncoming front, lowering the actual heat flow to the sample.

Additionally, the solid char surface shut off the blazing channel, preventing oxygen from diffusing to the hot gases within the nanocomposites and slowing the pace of combustion [151]. Samples with 1 wt.% of GNPs showed porous char, but those with higher filler and compatibilizer concentrations appeared uniform and solid due to reaching the percolation threshold [151, 153]. In addition, they discovered that although there was a discernible change in the composites PHRR depending on the percentage of the added GNPs, the inclusion of PP-g-MA did not significantly increase the combustion response of the PP composites with the same GNP amount [151].

EG and a PDPFDE were used to provide a synergistic fire-retardant effect on PP by Cheng et al. [154]. The barrier, as seen in Fig. 7.8, prevented heat transfer between the flame zone and the combustion matrix, slowing the pyrolysis of the PP to some extent as a result of the EG's rapid expansion into char layers. However, a major "popcorn effect" phenomenon that results in the development of ineffective loose chars as physical barriers to heat, and oxygen has a detrimental impact on the fire retardancy of PP composites. The LOI values of all PP composites increase to some extent when the addition of flame-retardant is 25 wt.%. The LOI value of PP/3PP-g-MA/25EG (PP1) rises to 22.6% with the addition of only 25 wt.% EG but failed the UL-94 test. According to Fig. 7.8, the flame of the PP/3PP-g-MA/22.5EG/2.5PDPFDE (PP2) composite is significantly smaller than the PP/3PP-g-MA/25EG (PP1) composite, most likely as a result of the catalytic charring effect of the PDPFDE [155], this aids in producing a thicker layer of char on the sample's surface, slowing the rate at which flames spread, and lessening the "popcorn effect" of EG during combustion. A sufficient compact char layer cannot be produced with less PDPFDE, which results in the product failing the UL-94 rating. PP/3PP-g-MA/20EG/5PDPFDE (PP3) may pass the UL-94 V-0 grade and get a LOI value of 28.8% by combining 20 wt.% EG and 5 wt.% PDPFDE. Additionally, Fig. 7.8 clearly showed that the PP3 was self-extinguishing after ignition, showing that a dense and high-quality char layer was generated quickly on the surface of the sample and that it afterward served as an effective barrier to heat and oxygen [156]. It is important to note that EG's "popcorn effect" disappeared. PP/3PP-g-MA/17.5EG/7.5PDPFDE (PP4) passed the UL-94 V-0 rating, and the self-extinguishing time increased to 1.3 s when the loading of PDPFDE was increased to 7.5 wt.%. The UL-94 rating dropped to V-1 or below when PDPFDE was added in amounts of 10 or 12.5 wt.%, and the "popcorn effect" of EG returned. It should be noted that as the loading ratios of PDPFDE increased, the LOI values and UL-94 ratings of both PP/3PP-g-MA/15EG/10PDPFDE (PP5) and PP/3PP-g-MA/12.5EG/12.5PDPFDE (PP6) tend to decline. There is an ideal addition ratio between PDPFDE and EG; hence, a higher PDPFDE concentration cannot result in superior fire retardancy. In other words, EG and PDPFDE have a clear synergistic fire retarding effect. The fire retardancy of PP composites diminished as a result of the excessive PDPFDE incorporation because of its catalytic degrading impact, which dominated and worked against the establishment of a compact char layer. While keeping the overall loading amount constant, the effects of PDPFDE with various amounts on the flame resistance of PP/EG/PDPFDE blends were investigated. The LOI value of PP2 increased to 25.7% when EG was substituted with 2.5 wt.% PDPFDE.

More recently, Li and colleagues added EG to PP composite as a flame-retardant additive while being encased in a poly-siloxane (Poly-DDPM) (Si@EG) as per Fig. 7.9 [157]. They noted that the control PP sample burned quickly after ignition and burned out in 600 s; during the quick combustion, a significant amount of heat was intensely produced. The pHRR value was 548 kW/m^2, and a colossal peak in the HRR curve appeared at 155 s. 129 MJ/m^2 was the THR. Only 10% of the original residue remained after burning. The TOC during the entire burning process was 83 g. Following the addition of 25% EG or Si@EG, the heat release was significantly reduced, as seen by the longer burning time of more than 1200 s less concentrated heat release, increased char, and lower TOC. Both PP/EG and PP/Si@EG have comparable pHRRs. The THR of PP/Si@EG, however, decreased to 76 MJ/m^2 from 98 MJ/m^2 in PP/EG. Last but not least, there was a larger %age of residue in PP/Si@EG 35% than in PP/EG. Both PP/EG and PP/Si@EG used significantly less oxygen during the entire burning process than the control PP, consuming just 62 g and 48 g O_2, respectively. In contrast, PP/Si@EG tended to generate less heat during combustion than PP/EG. It was discovered that the burning of the material was regulated with less combustible products when PP was protected by EG or Si@EG. Due to the "cage-like" framework of the small char Si@EG generated, it exhibited better than EG.

They further analyzed the morphology as in Fig. 7.10 and FTIR as per Fig. 7.11 of the char to expose the formation and propagation of the char framework and reveal the poly-DDPM pyrolysis during combustion. Composite samples were used to make the SEM samples after being exposed to artificial ignition for 0 s, 15 s, 25 s, and 35 s. After only 15 s of burning, the PP/EG char (Fig. 7.10a–b) clearly revealed a cellular structure with cracks and pores. On the other hand, the developing char of the PP/Si-EG sample (Fig. 7.10e–f) had a consistent layered shape. It was interesting to note that after 25 s of burning, there was some partially deteriorated PP matrix visible between the char layer of the PP/Si@EG sample (Fig. 7.10g). The poly-DDPM coating was said to increase the compatibility between EG and PP matrix, and the extended char was said to be able to protect PP matrix from further combustion. The char discrepancies between EG and Si@EG could be seen after 35 s of burning. EG was developed into a "worm-like" char with significant porosity features, while Si-EG had dense and compact borders and was intumescent to columnar char. It was

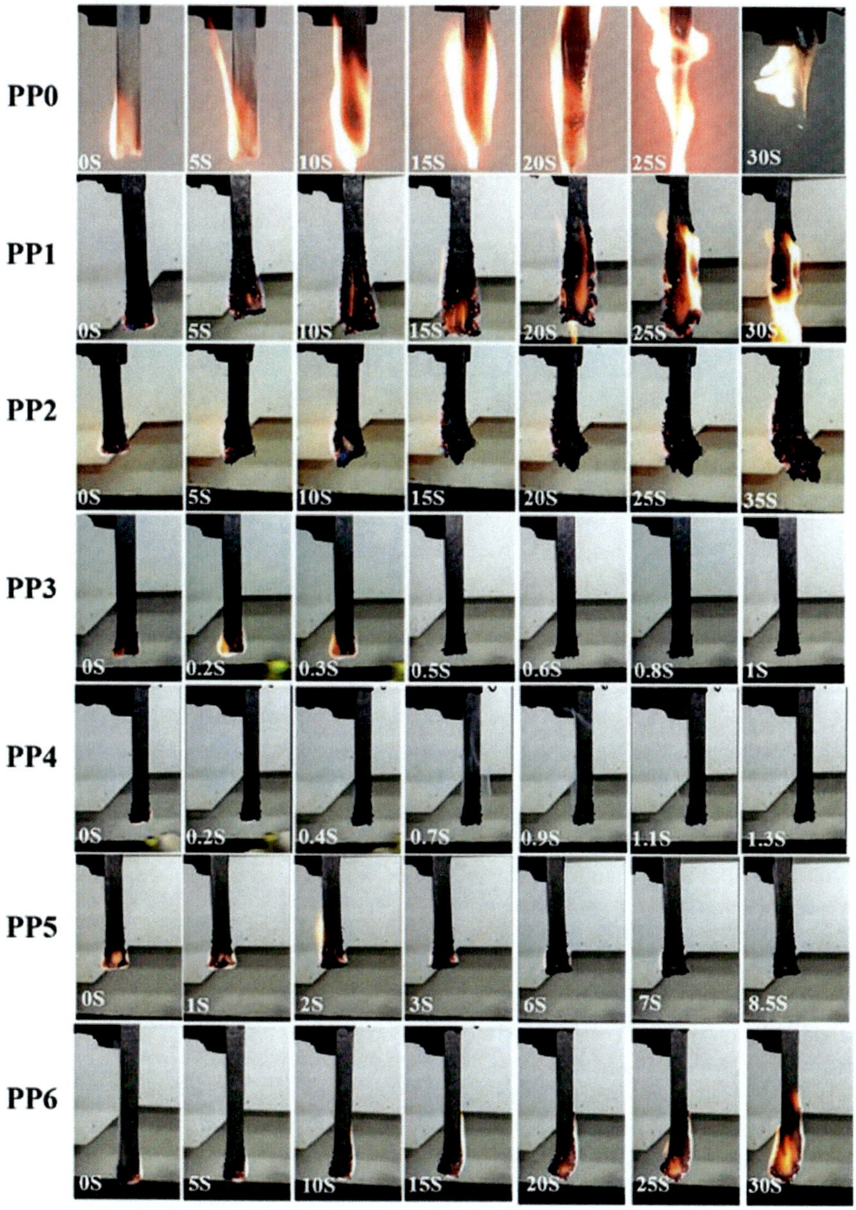

Fig. 7.8 Video screenshots of PP composites in the course of the UL-94 tests. Reproduced with permission from Cheng et al. [154], Copyright 2020, Elsevier Science Ltd.

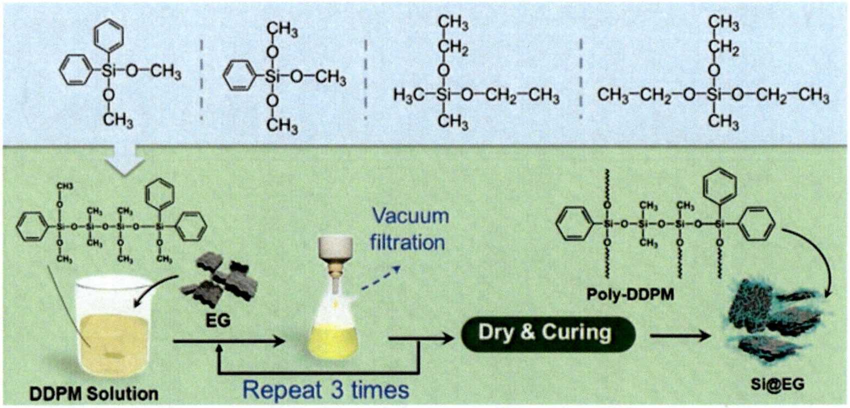

Fig. 7.9 Preparation schematic of Si@EG. Reproduced with permission from Li et al. [157], Copyright 2022. Elsevier Science Ltd.

hypothesized that melt and degraded PP matrix were covered and protected in the layer of extended char layers as a result of poly-DDPM altering the polarity of the EG surface. The SEM morphological evaluation indicated that poly-DDPM coating of EG had the potential to improve the compatibility of the PP matrix with EG particles both before and during combustion. Additionally, the Si-EG expanding char was observed to be faultless and to have uniform pores that are smaller than those of the EG system.

As the burning period was extended, the intensity of the bonds (Fig. 7.11) between Si–Ph at 1430 and 1128 cm^{-1}, Si–O–C at 802 and 842 cm^{-1}, and Si–C at 802, 842, and 700 cm^{-1} reduced, indicating some Si–C groups in poly-DDPM were gradually broken down. Simultaneously, the Si–O–Si peak's intensity at 1079 cm^{-1} increased, which explains how silicone changed into inorganic amorphous silica [157].

At the same time, Gao et al. [158] investigated the flame-retardant and combustion characteristics of PP composites treated with a solution of EG, nickel hydroxystannate, and pentaerythritol. PP/2PER displays a LOI value of 17.8% and an unclassified UL-94 test grade. The PP/2PER/30NHS sample exhibited a LOI value of 20.5%, which was slightly higher than that of the PP/2PER sample, but still failed the UL-94 test. This was due to the addition of 30 phr NHS and 2 phr PER into the PP matrix. PP/2PER/30EG sample had a LOI value of 27.3%, which was 9.5% points greater than that of PP/2PER, and it could achieve the UL-94 V-1 rating when 30 phr EG and 2 phr PER were added to the PP matrix. On the other hand, PP/2PER/28EG/2NHS exhibits a LOI value of 28.0% yet passed the UL-94 V-0 grade. This finding suggested that switching from 2 phr EG to 2 phr NHS helped pass the vertical combustion test.

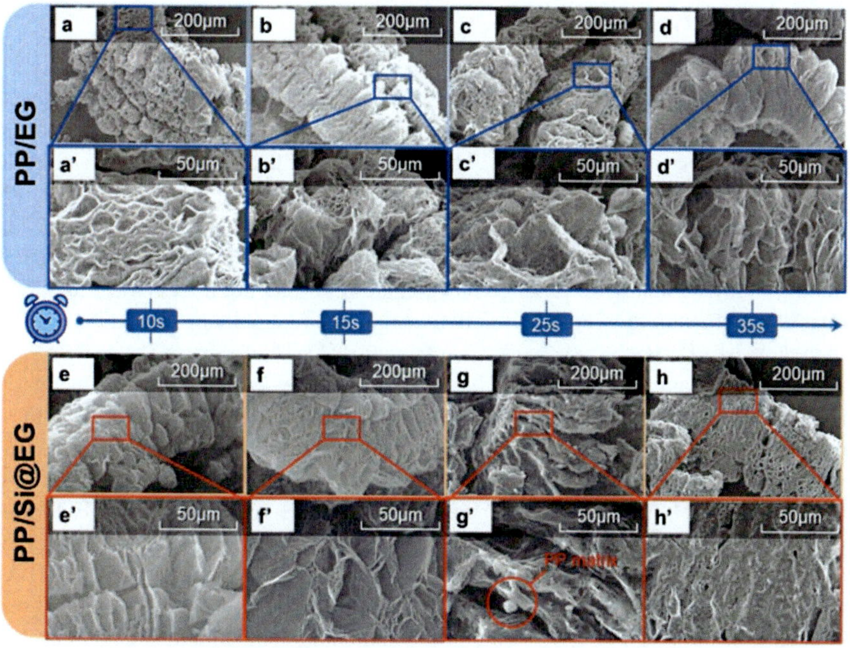

Fig. 7.10 SEM of PP/EG (a–d, a'–d') and PP/Si@EG (e–h, e'–h') after 10 s, 15 s, 25 s, and 30 s combustion. Reproduced with permission from Li et al. [157], Copyright 2022, Elsevier Science Ltd.

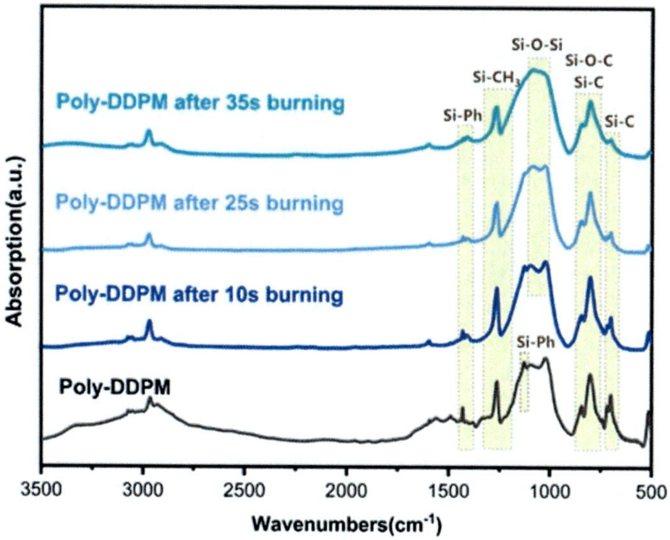

Fig. 7.11 FTIR spectra of poly-DDPM after different combustion times. Reproduced with permission from Li et al. [157]. Copyright 2022, Elsevier Science Ltd.

7.2.3 *Polystyrene (PS)*

GOs with different oxidation degrees (Table 7.2) [159] and GNPs have been investigated by Han and colleagues. They melt blended PS with GO having 7 different oxidation degrees and GNPs [159]. The study showed that the nanocomposites had a PHRR that was significantly lower than that of the virgin polymer. PS/GO1, PS/GO2, PS/GO3, and PS/graphene nanocomposites' PHRR values were, respectively, 39, 38, 32, and 47% lower and all values were lower than those of virgin PS. These findings showed that the high degree of oxidation of GO did not enhance its flame-retardant characteristics, and it was obvious that adding graphene to neat PS resulted in the lowest PHRR.

Flame-resistant PS composites have been made by Zhu et al. [160] using EG and AP. PS is a highly flammable polymer with a LOI value of just 17% and an unsatisfactory result in UL-94 tests. However, the LOI value of PS composites did slightly rise with the addition of EG or AP alone. For instance, when the EG addition content was 30%, the LOI value increased to 25%. However, the PS/30EG failed to get the UL-94 V-0 grade. On the other hand, the PS/25AP composite received a LOI value of 25.6% and a UL-94 V-0 grade. Every test sample with an EG/AP ratio of 1:1, 1:2, 1:3, 1:4, 2:1, or 3:1 containing 20 wt.% of both EG and AP was given a V-0 rating. The UL-94 V-0 grade, on the other hand, could not be passed by PS composites with just 20% EG or AP alone. The PS/10EG/5AP still receives a UL-94 V-0 rating and has a LOI value of 25.5%, even with a 15 wt.% FR content. These findings suggested that EG and AP work together synergistically to provide PS composites with their superior flame retardancy.

In a different study, Hu and colleagues investigated functionalizing GO by grafting a phosphonate derivative and a hyper-branched flame-retardant based on N-aminoethyl piperazine to lessen the flammability and toxicity of PS [161]. They discovered that the addition of 0.1% FGO significantly increased the onset of degradation of PS-FGO0.1 nanocomposites compared to pure PS, which was explained by the interaction of the physical barrier with the layers of FGO's ability to trap oxygen molecules and free radicals in the presence of air. Due to the flame-retardants' thermal instability, they were grafted onto the surface of graphene, which reduced the rate at which PS nanocomposites degraded at high GO levels. The PS-FGO nanocomposites gradually started to appear as the FGO content increased. They further looked at the influence of the hyper-branched structure on the flame retardancy of PS and observed a gradual decrease in the PHRR by approximately 39%. Additionally, FGO raises the temperature at the maximum heat release rate, which is in line with the

Table 7.2 Composition of GO1, GO2, and GO3, as well as statistics on elements

Sample	Composition
GO1	$C_8\,O_{2.1}\,H_{1.2}$
GO2	$C_8\,O_{2.8}\,H_{1.5}$
GO3	$C_8\,O_{3.7}\,H_{1.9}$

improved thermal stability seen from TGA. They offered two explanations for this: first, the FGO layers were dispersed throughout the PS matrix, acting as an excellent physical barrier during combustion, preventing fuels from entering the fire, capturing oxygen-free radicals, and preventing the degradation of the polymer matrix's chain; and second, the degradation of hyper-branched flame-retardants during combustion encouraged the appearance of char on the surface of burning material, reinforcing the barrier effect.

To learn more about how EG and AP affect PS's fire behavior, Zhu and colleagues [160] showed that PS nanocomposites with 30 wt.% EG had an enhanced LOI from 17 to 25% while still failing to meet the UL-94 V-0 rating. In comparison, when 25 wt.% of AP is added to PS, the composite received a LOI value of 25.6% and passed the UL-94 V-0 classification. When both AP and EG are added together to PS, all of the testing samples passed the UL-94 V-0 rating. The char remnants following the LOI test were examined by SEM in order to gain an understanding of the microstructure and purpose of the protective chars. On the surface of the "worm-like" char that formed during the burning of EG (Fig. 7.12a-2), they discovered pores which could carry the heat and oxygen required for polymer combustion. Figure 7.12c-2 depicted a "worm-like" char surface that was covered by char produced by the breakdown of AP, although this char surface was considerably more compact. The interaction of the flame-retardants caused an adhesive residue structure to form, which prevented heat and mass from being transferred from the flame to the supporting materials, which is what caused this morphology. Additionally, EG/APP displayed the specific char residue morphology. They added that in the gaseous phase, AP produces gaseous phosphorus-containing substances like PH_3 or/and PO_2^-, which can act to generate P•, PO•, and so on to capture the highly reactive radicals like HO• and H• generated during combustion and therefore efficiently put out the fire.

The combustion performance results showed a reduction in the PHRR from 730 to 163 kW/m^2 with the addition of 10 wt.% EG and 5 wt.% AP. The cone calorimeter test revealed no char residue for PS. However, the addition of both flame-retardants significantly increased the amount of char residue from 0 to 38.6%.

7.2.4 PE

The flame retardancy and thermo-oxidative stability of LDPE composites containing ATH, RP, and EG were investigated by Thi and coworkers [162]. To increase their compatibility with LDPE, the surfaces of ATH and RP were treated using MgST and poly(methylhydrosiloxane). They discovered that neat LDPE had a comparatively low LOI value of 17%, was combustible, and burned quickly. The inclusion of flame-retardants like ATH, RP, or EG enhanced the flame retardancy of LDPE. The filler with the highest flame-retardant impact on LDPE was RP. As shown in Table 7.3, the addition of 15 wt.% RP raises the LOI value of LDPE from 17.1 to 20.2%, while the LOI values of the composites loading 15 wt.% ATH and EG only reach 18.0 and 19.7%, respectively. LDPE was more fire-resistant when combined with ATH, RP,

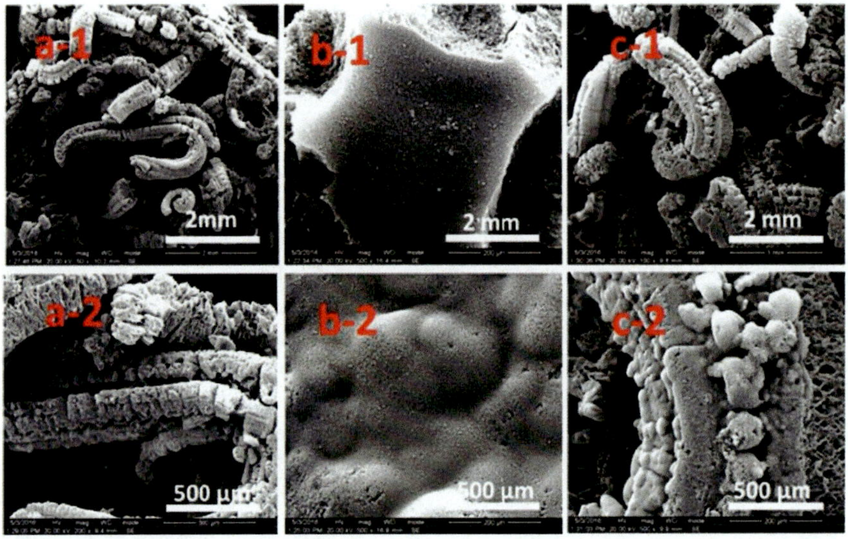

Fig. 7.12 SEM images of char residue from PS/15EG (**a-1**, **a-2**), PS/15AP (**b-1**, **b-2**), and PS/10EG/5AP (**c-1**, **c-2**) after LOI tests. Reproduced with permission from Zhu et al. [160]. Copyright 2018, Elsevier Science Ltd.

and EG; however, the flame-retardant quality of the composites with only one additive was not very good. At the same time, the mixtures of two flame-retardants with a total content of 15% in the LDPE matrix demonstrated superior flame retardancy compared to single flame-retardant systems with the same filler %age. LDPE had a poor flame-retardant impact in the composites with two fillers when ATH and EG were combined, whereas LDPE had a strong flame-retardant effect when RP and EG were combined. The composite with 10 wt.% ATH and 5 wt.% EG, in particular, failed the UL-94-V rating with a comparatively low LOI value of 21.1%, whereas the composite with 10 wt.% RP and 5 wt.% EG received the V-0 classification and a 24.1% LOI value. A significant improvement in the flame-retardant property of the LDPE composite was recorded when the three flame-retardants were added simultaneously. In particular, the composite obtained a V-0 classification and LOI value of 25.4% when 15 wt.% of the fillers were combined with an ATH: RP: EG weight ratio of 1:1:1.

In another study, Hu and colleagues [163] have looked at functionalizing GO by a hyper-branched flame-retardant (Fig. 7.13), using N-aminoethyl piperazine and di(acryloyloxyethyl)methylphosphonate and incorporating it in XLPE to make the matrix more flame-resistant. They observed that XLPE burns quickly after ignition, with a PHRR OF 1741 kW/m^2. The PHRR decreased with increasing content of FGO from 1671 kW/m^2 at 0.5 wt. % to 1241 kW/m^2 at 3 wt.% of functionalized GO. The addition of GO marginally enhanced the THR of XLPE composites due to GO's strong heat conduction. In addition, when the amount of GO was increased, the THR value gradually decreased. For the XLPE-FGO sample, however, the THR value

Table 7.3 An overview of the outcomes for polymer composite flame-retardants based on graphene

Graphene-based nanocomposite	TTI (s)	Flame out (s)	pkHHR (kW/m^2)	THR (MJ/m^2)	Residue mass (%)	UL-94	LOI (%)	ASEA (m^2/kg)	AMLR (g/s)	Heat flux (kW/m^2)	References
Neat PP	51 ± 8	207 ± 3	1218 ± 19	106 ± 1	0.0	–	–	–	–	35	[151]
PP/1GNPs	37 ± 2	207 ± 7	1001 ± 70	105 ± 1	1.0 ± 0.1	–	–	–	–		
PP/3GNPs	32 ± 3	219 ± 3	907 ± 13	105 ± 1	2.4 ± 0.1	–	–	–	–		
PP/3GNPs/3PP-g-MA	29 ± 5	216 ± 3	881 ± 42	104 ± 1	2.7 ± 0.1	–	–	–	–		
PP-FR	–	–	1909 ± 76	97 ± 2	–	–	–	–	–	50	[145]
PP-FR/5TRGO	–	–	489 ± 20	86 ± 1	–	–	–	–	–		
PP-FR/5MLG 250	–	–	612 ± 25	88 ± 1	–	–	–	–	–		
PP-FR/5CB	–	–	867 ± 36	92 ± 1	–	–	–	–	–		
PP-FR/5CNT	–	–	1027 ± 67	92 ± 1	–	–	–	–	–		
PP-FR/5EG 40	–	–	1048 ± 40	93 ± 2	–	–	–	–	–		
PP	–	–	548 ± 15	129 ± 5	4.5 ± 1	Failed	17.8	–	–	35	[157]
PP/EG	–	–	121 ± 19	98 ± 2	30 ± 3	Failed	24.7	–	–		
PP/Si-EG	–	–	125 ± 17	76 ± 1	35 ± 3	V0	25.6	–	–		
PP/2PER	–	–	992.8 ± 41	133.3 ± 2.3	0	Failed	17.8	–	–	50	[158]
PP/2PER/30NHS	–	–	843.3 ± 20	112.2 ± 1.2	13.8 ± 0.3	Failed	20.5	–	–		
PP/2PER/30EG	–	–	120.3 ± 15	25.3 ± 3.2	65.8 ± 0.5	V1	27.3	–	–		
PP/2PER/28EG/2NHS	–	–	77.0 ± 9	17.8 ± 2.4	75.1 ± 0.3	V0	28.0	–	–		
PP/3PP-g-MA	54 ± 1	–	861 ± 28	152 ± 1	0	Failed	17.0	–	–	35	[154]

(continued)

Table 7.3 (continued)

Graphene-based nanocomposite	TTI (s)	Flame out (s)	pkHHR (kW/m^2)	THR (MJ/m^2)	Residue mass (%)	UL-94	LOI (%)	ASEA (m^2/kg)	AMLR (g/s)	Heat flux (kW/m^2)	References
PP/3PP-g-MA/ 25EG	43 ± 1	–	183 ± 5.8	95 ± 1	31 ± 1.2	Failed	22.6	–	–		
PP/3PP-g-MA/ 22.5EG/ 2.5PDPFDE	37 ± 1	–	149 ± 7.1	90 ± 1	35 ± 0.8	Failed	25.7	–	–		
PP/3PP-g-MA/ 20EG/5PDPFDE	41 ± 0	–	156 ± 3.2	86 ± 1	34 ± 0.5	V0	28.8	–	–		
PP/3PP-g-MA/ 17.5EG/ 7.5PDPFDE	–	–	–	–	–	V0	28.0	–	–		
PP/3PP-g-MA/ 15EG/10PDPFDE	–	–	–	–	–	V1	27.1	–	–		
PP/3PP-g-MA/ 12.5EG/ 12.5PDPFDE	–	–	–	–	–	Failed	25.6	–	–		
PP	33	–	2011 ± 80	106 ± 4	0	–	19	–	–	50	[144]
PP/5.0CB	25		765 ± 30	103 ± 1	5.0	–	20.7	–	–		
PP/5.0MWNT	23	–	855 ± 34	100 ± 2	4.9	–	19.8	–	–		
PP/5.0EG40	24	–	1129 ± 73	105 ± 3	4.7	–	19.4	–	–		
PP/5.0EG60	24	–	1104 ± 45	105 ± 1	4.5	–	19.0	–	–		
PP/5.0MLG250	25	–	570 ± 22	96 ± 2	3.9	–	20.9	–	–		
PP/5.0TRGO	23	–	532 ± 21	97 ± 2	4.5	–	20.4	–	–		

(continued)

Table 7.3 (continued)

Graphene-based nanocomposite	TTI (s)	Flame out (s)	pkHHR (kW/m^2)	THR (MJ/m^2)	Residue mass (%)	UL-94	LOI (%)	ASEA (m^2/kg)	AMLR (g/s)	Heat flux (kW/m^2)	References
	–	–	–	–	–	–		–	–		
PP Neat	50	–	1025	110.8	–	NR	17.0	–	–	35	[149]
PP/25APP	–	–	–	–	–	NR	20.5	–	–		
PP/20.85APP/4.15CFA	–	–	–	–	–	V0	33.0	–	–		
PP/20APP/5CFA	–	–	–	–	–	V0	34.0	–	–		
PP/18.75APP/6.25CFA	–	–	–	–	–	V0	32.0	–	–		
PP/16.67APP/8.33CFA	–	–	–	–	–	V0	32.0	–	–		
PP/12.5APP/12.5CFA	–	–	–	–	–	V0	29.5	–	–		
PP/8.33APP/16.67CFA	–	–	–	–	–	V2	25.0	–	–		
PP/25CFA	–	–	–	–	–	NR	22.0	–	–		
PP/18.4APP/4.6CFA	–	–	–	–	–	V0	31.0	–	–		
PP/16APP/4CFA	–	–	–	–	–	V1	28.0	–	–		
PP/19.6APP/4.9CFA/0.5G	35	–	140	90.4	–	V0	32.0	–	–		

(continued)

Table 7.3 (continued)

Graphene-based nanocomposite	TTI (s)	Flame out (s)	pkHHR (kW/m^2)	THR (MJ/m^2)	Residue mass (%)	UL-94	LOI (%)	ASEA (m^2/kg)	AMLR (g/s)	Heat flux (kW/m^2)	References
PP/19.6APP/4.8CFA/1G	34	–	156	86.0	–	V0	28.0	–	–		
PP/18.4APP/4.6CFA/2G	36	–	262	94.4	–	V2	25.0	–	–		
PP Neat	45	–	1230	113.6	0.97	–		–	–	35	[190]
PP/GO	30	–	1105	97.5	3.14	–		–	–		
PP/FGO1	27	–	967	112.9	7.49	–		–	–		
PP/FGO2	35	–	829	92.0	7.33	–		–	–		
PP	184	–	–	–	–	NR	18.3	–	–	50	[191]
PP/10GNPs	350					NR	19.5	–	–		
PP/20GNPs	–	–	–	–	–	NR	20.6	–	–		
PP/30GNPs	–	–	–	–	–	NR	22.6	–	–		
PP/40GNPs	–	–	–	–	–	V0	23.4	–	–		
PP/50GNPs	–	–	–	–	–	V0	–				
PP Neat	45	–	1456	139.1	–	NR	17.0	–	–	35	[150]
PP/PA-APP/CFA	41	–	501	126.0	–	V0	35.0	–	–		
PP/PA-APP/CFA/0.5 RGO	39	–	434	125.1	–	V0	34.0	–	–		
PP/PA-APP/CFA/1 RGO	37	–	350	123.9	–	V0	32.0	–	–		

(continued)

Table 7.3 (continued)

Graphene-based nanocomposite	TTI (s)	Flame out (s)	pkHHR (kW/m^2)	THR (MJ/m^2)	Residue mass (%)	UL-94	LOI (%)	ASEA (m^2/kg)	AMLR (g/s)	Heat flux (kW/m^2)	References
PP/PA-APP/CFA/2RGO	37	–	397	125.4		V0	30.0	–	–		
PP/PA-APP/PA-APPG/CFA/0.5 RGO	38	–	401	117.3	–	V0	33.0	–	–		
PP/PA-APP/PA-APPG/CFA/1 RGO	34	–	290	117.1	–	V0	30.5	–	–		
PP/PA-APP/PA-APPG/CFA/2 RGO	31	–	464	125.8	–	V2	26.5	–	–		
PP Neat	42 ± 4	–	1242 ± 21	–	–	Failed	17.8	552 ± 14	0.049 ± 0.006	35	[146]
PP/2CNT	44 ± 2	–	538 ± 12	–	–	Failed	20.6	511 ± 15	0.048 ± 0.004		
PP/2RGO	43 ± 3	–	486 ± 14	–	–	Failed	20.1	474 ± 12	0.048 ± 0.004		
PP/1CNTs/1RGO	48 ± 2	–	465 ± 12			Failed	21.0	439 ± 12	0.046 ± 0.005		
PP/20IFR	53 ± 4	–	350 ± 11	–	–	V1	29.2	412 ± 13	0.037 ± 0.003		
PP/18IFR/2CNTs	64 ± 3	–	278 ± 11	–	–	V1	29.7	401 ± 13	0.035 ± 0.004		
PP/18IFR/2RGO	66 ± 4	–	245 ± 12	–	–	V1	30.6	396 ± 14	0.034 ± 0.004		
PP/18IFR/1CNTs/1RGO	82 ± 3	–	212 ± 8	–	–	V0	31.4	380 ± 12	0.032 ± 0.003		
PP	50	–	1044	101.4	–	–	–	–	–	35	[148]

(continued)

Table 7.3 (continued)

Graphene-based nanocomposite	TTI (s)	Flame out (s)	pkHHR (kW/m^2)	THR (MJ/m^2)	Residue mass (%)	UL-94	LOI (%)	ASEA (m^2/kg)	AMLR (g/s)	Heat flux (kW/m^2)	References
PP/0.5 FGO	40	–	892	104.1	–	–	–	–	–		
PP/1FGO	37	–	834	100.6	–	–	–	–	–		
PP/2FGO	33	–	739	98.7	–	–	–	–	–		
PP/1GO	33	–	979	108.2	–	–	–	–	–		
PP Neat	54	–	1199	97.8	–	–	–	–	–	50	[147]
PP/3PP-g-MA/20FR/	69	–	620	78.5	–	–	–	–	–		
PP/3PP-g-MA/18FR/2GO	64	–	473	79.0	–	–	–	–	–		
PP/3PP-g-MA/20FRs-FGO	66	–	397	73.9	–	–	–	–	–		
Neat LDPE	–	–	–	–	–	Failed	17.1	–	–	–	[162]
LDPE/15.75ATH	–	–	–	–	–	Failed	18.0	–	–		
LDPE/15.75RP	–	–	–	–	–	Failed	20.2	–	–		
LDPE/15EG	–	–	–	–	–	Failed	19.7	–	–		
LDPE/10.5ATH/5.25RP	–	–	–	–	–	V2	22.8	–	–		
LDPE/10.5ATH/5EG	–	–	–	–	–	Failed	21.1	–	–		
LDPE/5.25ATH/10.5RP	–	–	–	–	–	V2	22.6	–	–		

(continued)

Table 7.3 (continued)

Graphene-based nanocomposite	TTI (s)	Flame out (s)	pkHHR (kW/m^2)	THR (MJ/m^2)	Residue mass (%)	UL-94	LOI (%)	ASEA (m^2/kg)	AMLR (g/s)	Heat flux (kW/m^2)	References
LDPE/10.5RP/5EG	–	–	–	–	–	V0	24.1	–	–		
LDPE/5.25ATH/10EG	–	–	–	–	–	Failed	21.5	–	–		
LDPE/5.25RP/10EG	–	–	–	–	–	V0	23.2	–	–		
LDPE/5.25ATH/5.25RP/5EG	–	–	–	–	–	V0	25.4	–	–		
LDPE/5.25ATH/5.25RP/5EG/4POE	–	–	–	–	–	V0	25.4	–	–		
XLPE	66		1741.3	103.1	–	–	18.5	–	–	35	[163]
XLPE/0.5GO	51	–	1671.4	107.8	–	–	19.0	–	–		
XLPE/1GO	55	–	1595.8	106.5	–	–	19.0	–	–		
XLPE/3GO	65	–	1585.6	105.1	–	–	19.5	–	–		
XLPE/0.5FGO	66	–	1563	106.1	–	–	19.0	–	–		
XLPE/1FGO	62	–	1254	101.3	–	–	19.5	–	–		
XLPE/3FGO	52	–	1241	97.3	–	–	20.5	–	–		
PVC	–	–	289.4	40.6	8.1	V1	27.4	–	–	50	[176]
PVC3G	–	–	242.6	40.1	8.4	V1	27.6	–	–		
PVC5G	–	–	237.4	39.9	8.8	V1	28.0	–	–		

(continued)

Table 7.3 (continued)

Graphene-based nanocomposite	TTI (s)	Flame out (s)	pkHHR (kW/m^2)	THR (MJ/m^2)	Residue mass (%)	UL-94	LOI (%)	ASEA (m^2/kg)	AMLR (g/s)	Heat flux (kW/m^2)	References
PVC5Fe	–	–	211.6	28.1	15.2	V0	28.3	–	–		
PVC3G5Fe	–	–	150.8	24.2	16.0	V0	29.0	–	–		
PVC5G5Fe	–	–	112.0	20.0	16.3	V0	29.0	–	–		
Pure FPVC	15 ± 1	–	341 ± 11	57.3 ± 0.4	–	–	–	–	–	35	[177]
FPVC/1CGO	16 ± 1	–	211 ± 8	54.8 ± 0.5	–	–	–	–	–		
FPVC/2CGO	21 ± 1	–	189 ± 4	51.0 ± 0.7	–	–	–	–	–		
FPVC/3CGO	20 ± 1	–	169 ± 6	48.1 ± 0.5	–	–	–	–	–		
FPVC/2GO	13 ± 1	–	298 ± 13	61.4 ± 1.1	–	–	–	–	–		
FPVC/2CHS	22 ± 1	–	233 ± 8	59.0 ± 0.6	–	–	–	–	–		
Neat PVA	18 ± 2	–	373 ± 6	58 ± 0.6	–	–	–	590 ± 1	0.108 ± 0.012	35	[142]
PVA-MMT	20 ± 2	–	263 ± 7	58 ± 0.4	–	–	–	585 ± 1	0.092 ± 0.012		
PVA-MWNTs	24 ± 2	–	241 ± 8	52 ± 0.4	–	–	–	523 ± 1	0.090 ± 0.008		
PVA-G1	25 ± 3	–	214 ± 7	55 ± 0.5	–	–	–	574 ± 9	0.064 ± 0.008		
PVA-G2	29 ± 2	–	209 ± 6	47 ± 0.5	–	–	–	480 ± 9	0.059 ± 0.006		
PVA-G3	33 ± 2	–	190 ± 6	45 ± 0.3	–	–	–	356 ± 1	0.064 ± 0.005		
PVA-G4	39 ± 3	–	161 ± 7	44 ± 0.3	–	–	–	333 ± 8	0.043 ± 0.004		
PVA-G5	45 ± 3	–	133 ± 5	38 ± 0.5	–	–	–	319 ± 8	0.042 ± 0.004		
Neat PVA	18 ± 2	–	373 ± 6	58 ± 0.6	–	Failed	19.2	590 ± 1	0.108 ± 0.012	35	[143]
PVA/G1	25 ± 3	–	214 ± 7	55 ± 0.5	–	Failed	23.4	574 ± 9	0.064 ± 0.008		

(continued)

Table 7.3 (continued)

Graphene-based nanocomposite	TTI (s)	Flame out (s)	pkHHR (kW/m^2)	THR (MJ/m^2)	Residue mass (%)	UL-94	LOI (%)	ASEA (m^2/kg)	AMLR (g/s)	Heat flux (kW/m^2)	References
PVA/MPP10	48 ± 4	–	297 ± 9	38 ± 0.4	–	Failed	25.7	609 ± 2	0.082 ± 0.006		
PVA/G1/MPP10	53 ± 3	–	148 ± 5	36 ± 0.4	–	V0	29.6	403 ± 7	0.054 ± 0.005		
PVA/MPP20	64 ± 3	–	134 ± 4	19 ± 0.3	–	V0	31.7	754 ± 2	0.044 ± 0.004		
PVA/G1/MPP20	70 ± 4	–	114 ± 4	18 ± 0.2	–	V0	33.4	564 ± 1	0.040 ± 0.003		
PS	–	–	861	–	–	–	–	–	–	50	[159]
PS/GO1	–	–	525	–	–	–	–	–	–		
PS/GO2	–	–	536	–	–	–	–	–	–		
PS/GO3	–	–	587	–	–	–	–	–	–		
PS/G	–	–	454	–	–	–	–	–	–		
PS	–	–	840	–	–	–	–	–	–	–	[161]
PS-FGO0.1	–	–	750	–	–	–	–	–	–		
PS-FGO0.5	–	–	683	–	–	–	–	–	–		
PS-FGO1.0	–	–	620	–	–	–	–	–	–		
PS-FGO2.0	–	–	514	–	–	–	–	–	–		
PS	47	–	730	117	0	–	–	–	–	35	[160]
PS/15EG	32	–	208	85	43.3	–	–	–	–		
PS/15AP	31	–	254	87	15.6	–	–	–	–		
PS/10EG/5AP	23	–	163	71	38.6	–	–	–	–		
PS	50	–	1703	139	–	–	–	887	–	32	[192]

(continued)

Table 7.3 (continued)

Graphene-based nanocomposite	TTI (s)	Flame out (s)	pkHHR (kW/m^2)	THR (MJ/m^2)	Residue mass (%)	UL-94	LOI (%)	ASEA (m^2/kg)	AMLR (g/s)	Heat flux (kW/m^2)	References
PS/0.5FGO	30	–	1127	96	–	–	–	888	–		
PS/1FGO	25	–	1058	86	–	–	–	894	–		
PS/2FGO	20	–	908	92	–	–	–	1000	–		
PS/3FGO	20	–	805	93	–	–	–	1086	–		
PA6	70	–	1434	148.1	–	–	–	–	–	35	[187]
PA6/2G	79	–	1257	133.1	–	–	–	–	–		
PA6/2GNS-CO_3O_4	81	–	1282	141.1	–	–	–	–	–		
PA6/2GNS-NiO	75	–	1105	130.0	–	–	–	–	–		
PA 66	90 ± 10	–	481 ± 11	12.0 ± 0.4	–	–	20 ± 1	–	–	35	[189]
PA66-0.5GO-L/ 0.5GO-PA	55 ± 0	–	384 ± 17	13.0 ± 0.2	–	–	27 ± 0.5	–	–		
PA66-0.5GO-L/ 15PA	55 ± 3	–	378 ± 20	13.0 ± 0.1	–	–	24.5 ± 1	–	–		
PA66-0.5GO-L/ 2.5CS/05GO-PA	63 ± 0	–	394 ± 11	13.0 ± 0.2	–	–	26.5 ± 1	–	–		
PA66-0.5GO-L/ 2.5CS/15PA	58 ± 2	–	363 ± 5	13.0 ± 0.4	–	–	25.5 ± 1	–	–		
PA1010	82 ± 3	–	934 ± 2	156 ± 4.2	–	V2	24.2 ± 5.6	–	–	50	[193]
PA1010/2.5EG	43 ± 2	–	840 ± 9	184 ± 5.3	–	V2	25.2 ± 4.8	–	–		
PA1010/5EG	42 ± 2	–	647 ± 1	186 ± 4.7	–	V2	23.6 ± 8.9	–	–		

(continued)

Table 7.3 (continued)

Graphene-based nanocomposite	TTI (s)	Flame out (s)	pkHHR (kW/m^2)	THR (MJ/m^2)	Residue mass (%)	UL-94	LOI (%)	ASEA (m^2/kg)	AMLR (g/s)	Heat flux (kW/m^2)	References
PA1010/7.5EG	53 ± 3	–	405 ± 9	177 ± 2.6	–	V1	22.6 ± 6.5	–	–		
PA1010/10EG	50 ± 2	–	374 ± 6	176 ± 3.7	–	V1	22.1 ± 5.9	–	–		
EVA	57 ± 3	–	574 ± 2	124 ± 1.5	–	–	–	442 ± 1	0.057 ± 0.006	35	[194]
EVA/0.5CRGO	58 ± 2	–	451 ± 8	117 ± 1.5	–	–	–	418 ± 1	0.054 ± 0.005		
EVA/1CRGO	61 ± 2	–	394 ± 6	106 ± 1.2	–	–	–	386 ± 9	0.046 ± 0.004		
EVA/ 0.5CRGO-PPSPB	70 ± 4	–	381 ± 7	103 ± 1.0	–	–	–	426 ± 1	0.044 ± 0.003		
EVA/ 1CRGO-PPSPB	75 ± 4	–	309 ± 7	97 ± 1.2	–	–	–	405 ± 8	0.039 ± 0.002		

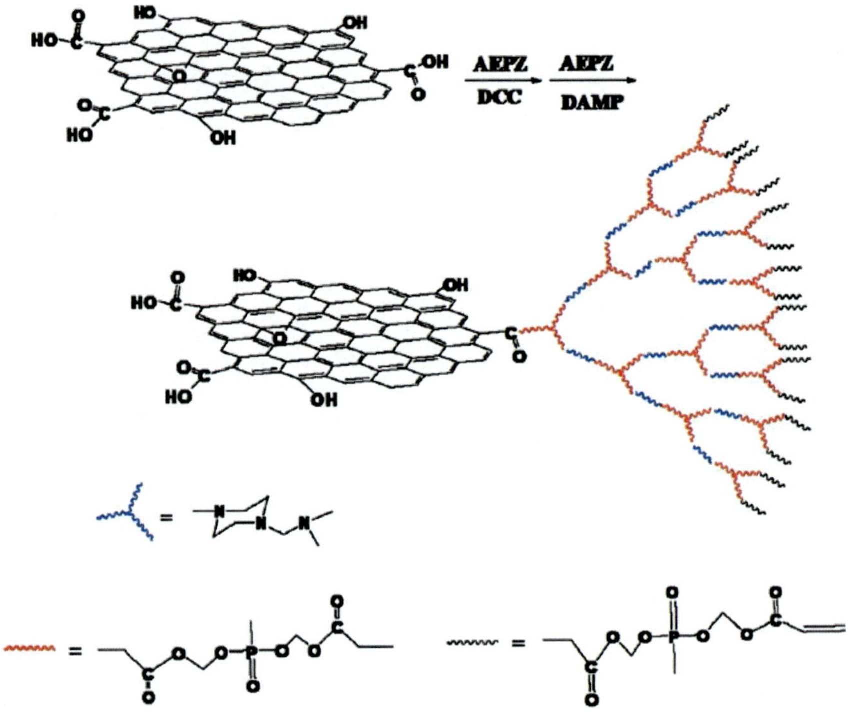

Fig. 7.13 Schematic of the fabrication of functionalized GO. Reproduced with permission from Hu et al. [163]. Copyright 2014, ACS Publications

declined noticeably as more FGO was applied, indicating that the functionalized GO improved the composites' flame resistance.

When they examined the LOI test findings, they discovered that pure XLPE had a LOI value of only 18.5%, but that when GO, and FGO were mixed into the XLPE, the LOI value increased. In contrast, the LOI of the XLPE-FGO sample in the same additive was marginally greater than the XLPE-GO sample. The self-extinguishing property provided the basis for the LOI test's measurement. The nanocomposites burned slowly when they ignited but were unable to extinguish themselves until the majority of the fuel burnt. As a result, XLPE-FGO nanocomposites had no impact on raising the LOI value while outperforming the competition in terms of flame-retardant efficiency.

7.2.5 Poly(Vinyl Chloride)

PVC is a sizable class of all-purpose plastic that is produced globally and has a wide range of uses, including those in construction, transportation, and communication [164]. However, due to the stiffness of its polymer chains, pure PVC resin is challenging to manufacture. Consequently, in order to satisfy the various requirements in most cases (e.g., wires, cables, wall, and floor coverings). To give PVC a good degree of flexibility and processability, plasticizers such as TOTM [165], DBP [166], and DOP [167] are employed [168]. Large amounts of plasticizers are incorporated into flexible PVC, which significantly reduces its flame-retardant properties because most plasticizers are easily FPVC [169]. Additionally, PVC materials have a propensity to generate a lot of smoke and harmful chemicals when burned [170]. Enhancing FPVC's flame-retardant and smoke suppression properties is, therefore, crucial. Phosphates make up the flame-retardants frequently utilized in FPVC [171], halogen-containing phosphates [172], antimony trioxide (Sb_2O_3) [173], metal hydroxides [174], zinc borate [175], etc. FPVC smoke increases when phosphates, phosphates containing halogens, and antimony trioxide are burned. In FPVC, metal hydroxides are less effective flame-retardants despite being environmentally friendly.

The flame retardancy and smoke suppression of PVC utilizing graphene and Fe_3O_4 have been investigated by Yao et al. [176]. They demonstrated that neat PVC burnt very fast after ignition with a sharp HRR peak appearing with a peak value of PHRR of 289.4 kW/m^2. The PHRR of PVC decreases with an increase in graphene from 3 wt.% to 5 wt.% at 242.6 and 237.4 kW/m^2, respectively. Compared to neat PVC, the PHRR of PVC composite with 5 wt.% of Fe_3O_4 reduced by 27% (to 211.5 kW/m^2), indicating that adding graphene or Fe_3O_4 nanoparticles alone was insufficient to significantly increase PVC's flame retardancy. However, the PHRR from the HRR plots of the ternary PVC5G5Fe composites revealed a clear reduction when compared with that of the binary PVC5G or PVC5Fe composite. For instance, PHRR = 150.8 kW/m^2 for PVC3G5Fe, which was reduced by 48% when compared to that of pure PVC. Interestingly, the ternary PVC5G5Fe composite HRR had the lowest peak value (PHRR = 112.0 kW/m^2, which is 61% lower than that of plain PVC).

In another work, Huang and colleagues [177] synthesized CHS-modified GO nanohybrid "CGO" through one-step in situ method to improve the flame retardancy of PVC. Due to the addition of significant amounts of the plasticizer DOP, the LOI of the pure FPVC was only 22.5%, and it received a failing grade on the UL-94 test, indicating that it is highly flammable. The LOI value of the FPVC composites with CGO increased consistently along with the increase in CGO content. The LOI value of sample FPVC/2CGO, which includes 2 wt.% CGO was higher than that of samples FPVC/2GO (at 23%) and FPVC/2CHS (at 25.5%) at the same dosage, demonstrating that CHS in combination with GO significantly reduces the flammability of FPVC composites. The three FPVC/CGO samples and sample FPVC/2CHS passed the UL-94 test with a V0 rating; however, sample FPVC/2GO received absolutely no rating. At the same time, the cone calorimeter test findings showed that the PHRR and THR in the FPVC/CGO composites generally decreased with an increase in the CGO content. The PHRR and THR values of sample FPVC/2CGO decreased by 44.6 and 10.5%, respectively, when compared to those of pure FPVC when the CGO content is 2 wt.%. Additionally, sample FPVC/2CGO's values were lower than those of samples FPVC/2GO and FPVC/2CHS. With the addition of CGO or CHS, the TTI of FPVC was extended, which was considered to be crucial for the fire rescue. They further explained that the shielding effect of the compact protective layer was primarily responsible for the decrease in the peak heat release rate of the FPVC/CGO composites during burning. Instead of the shielding effect, the addition of CGO results in a greater reduction of total heat release than the addition of GO or CHS at the same content due to the increased char yield and decreased effective heat of combustion (EHC, the ratio of THR to mass loss) [178, 179]. Since CHS nanoparticles were immobilized on the surface of GO, the material was more heat resistant and had a better shielding effect in FPVC than GO alone. The CHS from CGO tended to grab HCl formed from the deterioration of PVC resin to form metal chlorides ($CuCl_2$ and $SnCl_2$), effectively reducing the smoke [177, 180].

Digital images utilized to examine the morphology of the residual char of pure FPVC and FPVC composites after cone calorimeter experiments revealed that the combustion of pure FPVC and FPVC/2GO (Fig. 7.14a–b) produced little and brittle char. In contrast to the loose and glomerate-like chars left by sample FPVC/2CHS in Fig. 7.14c, the chars from FPVC/CGO composites (Fig. 7.14d–f) became increasingly compact as the CGO content rose [177].

Fig. 7.14 Digital photos of residual char of pure FPVC and FPVC composites, **a** Pure FPVC, **b** FPVC/2GO, **c** FPVC/2CH, **d** FPVC/1CGO, **e** FPVC/2CGO, and **f** FPVC/3CGO. Reproduced with permission from Huang et al. [177]. Copyright 2021, Elsevier Science Ltd.

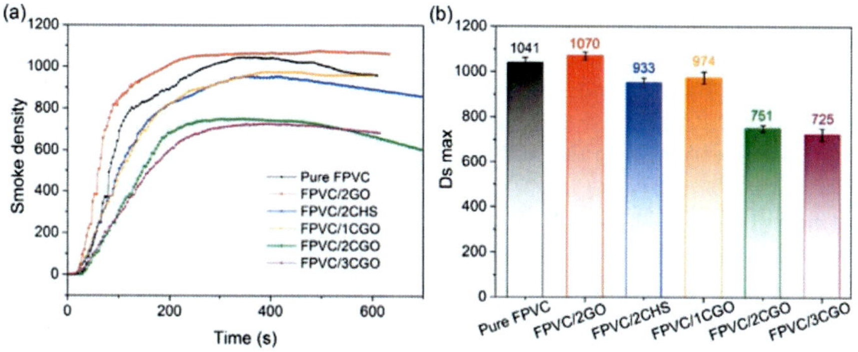

Fig. 7.15 Smoke density test results of pure FPVC and FPVC composites **a** Smoke density curves, **b** Maximum smoke density values. Reproduced with permission from Huang et al. [177]. Copyright 2021, Elsevier Science Ltd.

Another significant factor that was looked into was smoke density. Figure 7.15a depicts the smoke density curves for pure FPVC and FPVC composites, whereas Fig. 7.15b lists the highest smoke density values (Ds max). During the test, pure FPVC emitted a lot of smoke, and in the first 100 s of the test, the smoke density curve rose quickly, with the Ds max value reaching 1041. According to the smoke release in the cone calorimeter test, sample FPVC/2GO's smoke density was somewhat increased by the presence of GO. The Ds max value dropped to 933 for sample PVC/2CHS, and the smoke density curve was lower than for pure PFVC throughout the whole test. The smoke density curves were steadily lowered as the CGO content rose in FPVC/CGO composites. For instance, the Ds max value of the sample FPVC/2CGO fell by 27.9% when compared to pure FPVC. The findings on smoke density provided additional proof that CGO in FPVC effectively suppressed smoke [177].

7.2.6 Polyamide

Due to its remarkable characteristics, PAM has been utilized extensively as an engineering plastic [181]. PAM is a combustible material with a high heat release, so it needs to have its thermal and mechanical qualities improved if its use in advanced technologies is to be expanded. Additionally, it produces a significant amount of dense smoke and hazardous gaseous pollutants, whose inhalation is the primary cause of death in fires [182]. In order to reduce the overall fire dangers, it is necessary to take into account PAM's flammability, poisonous gases, and smoke. Even though halogen-containing flame-retardants are inexpensive and efficient for PAM, legislation prevents their further use because they burn with corrosive gases and black smoke [183]. PAM nanocomposites have been shown to be a cutting-edge, ecologically friendly, and successful technology to impart great mechanical qualities and increase flame resistance in recent years [184]. Clay [185] and layered double hydroxide [186] are examples of PAM/layered structures that have been extensively researched to improve thermal stability and flame retardancy.

In a study by Hong and colleagues [187], graphene nanosheets were coated with Co_3O_4 and NiO (referred to as GNS-Co_3O_4 and GNS-NiO) and melt-mixed with PA6 in order to improve the study of the fire-retardant qualities of the PA6. They discovered that pure PA6 ignited swiftly (70 s) and sharply (160 s) with a PHRR of 1434 kW/m^2. The addition of graphene to the PA6/graphene composite caused a delay in ignition of 79 s and a lowering in PHRR to 1257 kW/m^2, which was delayed by 233 s. It was believed that graphene served as physical barriers that retarded the growth of pyrolytic gases and transmitted the radiant heat flux to the sample, causing a considerable delay in PHRR. When the same quantity of GNS-NiO was added, the PHRR of the composite decreased by 22.9% in comparison with PA6, while the HRR for the composite containing 2 wt.% GNS-Co_3O_4 does not change significantly. However, the time to ignition was obviously delayed. After 800 s of exposure, the pure PA6 smoke density (Fig. 7.16) increased progressively from 0 to 320. The outcome showed that PA6 produced substantial smoke in an atmosphere with low oxygen levels. The smoke density of the composite fell after the addition of 2 wt.% graphene, only reading a maximum of 225. The barrier effect of graphene, which prevents volatile degradation products from escaping, may be to blame for the decrease in smoke production. In comparison with PA6/graphene, the composite incorporated with GNS-Co_3O_4 firstly produced comparable amounts of smoke. The PA6/GNS-NiO composite exhibited the greatest reduction in smoke density, proving that nickel oxide was more effective than cobalt oxide at reducing smoke emissions

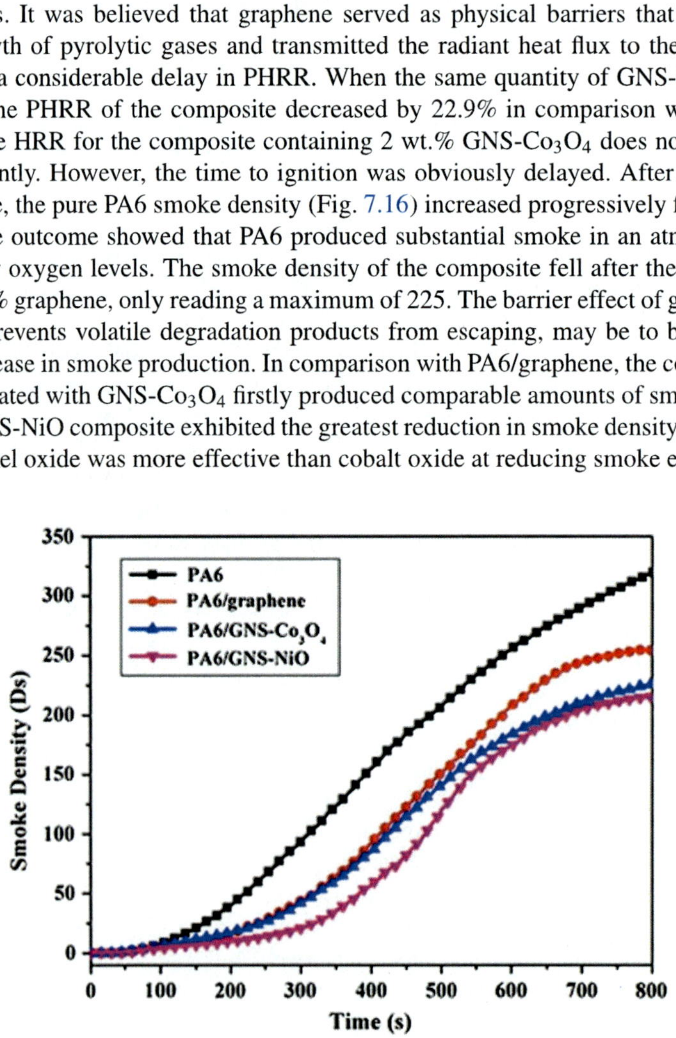

Fig. 7.16 Smoke density of PA6, PA6/graphene, PA6/GNS–Co_3O_4 and PA6/GNS–NiO composites. Reproduced with permission from Hong et al. [187], Copyright 2013, Elsevier Science Ltd.

due to its strong catalytic activity for the generation of coke [188]. Due to the catalytic carbonization impact of metal oxides and the physical barrier effect of graphene, the addition of metal oxides to graphene could further reduce the formation of smoke [187].

In order to create a graphene-based flame-retardant finishing on the surfaces of polyamide 66 fabric, Kundu et al. [189] functionalized new biomolecules, including phytic acid and lignin, with GO (see Figs. 7.17 and 7.18). To prevent the fabric's color and handling from being compromised, they concentrated on ensuring that the applied compounds were loaded minimally. To assess the function of a phosphorus molecule in a different form, they also employed pure phytic acid and compared it with GO-functionalized PAM. In addition, they combine lignin with pure chitosan, a novel charring agent and an intumescent additive, to create a hybrid charring system.

All of the treated fabric samples exhibited PHRR values that were lower and distinct from those of pure PA66, and they all arrived at those values over the course of various amounts of time. This might be a result of coatings that were applied

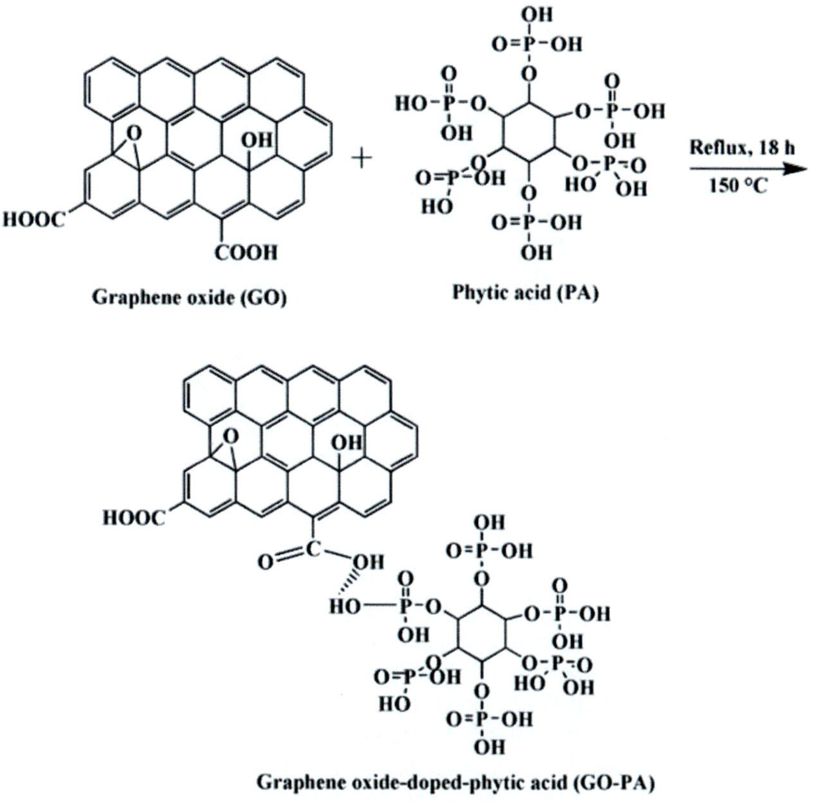

Fig. 7.17 Synthesis route of graphene oxide-doped-phytic acid (GO-PA). Reproduced with permission from Kundu et al. [189]. Copyright 2020, Elsevier Science Ltd.

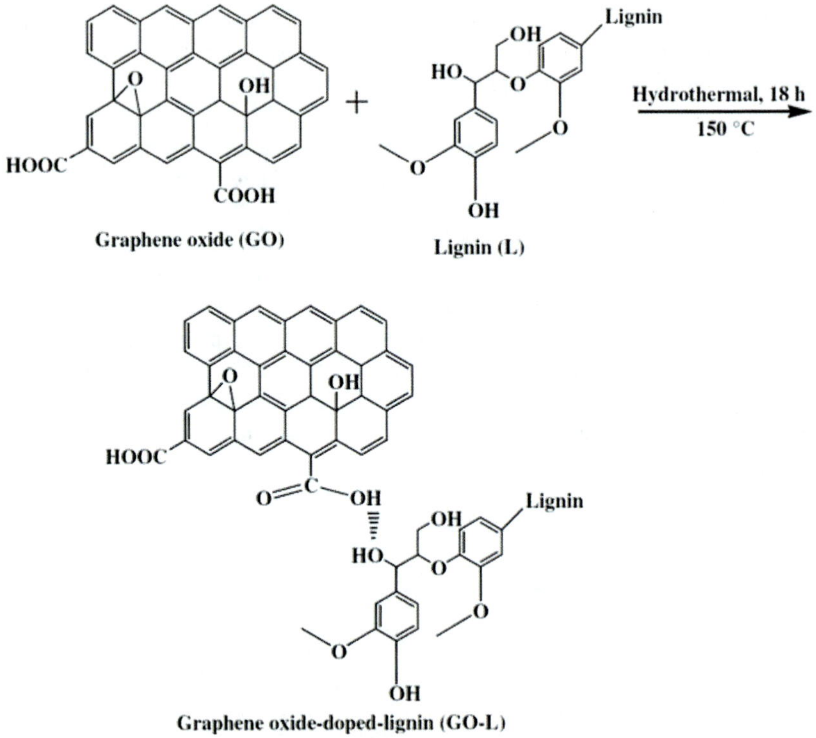

Fig. 7.18 Synthesis route of graphene oxide-doped-lignin (GO-L). Reproduced with permission from Kundu et al. [189]. Copyright 2020, Elsevier Science Ltd.

to fabric surfaces, and the different compositions of those coatings may have had an impact on the PHRR readings. The PA66-D fabric sample, which was treated with phytic acid (PA), CS, and GO-functionalized lignin (GO-L), has the highest reduction in PHRR value of all coated fabrics (i.e., 25%).

7.3 Summary

In conclusion, the application of graphene-bearing polymer composites in the realms of EMI shielding and flame-retardant materials marks a significant stride in material innovation. The remarkable properties of graphene, such as its high conductivity, mechanical strength, and thermal stability, synergistically enhance the performance of polymer matrices. This integration results in composite materials that excel in providing robust protection against unwanted electromagnetic interference, ensuring the seamless functionality of electronic devices. Simultaneously, these composites demonstrate exceptional flame-retardant characteristics, addressing critical safety concerns in industries where fire resistance is paramount. As we witness the successful deployment of graphene-bearing polymer composites in practical applications, it is evident that this technology holds the key to advancing the fields of electronics, aerospace, and beyond. The era of graphene-infused polymers has ushered in a new paradigm of materials engineering, where enhanced functionalities meet the stringent demands of our technologically driven world.

References

1. I. Pomestchenko, D. Stolyarov, E. Polyakova, I. Stolyarov, Thermoplastic polymer composites and methods for preparing, collecting, and tempering 3D printable materials and articles from same. Google Patents (2023)
2. D. Therriault, K. Chizari, Electrically conductive ink for solvent-cast 3D printing. Google Patents (2020)
3. L.W. Jen, F.Y.N., S. Jun, Electromagnetic wave shielding composite film (2014)
4. L. Yongyong, A kind of graphene-silicon electromagnetic shielding filler and electromagnetic screen coating (2018)
5. J.T. Orasugh, S.S. Ray, Functional and structural facts of effective electromagnetic interference shielding materials: a review. ACS Omega **8**(9), 8134–8158 (2023)
6. M. Taj, S.R. Manohara, B. Siddlingeshwar, N. Raghavendra, M. Faisal, U.V. Khadke, Anti-corrosion and electromagnetic interference shielding performance of bifunctional PEDOT-graphene nanocomposites. Diamond Relat. Mater. **132**, 109690 (2023)
7. P. Liu, Y. Huang, X. Zhang, Synthesis, characterization and excellent electromagnetic wave absorption properties of graphene/poly (3,4-ethylenedioxythiophene) hybrid materials with Fe3O4 nanoparticles. J. Alloy. Compd. **617**, 511–517 (2014)
8. P. Liu, Y. Huang, X. Zhang, Superparamagnetic NiFe2O4 particles on poly(3,4-ethylenedioxythiophene)–graphene: synthesis, characterization and their excellent microwave absorption properties. Compos. Sci. Technol. **95**, 107–113 (2014)
9. P. Liu, Y. Huang, X. Zhang, Preparation and excellent microwave absorption properties of ferromagnetic graphene/poly(3, 4-ethylenedioxythiophene)/CoFe2O4 nanocomposites. Powder Technol. **276**, 112–117 (2015)
10. J. Dalal, A. Gupta, S. Lather, K. Singh, S.K. Dhawan, A. Ohlan, Poly (3, 4-ethylene dioxythiophene) laminated reduced graphene oxide composites for effective electromagnetic interference shielding. J. Alloy. Compd. **682**, 52–60 (2016)
11. V.-T. Nguyen, B.K. Min, Y. Yi, S.J. Kim, C.-G. Choi, MXene(Ti3C2TX)/graphene/PDMS composites for multifunctional broadband electromagnetic interference shielding skins. Chem. Eng. J. **393**, 124608 (2020)
12. C. Xing, S. Zhu, Z. Ullah, X. Pan, F. Wu, X. Zuo, J. Liu, M. Chen, W. Li, Q. Li, L. Liu, Ultralight and flexible graphene foam coated with Bacillus subtilis as a highly efficient electromagnetic interference shielding film. Appl. Surf. Sci. **491**, 616–623 (2019)
13. H. Fang, H. Guo, Y. Hu, Y. Ren, P.-C. Hsu, S.-L. Bai, In-situ grown hollow Fe3O4 onto graphene foam nanocomposites with high EMI shielding effectiveness and thermal conductivity. Compos. Sci. Technol. **188**, 107975 (2020)
14. D. Han, Y.-H. Zhao, S.-L. Bai, W.C. Ping, High shielding effectiveness of multilayer graphene oxide aerogel film/polymer composites. RSC Adv. **6**(95), 92168–92174 (2016)
15. H. Liu, Y. Xu, J.-P. Cao, D. Han, Q. Yang, R. Li, F. Zhao, Skin structured silver/three-dimensional graphene/polydimethylsiloxane composites with exceptional electromagnetic interference shielding effectiveness. Compos. A Appl. Sci. Manuf. **148**, 106476 (2021)

16. S. Zhu, Q. Cheng, C. Yu, X. Pan, X. Zuo, J. Liu, M. Chen, W. Li, Q. Li, L. Liu, Flexible Fe3O4/graphene foam/poly dimethylsiloxane composite for high-performance electromagnetic interference shielding. Compos. Sci. Technol. **189**, 108012 (2020)
17. W. Gao, N. Zhao, T. Yu, J. Xi, A. Mao, M. Yuan, H. Bai, C. Gao, High-efficiency electromagnetic interference shielding realized in nacre-mimetic graphene/polymer composite with extremely low graphene loading. Carbon **157**, 570–577 (2020)
18. Z. Chen, C. Xu, C. Ma, W. Ren, H.-M. Cheng, Lightweight and flexible graphene foam composites for high-performance electromagnetic interference shielding. Adv. Mater. **25**(9), 1296–1300 (2013)
19. L.J. Romasanta, M. Hernández, M.A. López-Manchado, R. Verdejo, Functionalised graphene sheets as effective high dielectric constant fillers. Nanoscale Res. Lett. **6**(1), 508 (2011)
20. N. Gill, V. Gupta, M. Tomar, A.L. Sharma, O.P. Pandey, D.P. Singh, Improved electromagnetic shielding behaviour of graphene encapsulated polypyrrole-graphene nanocomposite in X-band. Compos. Sci. Technol. **192**, 108113 (2020)
21. T. Lin, H. Yu, Y. Wang, L. Wang, S.Z. Vatsadze, X. Liu, Z. Huang, S. Ren, M.A. Uddin, B.U. Amin, S. Fahad, Polypyrrole nanotube/ferrocene-modified graphene oxide composites: from fabrication to EMI shielding application. J. Mater. Sci. **56**(32), 18093–18115 (2021)
22. P.R. Modak, D.V. Nandanwar, S.B. Kondawar, in *Electromagnetic Interference Shielding Effectiveness of Graphene Based Conducting Polymer Nanocomposites*. NAC 2019: Proceedings of the 2nd International Conference on Nanomaterials and Advanced Composites (Springer, 2020), pp. 31–40
23. A. Nazir, H. Yu, L. Wang, Y. He, Q. Chen, B.U. Amin, D. Shen, Electromagnetic interference shielding properties of ferrocene-based polypyrrole/carbon material composites. Appl. Phys. A **126**(9), 749 (2020)
24. P. Sambyal, S.K. Dhawan, P. Gairola, S.S. Chauhan, S.P. Gairola, Synergistic effect of polypyrrole/BST/RGO/Fe3O4 composite for enhanced microwave absorption and EMI shielding in X-Band. Curr. Appl. Phys. **18**(5), 611–618 (2018)
25. H. Wadhwa, S. Mahendia, S. Kumar, in *Microwave Synthesized Graphene-PAni Nanocomposites for EMI Shielding*. AIP Conference Processing (AIP Publishing LLC, 2019), p. 030123
26. K. Cheng, H. Li, M. Zhu, H. Qiu, J. Yang, In situ polymerization of graphene-polyaniline@polyimide composite films with high EMI shielding and electrical properties. RSC Adv. **10**(4), 2368–2377 (2020)
27. S. Khasim, Polyaniline-graphene nanoplatelet composite films with improved conductivity for high performance X-band microwave shielding applications. Results Phys. **12**, 1073–1081 (2019)
28. M. Hamidinejad, B. Zhao, A. Zandieh, N. Moghimian, T. Filleter, C.B. Park, Enhanced electrical and electromagnetic interference shielding properties of polymer-graphene nanoplatelet composites fabricated via supercritical-fluid treatment and physical foaming. ACS Appl. Mater. Interfaces. **10**(36), 30752–30761 (2018)
29. E.H. Awad, K.F. El-Nemr, M.M. Atta, A. Abdel-Hakim, A. Sharaf, Electromagnetic interference shielding efficiency of irradiated wood-plastic composites based on graphene oxide nanoparticles. Radiat. Phys. Chem. **203**, 110629 (2023)
30. D.-X. Yan, H. Pang, L. Xu, Y. Bao, P.-G. Ren, J. Lei, Z.-M. Li, Electromagnetic interference shielding of segregated polymer composite with an ultralow loading of in situ thermally reduced graphene oxide. Nanotechnology **25**(14), 145705 (2014)
31. J. Jing, Y. Xiong, S. Shi, H. Pei, Y. Chen, P. Lambin, Facile fabrication of lightweight porous FDM-Printed polyethylene/graphene nanocomposites with enhanced interfacial strength for electromagnetic interference shielding. Compos. Sci. Technol. **207**, 108732 (2021)
32. H. Zhu, Y. Yang, H. Duan, G. Zhao, Y. Liu, Electromagnetic interference shielding polymer composites with magnetic and conductive FeCo/reduced graphene oxide 3D networks. J. Mater. Sci.: Mater. Electron. **30**(3), 2045–2056 (2019)
33. H. Ji, J. Li, J. Zhang, Y. Yan, Remarkable microwave absorption performance of ultralight graphene-polyethylene glycol composite aerogels with a very low loading ratio of graphene. Compos. A Appl. Sci. Manuf. **123**, 158–169 (2019)

34. D.-X. Yan, H. Pang, B. Li, R. Vajtai, L. Xu, P.-G. Ren, J.-H. Wang, Z.-M. Li, Structured reduced graphene oxide/polymer composites for ultra-efficient electromagnetic interference shielding. Adv. Func. Mater. **25**(4), 559–566 (2015)
35. Y. Chen, Y. Wang, H.-B. Zhang, X. Li, C.-X. Gui, Z.-Z. Yu, Enhanced electromagnetic interference shielding efficiency of polystyrene/graphene composites with magnetic Fe3O4 nanoparticles. Carbon **82**, 67–76 (2015)
36. F. Shahzad, S.H. Lee, S.M. Hong, C.M. Koo, Segregated reduced graphene oxide polymer composite as a high performance electromagnetic interference shield. Res. Chem. Intermed. **44**(8), 4707–4719 (2018)
37. D. Skoda, J. Vilcakova, R.S. Yadav, B. Hanulikova, T. Capkova, M. Jurca, M. Urbanek, P. Machac, L. Simonikova, J. Antos, I. Kuritka, Nickel nanoparticle–decorated reduced graphene oxide via one-step microwave-assisted synthesis and its lightweight and flexible composite with polystyrene-block-poly(ethylene-ran-butylene)-block-polystyrene polymer for electromagnetic wave shielding application. Adv. Compos. Hybrid Mater. **6**(3), 113 (2023)
38. Y. Guo, L. Pan, X. Yang, K. Ruan, Y. Han, J. Kong, J. Gu, Simultaneous improvement of thermal conductivities and electromagnetic interference shielding performances in polystyrene composites via constructing interconnection oriented networks based on electrospinning technology. Compos. A Appl. Sci. Manuf. **124**, 105484 (2019)
39. F. Shahzad, S. Yu, P. Kumar, J.-W. Lee, Y.-H. Kim, S.M. Hong, C.M. Koo, Sulfur doped graphene/polystyrene nanocomposites for electromagnetic interference shielding. Compos. Struct. **133**, 1267–1275 (2015)
40. N. Bagotia, H. Mohite, N. Tanaliya, D.K. Sharma, A comparative study of electrical, EMI shielding and thermal properties of graphene and multiwalled carbon nanotube filled polystyrene nanocomposites. Polym. Compos. **39**(S2), E1041–E1051 (2018)
41. S. Maiti, N.K. Shrivastava, S. Suin, B.B. Khatua, Polystyrene/MWCNT/graphite nanoplate nanocomposites: efficient electromagnetic interference shielding material through graphite nanoplate–MWCNT–graphite nanoplate networking. ACS Appl. Mater. Interfaces. **5**(11), 4712–4724 (2013)
42. L. Wei, W. Zhang, J. Ma, S.-L. Bai, Y. Ren, C. Liu, D. Simion, J. Qin, π-π stacking interface design for improving the strength and electromagnetic interference shielding of ultrathin and flexible water-borne polymer/sulfonated graphene composites. Carbon **149**, 679–692 (2019)
43. J. Zhou, C. Liu, L. Xia, L. Wang, C. Qi, G. Zhang, Z. Tan, B. Ren, B. Yuan, Bridge-graphene connecting polymer composite with a distinctive segregated structure for simultaneously improving electromagnetic interference shielding and flame-retardant properties. Colloids Surf., A **661**, 130853 (2023)
44. B. Lee, U. Hwang, J. Kim, S.-H. Kim, K. Choi, I.-K. Park, C. Choi, J. Suhr, J.-D. Nam, Highly dispersed graphene nanoplatelets in polypropylene composites by employing high-shear stress for enhanced dielectric properties and frequency-selective electromagnetic interference shielding capability. Compos. Commun. **37**, 101409 (2023)
45. Y. Wu, Z. Wang, X. Liu, X. Shen, Q. Zheng, Q. Xue, J.-K. Kim, Ultralight graphene foam/conductive polymer composites for exceptional electromagnetic interference shielding. ACS Appl. Mater. Interfaces. **9**(10), 9059–9069 (2017)
46. G. George, S.M. Simon, V. Prakashan, M. Sajna, M. Faisal, R. Wilson, A. Chandran, P. Biju, C. Joseph, N. Unnikrishnan, Green and facile approach to prepare polypropylene/in situ reduced graphene oxide nanocomposites with excellent electromagnetic interference shielding properties. RSC Adv. **8**(53), 30412–30428 (2018)
47. A. Kaushal, V. Singh, Effect of filler loading on the shielding of electromagnetic interference of reduced graphene oxide reinforced polypropylene nanocomposites prepared via a twin-screw extruder. J. Mater. Sci.: Mater. Electron. **31**(24), 22162–22170 (2020)
48. R.S. Yadav, Anju, T. Yadav, I. Kuřitka, J. Vilčáková, D. Škoda, P. Urbánek, M. Machovský, M. Masař, M. Urbánek, L. Kalina, J. Havlica, Excellent, lightweight and flexible electromagnetic interference shielding nanocomposites based on polypropylene with MnFe2O4 spinel ferrite nanoparticles and reduced graphene oxide. Nanomaterials [Online] (2020)

49. R.S. Yadav, I. Kuřitka, J. Vilcakova, M. Machovsky, D. Skoda, P. Urbánek, M. Masař, M. Jurča, M. Urbánek, L. Kalina, J. Havlica, NiFe2O4 nanoparticles synthesized by dextrin from corn-mediated sol-gel combustion method and its polypropylene nanocomposites engineered with reduced graphene oxide for the reduction of electromagnetic pollution. ACS Omega **4**(26), 22069–22081 (2019)
50. K. Raagulan, J. Ghim, R. Braveenth, K.Y. Chai, B.M. Kim, Improving the EMI shielding of graphene oxide (GNO)-coated glass-fiber–GNO–MA-grafted polypropylene (PP) composites and nylon 1D–2D nanocomposite foams. RSC Adv. **12**(24), 15316–15328 (2022)
51. V.K. Singh, A. Shukla, M.K. Patra, L. Saini, R.K. Jani, S.R. Vadera, N. Kumar, Microwave absorbing properties of a thermally reduced graphene oxide/nitrile butadiene rubber composite. Carbon **50**(6), 2202–2208 (2012)
52. Y. Zhan, J. Wang, K. Zhang, Y. Li, Y. Meng, N. Yan, W. Wei, F. Peng, H. Xia, Fabrication of a flexible electromagnetic interference shielding Fe3O4@ reduced graphene oxide/natural rubber composite with segregated network. Chem. Eng. J. **344**, 184–193 (2018)
53. A.A. Al-Ghamdi, A.A. Al-Ghamdi, Y. Al-Turki, F. Yakuphanoglu, F. El-Tantawy, Electromagnetic shielding properties of graphene/acrylonitrile butadiene rubber nanocomposites for portable and flexible electronic devices. Compos. B Eng. **88**, 212–219 (2016)
54. Y. Zhan, J. Wang, K. Zhang, Y. Li, Y. Meng, N. Yan, W. Wei, F. Peng, H. Xia, Fabrication of a flexible electromagnetic interference shielding Fe3O4@reduced graphene oxide/natural rubber composite with segregated network. Chem. Eng. J. **344**, 184–193 (2018)
55. G. Wang, X. Liao, J. Yang, W. Tang, Y. Zhang, Q. Jiang, G. Li, Frequency-selective and tunable electromagnetic shielding effectiveness via the sandwich structure of silicone rubber/graphene composite. Compos. Sci. Technol. **184**, 107847 (2019)
56. D. Liu, Q.-Q. Kong, H. Jia, L.-J. Xie, J. Chen, Z. Tao, Z. Wang, D. Jiang, C.-M. Chen, Dual-functional 3D multi-wall carbon nanotubes/graphene/silicone rubber elastomer: thermal management and electromagnetic interference shielding. Carbon **183**, 216–224 (2021)
57. Y. Wang, Y. Guan, D. Liao, Y. He, S. Li, L. Zhou, C. Yu, Y. Chen, Y. Liu, H. Liu, Fabrication of cellulose nanofiber/reduced graphene oxide/nitrile rubber flexible films using pickering emulsion technology for electromagnetic interference shielding and piezoresistive sensor. Macromol. Mater. Eng. **306**(6), 2100070 (2021)
58. A. Nasr Esfahani, A. Katbab, A. Taeb, L. Simon, M.A. Pope, Correlation between mechanical dissipation and improved X-band electromagnetic shielding capabilities of amine functionalized graphene/thermoplastic polyurethane composites. Eur. Polymer J. **95**, 520–538 (2017)
59. S.-T. Hsiao, C.-C.M. Ma, H.-W. Tien, W.-H. Liao, Y.-S. Wang, S.-M. Li, Y.-C. Huang, Using a non-covalent modification to prepare a high electromagnetic interference shielding performance graphene nanosheet/water-borne polyurethane composite. Carbon **60**, 57–66 (2013)
60. Y. Li, B. Shen, D. Yi, L. Zhang, W. Zhai, X. Wei, W. Zheng, The influence of gradient and sandwich configurations on the electromagnetic interference shielding performance of multilayered thermoplastic polyurethane/graphene composite foams. Compos. Sci. Technol. **138**, 209–216 (2017)
61. B. Shen, Y. Li, W. Zhai, W. Zheng, Compressible graphene-coated polymer foams with ultralow density for adjustable electromagnetic interference (EMI) shielding. ACS Appl. Mater. Interfaces. **8**(12), 8050–8057 (2016)
62. S.-C. Lin, C.-C.M. Ma, S.-T. Hsiao, Y.-S. Wang, C.-Y. Yang, W.-H. Liao, S.-M. Li, J.-A. Wang, T.-Y. Cheng, C.-W. Lin, R.-B. Yang, Electromagnetic interference shielding performance of waterborne polyurethane composites filled with silver nanoparticles deposited on functionalized graphene. Appl. Surf. Sci. **385**, 436–444 (2016)
63. Q. Jiang, X. Liao, J. Li, J. Chen, G. Wang, J. Yi, Q. Yang, G. Li, Flexible thermoplastic polyurethane/reduced graphene oxide composite foams for electromagnetic interference shielding with high absorption characteristic. Compos. A Appl. Sci. Manuf. **123**, 310–319 (2019)

64. V. Eswaraiah, V. Sankaranarayanan, S. Ramaprabhu, Functionalized graphene–PVDF foam composites for EMI shielding. Macromol. Mater. Eng. **296**(10), 894–898 (2011)
65. S. Anand, S. Pauline, Electromagnetic interference shielding properties of BaCo2Fe16O27 nanoplatelets and RGO reinforced PVDF polymer composite flexible films. Adv. Mater. Interfaces **8**(3), 2001810 (2021)
66. J. Wang, H. Li, Z. Wang, D. Xin, J. Luo, S. Bai, H. Zhou, Improved electromagnetic interference shielding properties of poly (vinylidene fluoride) composites based on carbon nanotubes and graphene nanoplatelets. Polym. Compos. **43**(10), 6966–6974 (2022)
67. Q. Qi, L. Ma, B. Zhao, S. Wang, X. Liu, Y. Lei, C.B. Park, An effective design strategy for the sandwich structure of PVDF/GNP-Ni-CNT composites with remarkable electromagnetic interference shielding effectiveness. ACS Appl. Mater. Interfaces. **12**(32), 36568–36577 (2020)
68. R. Yang, Y. Zhou, Y. Ren, D. Xu, L. Guan, X. Guo, R. Zhang, B. Zhao, Promising PVDF-CNT-graphene-NiCo chains composite films with excellent electromagnetic interference shielding performance. J. Alloy. Compd. **908**, 164538 (2022)
69. B. Zhao, C. Zhao, M. Hamidinejad, C. Wang, R. Li, S. Wang, K. Yasamin, C.B. Park, Incorporating a microcellular structure into PVDF/graphene–nanoplatelet composites to tune their electrical conductivity and electromagnetic interference shielding properties. J. Mater. Chem. C **6**(38), 10292–10300 (2018)
70. J. Fang, C. Chen, H. Qi, J. Zhang, X. Hou, L. Pan, X. Wang, Flexible multilayered poly(vinylidene fluoride)/graphene-poly(vinylidene fluoride) films for efficient electromagnetic shielding. ACS Appl. Nano Mater. **6**(8), 6858–6868 (2023)
71. J. Ling, W. Zhai, W. Feng, B. Shen, J. Zhang, W.G. Zheng, Facile preparation of lightweight microcellular polyetherimide/graphene composite foams for electromagnetic interference shielding. ACS Appl. Mater. Interfaces **5**(7), 2677–2684 (2013)
72. B. Shen, W. Zhai, M. Tao, J. Ling, W. Zheng, Lightweight, multifunctional polyetherimide/graphene@Fe3O4 composite foams for shielding of electromagnetic pollution. ACS Appl. Mater. Interfaces. **5**(21), 11383–11391 (2013)
73. H. Yang, Z. Yu, P. Wu, H. Zou, P. Liu, Electromagnetic interference shielding effectiveness of microcellular polyimide/in situ thermally reduced graphene oxide/carbon nanotubes nanocomposites. Appl. Surf. Sci. **434**, 318–325 (2018)
74. Y. Meng, S. Sharma, J.S. Chung, W. Gan, S.H. Hur, W.M. Choi, Enhanced electromagnetic interference shielding properties of immiscible polyblends with selective localization of reduced graphene oxide networks. Polymers [Online] (2022)
75. S. Mondal, A. Kumar, Multifunctional silanized silica nanoparticle functionalized graphene oxide: polyetherimide composite film for EMI shielding applications. J. Mater. Sci.: Mater. Electron. **29**(16), 14122–14131 (2018)
76. Y. Li, X. Pei, B. Shen, W. Zhai, L. Zhang, W. Zheng, Polyimide/graphene composite foam sheets with ultrahigh thermostability for electromagnetic interference shielding. RSC Adv. **5**(31), 24342–24351 (2015)
77. S. Kim, J.-S. Oh, M.-G. Kim, W. Jang, M. Wang, Y. Kim, H.W. Seo, Y.C. Kim, J.-H. Lee, Y. Lee, J.-D. Nam, Electromagnetic interference (EMI) transparent shielding of reduced graphene oxide (RGO) interleaved structure fabricated by electrophoretic deposition. ACS Appl. Mater. Interfaces. **6**(20), 17647–17653 (2014)
78. P. Sawai, P.P. Chattopadhaya, S. Banerjee, Synthesized reduce graphene oxide (rGO) filled polyetherimide based nanocomposites for EMI shielding applications. Mater. Today: Proc. **5**(3, Part 3), 9989–9999 (2018)
79. H. Luo, J. Xie, L. Xiong, Y. Zhu, Z. Yang, Y. Wan, Fabrication of flexible, ultra-strong, and highly conductive bacterial cellulose-based paper by engineering dispersion of graphene nanosheets. Compos. B Eng. **162**, 484–490 (2019)
80. C. Wan, J. Li, Graphene oxide/cellulose aerogels nanocomposite: preparation, pyrolysis, and application for electromagnetic interference shielding. Carbohyd. Polym. **150**, 172–179 (2016)

81. L. Li, Z. Ma, P. Xu, B. Zhou, Q. Li, J. Ma, C. He, Y. Feng, C. Liu, Flexible and alternant-layered cellulose nanofiber/graphene film with superior thermal conductivity and efficient electromagnetic interference shielding. Compos. A Appl. Sci. Manuf. **139**, 106134 (2020)
82. W. Yang, Y. Zhang, T. Liu, R. Huang, S. Chai, F. Chen, Q. Fu, Completely green approach for the preparation of strong and highly conductive graphene composite film by using nanocellulose as dispersing agent and mechanical compression. ACS Sustain. Chem. Eng. **5**(10), 9102–9113 (2017)
83. G. Han, Z. Ma, B. Zhou, C. He, B. Wang, Y. Feng, J. Ma, L. Sun, C. Liu, Cellulose-based Ni-decorated graphene magnetic film for electromagnetic interference shielding. J. Colloid Interface Sci. **583**, 571–578 (2021)
84. Y. Huo, D. Guo, J. Yang, Y. Chang, C. Mu, A. Nie, B. Wang, J. Xiang, K. Zhai, T. Xue, F. Wen, Flexible graphene/bacterial celluloses Janus structure film with excellent electromagnetic interference shielding and Joule heating performance. Mater. Chem. Phys. **287**, 126318 (2022)
85. M. Li, F. Han, S. Jiang, M. Zhang, Q. Xu, J. Zhu, A. Ge, L. Liu, Lightweight cellulose nanofibril/reduced graphene oxide aerogels with unidirectional pores for efficient electromagnetic interference shielding. Adv. Mater. Interfaces **8**(24), 2101437 (2021)
86. Z. Chen, Q. Zhang, W. Meng, Z. Wang, X. Han, J. Pu, Nickel-reduced graphene oxide-cellulose nanofiber composite papers for electromagnetic interference shielding. BioResources **15**(1), 814–824 (2020)
87. W. Yang, Z. Zhao, K. Wu, R. Huang, T. Liu, H. Jiang, F. Chen, Q. Fu, Ultrathin flexible reduced graphene oxide/cellulose nanofiber composite films with strongly anisotropic thermal conductivity and efficient electromagnetic interference shielding. J. Mater. Chem. C **5**(15), 3748–3756 (2017)
88. J. Li, S. Zhang, L. Wang, X. Liu, In Situ growth of Fe3O4 nanoparticles in poly(arylene ether nitrile)/graphene/carbon nanotube foams for electromagnetic interference shielding. ACS Appl. Nano Mater. **6**(9), 7802–7813 (2023)
89. S. Zhang, J. Ye, X. Liu, Constructing conductive network using 1D and 2D conductive fillers in porous poly(aryl ether nitrile) for EMI shielding. Colloids Surf. A **656**, 130414 (2023)
90. S. Zhang, H. Sun, T. Lan, Z. Bai, X. Liu, Facile preparation of graphene film and sandwiched flexible poly(arylene ether nitrile)/graphene composite films with high EMI shielding efficiency. Compos. A Appl. Sci. Manuf. **154**, 106777 (2022)
91. Dandapani, K. Devendra, Revannasiddappa, K.R. Vishnu, Thermal stability and electromagnetic interference of epoxy-graphene/hybrid composite materials. Mater. Today: Proc. **66**, 1664–1670 (2022)
92. Q. Gao, G. Zhang, Y. Zhang, X. Fan, Z. Wang, S. Zhang, R. Xiao, F. Huang, X. Shi, J. Qin, Absorption dominated high-performance electromagnetic interference shielding epoxy/functionalized reduced graphene oxide/Ni-chains microcellular foam with asymmetric conductive structure. Compos. Sci. Technol. **223**, 109419 (2022)
93. J. Li, G. Zhang, X. Fan, Q. Gao, H. Zhang, J. Qin, X. Shi, X. Fang, Microcellular epoxy/reduced graphene oxide/multi-walled carbon nanotube nanocomposite foams for electromagnetic interference shielding. Appl. Surf. Sci. **552**, 149232 (2021)
94. S. Chhetri, P. Samanta, N.C. Murmu, S.K. Srivastava, T. Kuila, Electromagnetic interference shielding and thermal properties of non-covalently functionalized reduced graphene oxide/epoxy composites. AIMS Mater. Sci **4**(1), 61–74 (2016)
95. C. Liang, P. Song, H. Qiu, Y. Zhang, X. Ma, F. Qi, H. Gu, J. Kong, D. Cao, J. Gu, Constructing interconnected spherical hollow conductive networks in silver platelets/reduced graphene oxide foam/epoxy nanocomposites for superior electromagnetic interference shielding effectiveness. Nanoscale **11**(46), 22590–22598 (2019)
96. A.F. Ahmad, S. Ab Aziz, Z. Abbas, S.J. Obaiys, A.M. Khamis, I.R. Hussain, M.H. Zaid, Preparation of a chemically reduced graphene oxide reinforced epoxy resin polymer as a composite for electromagnetic interference shielding and microwave-absorbing applications. Polymers [Online] (2018)
97. X. Bai, Y. Zhai, Y. Zhang, Green approach to prepare graphene-based composites with high microwave absorption capacity. J. Phys. Chem. C **115**(23), 11673–11677 (2011)

98. J. Xu, R. Chen, Z. Yun, Z. Bai, K. Li, S. Shi, J. Hou, X. Guo, X. Zhang, J. Chen, Lightweight epoxy/cotton fiber-based nanocomposites with carbon and Fe3O4 for electromagnetic interference shielding. ACS Omega **7**(17), 15215–15222 (2022)
99. Y. Pan, B. Yang, N. Jia, Y. Yang, Y. Wang, N. Zhang, R. Xia, J. Qian, S. Wang, Y. Tu, Y. Shi, Y. Fang, Polymethyl methacrylate (PMMA) nanocomposites containing graphene nanoplatelets decorated with nickel nanoparticles for electromagnetic interference (EMI) shielding and thermal management applications. Macromol. Mater. Eng. **307**(9), 2200220 (2022)
100. S. Acharya, J. Ray, T. Patro, P. Alegaonkar, S. Datar, Microwave absorption properties of reduced graphene oxide strontium hexaferrite/poly (methyl methacrylate) composites. Nanotechnology **29**(11), 115605 (2018)
101. H.-B. Zhang, W.-G. Zheng, Q. Yan, Z.-G. Jiang, Z.-Z. Yu, The effect of surface chemistry of graphene on rheological and electrical properties of polymethylmethacrylate composites. Carbon **50**(14), 5117–5125 (2012)
102. F. Sharif, M. Arjmand, A.A. Moud, U. Sundararaj, E.P.L. Roberts, Segregated hybrid poly(methyl methacrylate)/graphene/magnetite nanocomposites for electromagnetic interference shielding. ACS Appl. Mater. Interfaces **9**(16), 14171–14179 (2017)
103. M.C. Vu, D. Mani, J.-B. Kim, T.-H. Jeong, S. Park, G. Murali, I. In, J.-C. Won, D. Losic, C.-S. Lim, S.-R. Kim, Hybrid shell of MXene and reduced graphene oxide assembled on PMMA bead core towards tunable thermoconductive and EMI shielding nanocomposites. Compos. A Appl. Sci. Manuf. **149**, 106574 (2021)
104. H. Zhang, G. Zhang, Q. Gao, M. Zong, M. Wang, J. Qin, Electrically electromagnetic interference shielding microcellular composite foams with 3D hierarchical graphene-carbon nanotube hybrids. Compos. A Appl. Sci. Manuf. **130**, 105773 (2020)
105. I. Ibrahim Lakin, Z. Abbas, R.S. Azis, I. Abubakar Alhaji, Complex permittivity and electromagnetic interference shielding effectiveness of OPEFB fiber-polylactic acid filled with reduced graphene oxide. Materials [Online] (2020)
106. S. Shi, M. Dai, X. Tao, F. Wu, J. Sun, Y. Chen, 3D printed polylactic acid/graphene nanocomposites with tailored multifunctionality towards superior thermal management and high-efficient electromagnetic interference shielding. Chem. Eng. J. **450**, 138248 (2022)
107. S. Kashi, R.K. Gupta, T. Baum, N. Kao, S.N. Bhattacharya, Morphology, electromagnetic properties and electromagnetic interference shielding performance of poly lactide/graphene nanoplatelet nanocomposites. Mater. Des. **95**, 119–126 (2016)
108. Y. Wang, P. Wang, Z. Du, C. Liu, C. Shen, Y. Wang, Electromagnetic interference shielding enhancement of poly(lactic acid)-based carbonaceous nanocomposites by poly(ethylene oxide)-assisted segregated structure: a comparative study of carbon nanotubes and graphene nanoplatelets. Adv. Compos. Hybrid Mater. **5**(1), 209–219 (2022)
109. S. Shi, Z. Peng, J. Jing, L. Yang, Y. Chen, 3D printing of delicately controllable cellular nanocomposites based on polylactic acid incorporating graphene/carbon nanotube hybrids for efficient electromagnetic interference shielding. ACS Sustain. Chem. Eng. **8**(21), 7962–7972 (2020)
110. S. Shi, Z. Peng, J. Jing, L. Yang, Y. Chen, R. Kotsilkova, E. Ivanov, Preparation of highly efficient electromagnetic interference shielding polylactic acid/graphene nanocomposites for fused deposition modeling three-dimensional printing. Ind. Eng. Chem. Res. **59**(35), 15565–15575 (2020)
111. P. Bhawal, S. Ganguly, T.K. Das, S. Mondal, S. Choudhury, N.C. Das, Superior electromagnetic interference shielding effectiveness and electro-mechanical properties of EMA-IRGO nanocomposites through the in-situ reduction of GO from melt blended EMA-GO composites. Compos. B Eng. **134**, 46–60 (2018)
112. V. Shukla, S.K. Srivastava, Reduced graphene oxide/PdNi/poly(ethylene-co-vinyl acetate) nanocomposites for electromagnetic interference shielding. Mater. Chem. Phys. **276**, 125418 (2022)
113. N. Bagotia, D.K. Sharma, Exfoliated graphene lead to extraordinary electromagnetic interference shielding of polycarbonate/ethylene methyl acrylate blend nanocomposites. J. Polym. Res. **30**(6), 190 (2023)

114. N. Bagotia, V. Choudhary, D.K. Sharma, Synergistic effect of graphene/multiwalled carbon nanotube hybrid fillers on mechanical, electrical and EMI shielding properties of polycarbonate/ethylene methyl acrylate nanocomposites. Compos. B Eng. **159**, 378–388 (2019)
115. S. Ganguly, S. Ghosh, P. Das, T.K. Das, S.K. Ghosh, N.C. Das, Poly(N-vinylpyrrolidone)-stabilized colloidal graphene-reinforced poly(ethylene-co-methyl acrylate) to mitigate electromagnetic radiation pollution. Polym. Bull. **77**(6), 2923–2943 (2020)
116. N. Bagotia, V. Choudhary, D.K. Sharma, Superior electrical, mechanical and electromagnetic interference shielding properties of polycarbonate/ethylene-methyl acrylate-in situ reduced graphene oxide nanocomposites. J. Mater. Sci. **53**(23), 16047–16061 (2018)
117. L. Yang, Y. Chen, M. Wang, S. Shi, J. Jing, Fused deposition modeling 3D printing of novel poly(vinyl alcohol)/graphene nanocomposite with enhanced mechanical and electromagnetic interference shielding properties. Ind. Eng. Chem. Res. **59**(16), 8066–8077 (2020)
118. A.A. Khodiri, M.Y. Al-Ashry, A.G. El-Shamy, Novel hybrid nanocomposites based on polyvinyl alcohol/graphene/magnetite nanoparticles for high electromagnetic shielding performance. J. Alloy. Compd. **847**, 156430 (2020)
119. K. Manna, S.K. Srivastava, V. Mittal, Role of enhanced hydrogen bonding of selectively reduced graphite oxide in fabrication of poly(vinyl alcohol) nanocomposites in water as EMI shielding material. J. Phys. Chem. C **120**(30), 17011–17023 (2016)
120. Y. Yao, S. Jin, M. Wang, F. Gao, B. Xu, X. Lv, Q. Shu, MXene hybrid polyvinyl alcohol flexible composite films for electromagnetic interference shielding. Appl. Surf. Sci. **578**, 152007 (2022)
121. S.K. Marka, B. Sindam, K.J. Raju, V.V. Srikanth, Flexible few-layered graphene/poly vinyl alcohol composite sheets: synthesis, characterization and EMI shielding in X-band through the absorption mechanism. RSC Adv. **5**(46), 36498–36506 (2015)
122. X.-Z. Jin, Z.-Y. Yang, C.-H. Huang, J.-H. Yang, Y. Wang, PEDOT:PSS/MXene/PEG composites with remarkable thermal management performance and excellent HF-band and X-band electromagnetic interference shielding efficiency for electronic packaging. Chem. Eng. J. **448**, 137599 (2022)
123. N. Zhang, Z. Wang, R. Song, Q. Wang, H. Chen, B. Zhang, H. Lv, Z. Wu, D. He, Flexible and transparent graphene/silver-nanowires composite film for high electromagnetic interference shielding effectiveness. Sci. Bull. **64**(8), 540–546 (2019)
124. Y. Huo, B. Wang, A. Nie, C. Mu, J. Xiang, K. Zhai, T. Xue, F. Wen, Flexible aramid nanofiber/bacterial cellulose/graphene papers with nickel nanoparticles for enhanced electromagnetic interference shielding and joule heating performance. ACS Appl. Nano Mater. **5**(4), 5589–5598 (2022)
125. M. Das, P.P. Sethy, B. Sundaray, EMI shielding performance of graphene oxide reinforced polyaniline/polystyrene solution cast thin films. Synth. Met. **296**, 117369 (2023)
126. G.P. Abhilash, D. Sharma, S. Bose, C. Shivakumara, PANI-wrapped BaFe12O19 and SrFe12O19 with rGO composite materials for electromagnetic interference shielding applications. Heliyon **9**(3), e13648 (2023)
127. A. Nasir, A. Raza, M. Tahir, T. Yasin, M. Nadeem, B. Ahmad, Synthesis and study of polyaniline grafted graphene oxide nanohybrids. Mater. Res. Bull. **157**, 112006 (2023)
128. X. Zhang, X. Zhang, M. Yang, S. Yang, H. Wu, S. Guo, Y. Wang, Ordered multilayer film of (graphene oxide/polymer and boron nitride/polymer) nanocomposites: an ideal EMI shielding material with excellent electrical insulation and high thermal conductivity. Compos. Sci. Technol. **136**, 104–110 (2016)
129. P. Rani, M. Basheer Ahamed, K. Deshmukh, Dielectric and electromagnetic interference shielding performance of graphene nanoplatelets and copper oxide nanoparticles reinforced polyvinylidenefluoride/poly(3,4-ethylenedioxythiophene)-block-poly (ethylene glycol) blend nanocomposites. Synth. Met. **282**, 116923 (2021)
130. N. Jia, B. Yang, X. Wang, Y. Zuo, P. Chen, R. Xia, J. Miao, Z. Zheng, J. Qian, Y. Ke, W. Zhang, Y. Pan, Composites of an epoxy resin (EP)/PVDF/NiCo-graphene nanosheet (GNS) for electromagnetic shielding. ACS Appl. Nano Mater. **6**(9), 7731–7744 (2023)

131. M. Xiang, H. Niu, S. Qin, R. Yang, W. Lin, S. Zhou, Z. Yang, S. Dong, Modification of graphene by polypyrrole and ionic liquids for dual-band electromagnetic interference shielding hydrogels. J. Mater. Sci. **57**(24), 10983–10996 (2022)
132. C. Li, J. Guo, P. Xu, W. Hu, J. Lv, B. Shi, Z. Zhang, R. Li, Facile preparation of superior compressibility and hydrophobic reduced graphene oxide@cellulose nanocrystals/EPDM composites for highly efficient oil/organic solvent adsorption and enhanced electromagnetic interference shielding. Sep. Purif. Technol. **307**, 122775 (2023)
133. P. Rani, M.B. Ahamed, K. Deshmukh, Structural, dielectric and EMI shielding properties of polyvinyl alcohol/chitosan blend nanocomposites integrated with graphite oxide and nickel oxide nanofillers. J. Mater. Sci.: Mater. Electron. **32**(1), 764–779 (2021)
134. A. Sadeghi, R. Moeini, J.K. Yeganeh, Highly conductive PP/PET polymer blends with high electromagnetic interference shielding performances in the presence of thermally reduced graphene nanosheets prepared through melt compounding. Polym. Compos. **40**(S2), E1461–E1469 (2019)
135. P. Das, A.B. Deoghare, S. Ranjan Maity, Synergistically improved thermal stability and electromagnetic interference shielding effectiveness (EMI SE) of in-situ synthesized polyaniline/sulphur doped reduced graphene oxide (PANI/S-RGO) nanocomposites. Ceram. Int. **48**(8), 11031–11042 (2022)
136. J. Zhang, Y. Qi, Y. Zhang, J. Duan, B. Liu, B. Liu, Z. Sun, Y. Xu, W. Hu, N. Zhang, Lignin based flexible electromagnetic shielding pu synergized with graphite. Fibers Polym. **22**(1), 1–8 (2021)
137. A.F. Ahmad, S.A. Aziz, Z. Abbas, S.J. Obaiys, K.A. Matori, M.H. Zaid, H.K. Raad, U.S. Aliyu, Chemically reduced graphene oxide-reinforced poly(lactic acid)/poly(ethylene glycol) nanocomposites: preparation, characterization, and applications in electromagnetic interference shielding. Polymers [Online] (2019)
138. S.-T. Hsiao, C.-C.M. Ma, H.-W. Tien, W.-H. Liao, Y.-S. Wang, S.-M. Li, C.-Y. Yang, S.-C. Lin, R.-B. Yang, Effect of covalent modification of graphene nanosheets on the electrical property and electromagnetic interference shielding performance of a water-borne polyurethane composite. ACS Appl. Mater. Interfaces **7**(4), 2817–2826 (2015)
139. W. Yang, H. Bai, B. Jiang, C. Wang, W. Ye, Z. Li, C. Xu, X. Wang, Y. Li, Flexible and densified graphene/waterborne polyurethane composite film with thermal conducting property for high performance electromagnetic interference shielding. Nano Res. **15**(11), 9926–9935 (2022)
140. S. Ganguly, N. Kanovsky, P. Das, A. Gedanken, S. Margel, Photopolymerized thin coating of polypyrrole/graphene nanofiber/iron oxide onto nonpolar plastic for flexible electromagnetic radiation shielding, strain sensing, and non-contact heating applications. Adv. Mater. Interfaces **8**(23), 2101255 (2021)
141. M. Khan, A.N. Khan, A. Saboor, I.H. Gul, Investigating mechanical, dielectric, and electromagnetic interference shielding properties of polymer blends and three component hybrid composites based on polyvinyl alcohol, polyaniline, and few layer graphene. Polym. Compos. **39**(10), 3686–3695 (2018)
142. G. Huang, J. Gao, X. Wang, H. Liang, C. Ge, How can graphene reduce the flammability of polymer nanocomposites? Mater. Lett. **66**(1), 187–189 (2012)
143. G. Huang, H. Liang, Y. Wang, X. Wang, J. Gao, Z. Fei, Combination effect of melamine polyphosphate and graphene on flame retardant properties of poly(vinyl alcohol). Mater. Chem. Phys. **132**(2), 520–528 (2012)
144. B. Dittrich, K.-A. Wartig, D. Hofmann, R. Mülhaupt, B. Schartel, Flame retardancy through carbon nanomaterials: carbon black, multiwall nanotubes, expanded graphite, multi-layer graphene and graphene in polypropylene. Polym. Degrad. Stab. **98**(8), 1495–1505 (2013)
145. D. Hofmann, K.A. Wartig, R. Thomann, B. Dittrich, B. Schartel, R. Mülhaupt, Functionalized graphene and carbon materials as additives for melt-extruded flame retardant polypropylene. Macromol. Mater. Eng. **298**(12), 1322–1334 (2013)
146. G. Huang, S. Wang, P.A. Song, C. Wu, S. Chen, X. Wang, Combination effect of carbon nanotubes with graphene on intumescent flame-retardant polypropylene nanocomposites. Compos. A: Appl. Sci. Manuf. **59**, 18–25 (2014)

147. B. Yu, X. Wang, X. Qian, W. Xing, H. Yang, L. Ma, Y. Lin, S. Jiang, L. Song, Y. Hu, Functionalized graphene oxide/phosphoramide oligomer hybrids flame retardant prepared via in situ polymerization for improving the fire safety of polypropylene. RSC Adv. **4**(60), 31782–31794 (2014)
148. B. Yuan, H. Sheng, X. Mu, L. Song, Q. Tai, Y. Shi, K.M. Liew, Y. Hu, Enhanced flame retardancy of polypropylene by melamine-modified graphene oxide. J. Mater. Sci. **50**, 5389–5401 (2015)
149. B. Yuan, A. Fan, M. Yang, X. Chen, Y. Hu, C. Bao, S. Jiang, Y. Niu, Y. Zhang, S. He, H. Dai, The effects of graphene on the flammability and fire behavior of intumescent flame retardant polypropylene composites at different flame scenarios. Polym. Degrad. Stab. **143**, 42–56 (2017)
150. B. Yuan, Y. Sun, X. Chen, Y. Shi, H. Dai, S. He, Poorly-/well-dispersed graphene: abnormal influence on flammability and fire behavior of intumescent flame retardant. Compos. A Appl. Sci. Manuf. **109**, 345–354 (2018)
151. G. Trusiano, S. Matta, M. Bianchi, L.G. Rizzi, A. Frache, Evaluation of nanocomposites containing graphene nanoplatelets: mechanical properties and combustion behavior. Polym. Eng. Sci. **59**(10), 2062–2071 (2019)
152. K. Yang, M. Endoh, R. Trojanowski, R.P. Ramasamy, M.M. Gentleman, T.A. Butcher, M.H. Rafailovich, The thermo-mechanical response of PP nanocomposites at high graphene loading. Nanocomposites **1**(3), 126–137 (2015)
153. H. Quan, B.-Q. Zhang, Q. Zhao, R.K. Yuen, R.K. Li, Facile preparation and thermal degradation studies of graphite nanoplatelets (GNPs) filled thermoplastic polyurethane (TPU) nanocomposites. Compos. A Appl. Sci. Manuf. **40**(9), 1506–1513 (2009)
154. Z. Cheng, D. Liao, X. Hu, W. Li, C. Xie, H. Zhang, W. Yang, Synergistic fire retardant effect between expandable graphite and ferrocene-based non-phosphorus polymer on polypropylene. Polym. Degrad. Stab. **178**, 109201 (2020)
155. D.-J. Liao, Q.-K. Xu, R.W. McCabe, H.V. Babu, X.-P. Hu, N. Pan, D.-Y. Wang, T.R. Hull, Ferrocene-based nonphosphorus copolymer: synthesis, high-charring mechanism, and its application in fire retardant epoxy resin. Ind. Eng. Chem. Res. **56**(44), 12630–12643 (2017)
156. T. Tang, X. Chen, H. Chen, X. Meng, Z. Jiang, W. Bi, Catalyzing carbonization of polypropylene itself by supported nickel catalyst during combustion of polypropylene/clay nanocomposite for improving fire retardancy. Chem. Mater. **17**(11), 2799–2802 (2005)
157. J. Li, S. Wang, G. Zhang, H. Li, J. Sun, X. Gu, S. Zhang, Burning behavior analysis of polypropylene composite containing poly-siloxane encapsulated expandable graphite. Polym. Degrad. Stab. **202**, 110006 (2022)
158. Q. Gao, H. Zhao, X.-L. Zhou, F.-Y. Liu, Y.-H. Jiao, J.-X. Xie, H.-Q. Qu, J.-Z. Xu, H.-Y. Ma, Flame retardant, combustion and thermal degradation properties of polypropylene composites treated with the mixture of pentaerythritol, nickel hydroxystannate and expandable graphite. Polym. Degrad. Stab. **203**, 110084 (2022)
159. Y. Han, Y. Wu, M. Shen, X. Huang, J. Zhu, X. Zhang, Preparation and properties of polystyrene nanocomposites with graphite oxide and graphene as flame retardants. J. Mater. Sci. **48**, 4214–4222 (2013)
160. Z.-M. Zhu, W.-H. Rao, A.-H. Kang, W. Liao, Y.-Z. Wang, Highly effective flame retarded polystyrene by synergistic effects between expandable graphite and aluminum hypophosphite. Polym. Degrad. Stab. **154**, 1–9 (2018)
161. W. Hu, B. Yu, S.-D. Jiang, L. Song, Y. Hu, B. Wang, Hyper-branched polymer grafting graphene oxide as an effective flame retardant and smoke suppressant for polystyrene. J. Hazard. Mater. **300**, 58–66 (2015)
162. N.H. Thi, T.N. Nguyen, H.T. Oanh, N.T.T. Trang, D.Q. Tham, H.T. Nguyen, T. Van Nguyen, M.H. Hoang, Synergistic effects of aluminum hydroxide, red phosphorus, and expandable graphite on the flame retardancy and thermal stability of polyethylene. J. Appl. Polym. Sci. **138**(17), 50317 (2021)
163. W. Hu, J. Zhan, X. Wang, N. Hong, B. Wang, L. Song, A.A. Stec, T.R. Hull, J. Wang, Y. Hu, Effect of functionalized graphene oxide with hyper-branched flame retardant on flammability

and thermal stability of cross-linked polyethylene. Ind. Eng. Chem. Res. **53**(8), 3073–3083 (2014)
164. K. Shen, in *Overview of Flame Retardancy and Smoke Suppression of Flexible PVC*. Proceedings SPE Vinyltech Conference, Atlanta (2006)
165. Y.-T. Pan, D.-Y. Wang, One-step hydrothermal synthesis of nano zinc carbonate and its use as a promising substitute for antimony trioxide in flame retardant flexible poly (vinyl chloride). RSC Adv. **5**(35), 27837–27843 (2015)
166. P. Jia, G. Feng, C. Bo, L. Hu, X. Yang, L. Zhang, M. Zhang, Y. Zhou, A composition of phosphaphenanthrene groups-containing castor-oil-based phosphate plasticizer for PVC: synthesis, characterization and property. J. Ind. Eng. Chem. **60**, 192–205 (2018)
167. P. Jia, Y. Ma, H. Xia, M. Zheng, G. Feng, L. Hu, M. Zhang, Y. Zhou, Clean synthesis of epoxidized tung oil derivatives via phase transfer catalyst and thiol–ene reaction: a detailed study. ACS Sustain. Chem. Eng. **6**(11), 13983–13994 (2018)
168. V. Najafi, H. Abdollahi, Internally plasticized PVC by four different green plasticizer compounds. Eur. Polymer J. **128**, 109620 (2020)
169. L. Han, W. Wu, H. Qu, X. Han, A. Wang, Y. Jiao, J. Xu, Metallic ferrites as flame retardants and smoke suppressants in flexible poly (vinyl chloride). J. Therm. Anal. Calorim. **123**, 293–300 (2016)
170. S. He, W. Wu, M. Zhang, H. Han, Y. Jiao, H. Qu, J. Xu, Reduction in smoke emitted and fire hazard presented by flexible poly (vinyl chloride) through novel synthesis of SnO2 supported by activated carbon spheres. Polym. Adv. Technol. **29**(9), 2505–2514 (2018)
171. Y.-T. Pan, M. Castillo-Rodríguez, D.-Y. Wang, Mesoporous metal oxide/pyrophosphate hybrid originated from reutilization of water treatment resin as a novel fire hazard suppressant. Mater. Chem. Phys. **203**, 49–57 (2018)
172. C. Cullis, M. Hirschler, Q. Tao, Studies of the effects of phosphorus-nitrogen-bromine systems on the combustion of some thermoplastic polymers. Eur. Polymer J. **27**(3), 281–289 (1991)
173. Y. Li, L. Lv, W. Wang, J. Zhang, J. Lin, J. Zhou, M. Dong, Y. Gan, I. Seok, Z. Guo, Effects of chlorinated polyethylene and antimony trioxide on recycled polyvinyl chloride/acryl-butadiene-styrene blends: flame retardancy and mechanical properties. Polymer **190**, 122198 (2020)
174. H. Qu, C. Liu, W. Wu, L. Chen, J. Xu, Using cone calorimeter to study thermal degradation of flexible PVC filled with zinc ferrite and Mg (OH) 2. J. Therm. Anal. Calorim. **115**, 1081–1087 (2014)
175. H. Pi, S. Guo, Y. Ning, Mechanochemical improvement of the flame-retardant and mechanical properties of zinc borate and zinc borate–aluminum trihydrate-filled poly (vinyl chloride). J. Appl. Polym. Sci. **89**(3), 753–762 (2003)
176. K. Yao, J. Gong, N. Tian, Y. Lin, X. Wen, Z. Jiang, H. Na, T. Tang, Flammability properties and electromagnetic interference shielding of PVC/graphene composites containing Fe 3 O 4 nanoparticles. RSC Adv. **5**(40), 31910–31919 (2015)
177. Z. Huang, Z. Wang, Synthesis of a copper hydroxystannate modified graphene oxide nanohybrid and its high performance in flexible polyvinyl chloride with simultaneously improved flame retardancy, smoke suppression and mechanical properties. Polym. Degrad. Stab. **183**, 109425 (2021)
178. S. Pappalardo, P. Russo, D. Acierno, S. Rabe, B. Schartel, The synergistic effect of organically modified sepiolite in intumescent flame retardant polypropylene. Eur. Polymer J. **76**, 196–207 (2016)
179. B. Zirnstein, W. Tabaka, D. Frasca, D. Schulze, B. Schartel, Graphene/hydrogenated acrylonitrile-butadiene rubber nanocomposites: dispersion, curing, mechanical reinforcement, multifunctional filler. Polym. Test. **66**, 268–279 (2018)
180. H. Qu, W. Wu, J. Xie, J. Xu, Zinc hydroxystannate-coated metal hydroxides as flame retardant and smoke suppression for flexible poly vinyl chloride. Fire Mater. Int. J. **33**(4), 201–210 (2009)
181. S.V. Levchik, E.D. Weil, Combustion and fire retardancy of aliphatic nylons. Wiley Online Library (2000)

182. P. Gijsman, R. Steenbakkers, C. Fürst, J. Kersjes, Differences in the flame retardant mechanism of melamine cyanurate in polyamide 6 and polyamide 66. Polym. Degrad. Stab. **78**(2), 219–224 (2002)
183. M.A. Uddin, T. Bhaskar, J. Kaneko, A. Muto, Y. Sakata, T. Matsui, Dehydrohalogenation during pyrolysis of brominated flame retardant containing high impact polystyrene (HIPS-Br) mixed with polyvinylchloride (PVC). Fuel **81**(14), 1819–1825 (2002)
184. P. Kiliaris, C. Papaspyrides, Polymer/layered silicate (clay) nanocomposites: an overview of flame retardancy. Prog. Polym. Sci. **35**(7), 902–958 (2010)
185. X. Hao, G. Gai, J. Liu, Y. Yang, Y. Zhang, C.-W. Nan, Flame retardancy and antidripping effect of OMT/PA nanocomposites. Mater. Chem. Phys. **96**(1), 34–41 (2006)
186. D. Tabuani, S. Ceccia, G. Camino, Nylon-6 nanocomposites, study of the influence of the nanofiller nature on morphology and material properties. J. Polym. Sci., Part B: Polym. Phys. **47**(19), 1935–1948 (2009)
187. N. Hong, L. Song, T.R. Hull, A.A. Stec, B. Wang, Y. Pan, Y. Hu, Facile preparation of graphene supported Co3O4 and NiO for reducing fire hazards of polyamide 6 composites. Mater. Chem. Phys. **142**(2), 531–538 (2013)
188. B.M. Güell, I. Babich, L. Lefferts, K. Seshan, Steam reforming of phenol over Ni-based catalysts–A comparative study. Appl. Catal. B **106**(3–4), 280–286 (2011)
189. C.K. Kundu, Z. Li, X. Li, Z. Zhang, Y. Hu, Graphene oxide functionalized biomolecules for improved flame retardancy of Polyamide 66 fabrics with intact physical properties. Int. J. Biol. Macromol. **156**, 362–371 (2020)
190. B. Yuan, Y. Hu, X. Chen, Y. Shi, Y. Niu, Y. Zhang, S. He, H. Dai, Dual modification of graphene by polymeric flame retardant and Ni (OH) 2 nanosheets for improving flame retardancy of polypropylene. Compos. A Appl. Sci. Manuf. **100**, 106–117 (2017)
191. F. Seidi, E. Movahedifar, G. Naderi, V. Akbari, F. Ducos, R. Shamsi, H. Vahabi, M.R. Saeb, Flame retardant polypropylenes: a review. Polymers (Basel) **12**(8) (2020)
192. C. Bao, Y. Guo, B. Yuan, Y. Hu, L. Song, Functionalized graphene oxide for fire safety applications of polymers: a combination of condensed phase flame retardant strategies. J. Mater. Chem. **22**(43), 23057–23063 (2012)
193. D. Marset, E. Fages, E. Gonga, J. Ivorra-Martinez, L. Sánchez-Nacher, L. Quiles-Carrillo, Development and characterization of high environmentally friendly composites of bio-based polyamide 1010 with enhanced fire retardancy properties by expandable graphite. Polymers **14**(9), 1843 (2022)
194. G. Huang, S. Chen, S. Tang, J. Gao, A novel intumescent flame retardant-functionalized graphene: nanocomposite synthesis, characterization, and flammability properties. Mater. Chem. Phys. **135**(2), 938–947 (2012)

Chapter 8
Current Challenges and a Way Forward

8.1 Challenges and the Way Forward for Polymeric EMI Shielding Nanocomposite Materials

Numerous graphene-based application concepts have been demonstrated since the graphene revolution began in 2004. High-performance flame-retarding materials made of graphene-based nanofillers and their EMI shielding counterparts have achieved performance levels that are comparable to the best systems with years (decades) of research work behind them in only a few short years [1–3].

Advanced graphene-based nanofiller augmented polymer nanocomposites as well as their blends have special features that enable new and intriguing structural manufacturing applications. These frequently have low cost, excellent processability in polymers and solvents, and tunable chemical properties. "Graphene is possibly the single system where ideas from quantum field theory might lead to patented improvements," said Nobel laureate Frank Wilczek [4].

Research on graphene-bearing polymer nanocomposite materials has become the most important study area in the last ten years. Graphene-based nanofillers, however, face a number of obstacles before they can reach their full potential in the market, despite the fact that it has the ability to provide a wide range of solutions to many industrial areas in the long run. The sluggish adoption of graphene-based nanofillers by various industries is due to supply chain issues, including a lack of potential for scaling up, high costs, and poor quality of the created graphene-based nanofillers (Fig. 8.1). Even after gaining a fundamental grasp of polymeric nanocomposites based on graphene-based nanofillers, there are still a lot of unanswered concerns. Below is a list of some of the main difficulties.

The modification effectiveness in composites falls far short of expectations due to inconsistencies in critical material parameters, including the size, structural integrity, and purity of commercial graphene-based filler material systems [5].

The lack of rigorous, explicit standards for graphene-based nanofiller characterization that consider the material's physical characteristics, safety precautions, and

S. S. Ray et al., *Graphene-Bearing Polymer Composites*, Springer Series in Materials Science 340, https://doi.org/10.1007/978-3-031-51924-6_8

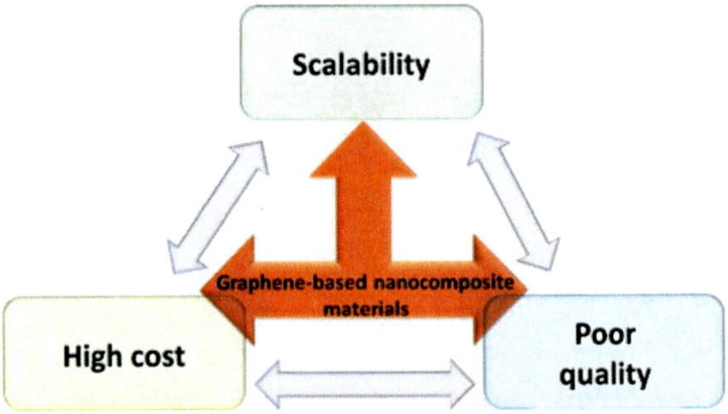

Fig. 8.1 Primary supply chain variables influencing the rapid use of GNP-based nanocomposite materials as EMI filters and flame-retardants

application-specific needs is of great concern [5]. There are currently few new scalable manufacturing procedures available for the mass manufacture of specialized items based on graphene, which is of great concern, especially for the two niches this monograph focuses on.

It is challenging to design graphene-based nanocomposite systems that can provide excellent fire retardants or EMI shield properties as well as foreseeable but also safe failure modes because existing models of graphene-modified (nano)composites for high-performance fire retardants or EMI shields are either rare or limited. To some extent, there is a lack of practically proven information on the interface characteristics between graphene-based nanofiller and polymer matrix under high loading circumstances and high-performance structural applications of graphene-based polymer composites. GNP-based materials' potential effects on human health and/or the environment are not fully understood [6].

There is currently a shortage of precise legislation and requirements, documented safety procedures, and experimental techniques for the disposal of the derived wastes.

In recent years, a breakthrough in some of the aforementioned restrictions has been made possible by state-of-the-art, economical synthesis techniques and the advent of multiple commercial graphene-based nanofiller producers. One of the main factors that businesses consider when deciding whether or not to utilize graphene in their products, like with any other product, is the cost versus performance. One of the primary obstacles to their introduction into the market has been regarded since their inception as the high price of graphene and related materials. However, over the past ten years, the market price of materials made of graphene has been consistently on a decline. For instance, graphene, a prominent graphene-based nanofiller supplier, claims that although while graphene is currently utilized for applications that other materials can't possibly support, the cost of high-quality sheet graphene will soon be lower than the cost of the primary competing materials [5]. Besides, not all applications necessitate a high-quality material, and the price of graphene is directly related to its quality. Because of this, it is crucial to understand the proper application area for each type of graphene. Additionally, the development of raw materials and catalysts by significant producers, including Hyperion USA, Bayer AG, as well as Arkema In France, the manufacturing of carbon-based NMTs, such as graphene, has significantly increased from grams to tons [5].

In order to create a successful polymer nanocomposite, we need to handle a number of important technological challenges, as we have seen thus far. First, these graphene-based nanocomposites must be processed to achieve uniform dispersion of the selected graphene-based reinforcing filler inside the selected polymer matrix. The case studies should make it obvious that measuring their dispersion and alignment, as well as the final nanocomposite's material properties consistently, is required if the graphene-based filler-bearing polymer composite is to be employed in manufacturing. Finding an effective, cost-effective production path is also essential. The study of polymer (nano)composites is seen to be relatively new, but it is advancing quickly on both the technological and business fronts. However, there are currently several new concerns, including commercial, technical, and regulatory ones, that possess the capacity to have a substantial impact on the field's future course.

8.1.1 Creation of Standardized Techniques for Describing Nanomaterials

In order to assess the quality as well as purity of their reinforcing nanofillers but also gauging how well they disperse in polymer matrices, different commercial vendors and research facilities employ distinct techniques [7]. This is a persistent problem when using nano-reinforcements as raw materials to create polymer nanocomposites. Therefore, how can a consumer trustfully compare products that come from several sources? We'll take a look at two relevant examples: measuring graphene-based nanofillers dispersion with microscopy and their purity using TGA. The purity of graphene-based nanofillers can be studied following the procedure reported for CNTs [8], seeing that both CNTs and GNPs are constituently made of graphene sheets. The average values of T_0, the sample's peak oxidation temperature, and M_{res}, , the amount of residual mass (or ash content) left after the TGA experiment, are indicators of the sample's purity (higher oxidation temperatures and lower residual ash contents are frequently used to demonstrate higher NT purity), while the standard deviations of T_0 and M_{res} are indicators of the sample's inhomogeneity between replicate experiments. However, the burning of the NTs during the TGA experiment could result in an abrupt release of heat, which would generate spikes in weight loss as a function of the temperatures that are recorded. Remember that in TGA, we evaluate the weight loss of the sample as a function of its temperature while heating the sample at a constant rate. The ash content (M_{res}) values obtained after combustion are lower, and the M_{res} standard deviation over repeated trials is higher. Due to smoke particulates being ejected from the sample pan, some samples may potentially be lost. Similar "burst expansion" is seen when GNP is characterized by TGA. The US National Institute of Standards and Technology, NIST, has created a recommended practice guide for conducting TGA of CNTs in order to overcome these issues and provide a standard test procedure that can be used to compare CNT samples with confidence [8]. Similarly, NIST has suggested a common microscopy procedure to

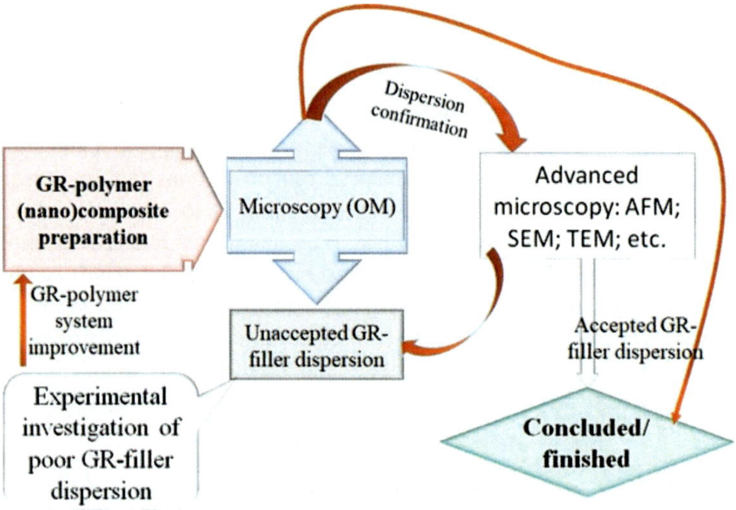

Fig. 8.2 Flowchart for evaluating the dispersion of reinforcing nanofillers in graphene-based polymer (nano)composite materials using the NIST-recommended microscopy procedure for SWCNTs, which is adaptable for graphene-based polymer composite systems

study SWCNTs, which is adaptable for graphene-based systems, too [9]. In this top-down method, samples are screened using more basic techniques, and where necessary, increasingly sophisticated techniques are added to confirm the dispersion of the nano-reinforcement in the matrix. This minimizes the practical amount of time and money needed for analysis. Sample preparation is typically the most challenging as well as time-consuming step in microscopic analysis. For efficacy, specimens are prepared for investigation as cast films and/or thin sections. Figure 8.2 is useful in summarizing the protocol.

8.1.2 Nanofillers but also Nanocomposites' Safety and Toxicity

We have already shown that the phrase "nanotechnology" can be confusing when referring to the technical elements of polymeric nanocomposites because it encompasses a wide range of techniques, materials, applications, and ideas that are all sized differently. We also mentioned that the physical characteristics of a certain substance (such as carbon, graphene, silicon, cellulose, or metals) differ at the nanoscale from those that exist at the macroscale. This could be true of both their bodily behavior and their material characteristics!

The impact of emerging nanofillers on health and safety is largely unknown as of this writing. Therefore, as nanofillers are employed more often in industry, there is an increasing danger of exposure to them by accident at work (ISO 2008; HSE 2013). Nanofiller exposure at work may happen accidentally: When creating NPs in non-enclosed systems, through manufacturing, when incorporating them into other materials (like polymer nanocomposites), when researching their properties and applications, when cleaning dust collection systems used to catch NPs, during improper disposal, and when they accidentally spill.

One of the first nations to regulate the use of nanomaterials is the UK. For instance, the BSI hired the KTN to conduct a study on the opportunities and obstacles for commercializing GNPs. The study also sought to determine how standards can be an important factor in accelerating the supply of GNPs and GNP-based products into the market. The research's objectives included examining and finding gaps in existing standards, identifying crucial areas where standards are required, and creating a roadmap for standards. The study included primary research, comprehensive community outreach via a survey, in-depth fact-finding secondary research, and primary research. A questionnaire created especially for the project was used to get input from the community. Online surveys, interactive surveys at workshops (69 respondents in two workshops), and one-on-one thorough interviews with top business leaders from 17 different organizations collectively yielded 129 detailed responses. The poll received responses from 60% of UK enterprises, 29% of universities and research institutions, and 11% of other organizations. The GNP sector is ready to accept standards, according to the report, as it works to foster a culture of trust and give potential OEMs along with end users guarantees and confidence. The following important requirements were noted.

(a) Standards are needed for graphene definition, production, characterization, and application-specific characteristics, among other things.
(b) Waste management and the production of graphene require health and safety regulations.
(c) To foster open supply chains, lower costs through economies of scale, and boost user and investor trust in the usage of graphene, standards must be developed and put into practice.
(d) To maximize the influence on the commercialization of graphene, new standards should be vigorously promoted to hasten corporate adoption.
(e) Graphene standards should be quickly adopted internationally to have a stronger impact.
(f) There is a need to spread knowledge of the current graphene-relevant standards, such as the nomenclature standard (ISO/TS 80004-13).

They suggested that:

a. The industry must actively promote current graphene-related standards by enlisting networks and industry/technological groups.
b. The following are priority areas for the creation of new graphene standards were considered: Production and manufacturing of graphene; descriptions of the different types, forms, and properties of graphene; information provided by suppliers about the material's characteristics and production process; data on the

material's properties that can be used to compare it to other materials; and the health and safety of producing and recycling graphene.

Another health and safety legislation that could greatly impact how nanomaterials are used in Europe in the future is the Registration, Evaluation, REACH law [10]. REACH, also known as EC 1907/2006, is a European Community Regulation regulating chemicals and their safe use. Chemical substance producers and importers are obligated by REACH to collect data on the qualities of the chemicals they offer in order to ensure their safe handling. The ECHA, in Helsinki has recorded this data in a consolidated database. REACH went into effect on June 1st, 2007, and its requirements will be implemented gradually over the next 11 years. A separate REACH registration was not necessary for nanoscale versions of known materials because they were not considered to be "novel" substances at the time of its debut (Gergely 2012). Whether NPTs will be treated differently from their microscale equivalents (e.g., GP versus GNP) is a crucial topic for NMTs.

8.1.3 Nanocomposites' Reuse and Recycling

Reduced waste along with the reuse or recycling of raw materials and finished goods at the end of their useful lives are becoming increasingly important in all sectors of industry and business, both from the perspective of customers and regulatory authorities. Let's look at two instances of industries that now use polymer nanocomposites: electronics, automotive, etc., for which EMI shields designed from graphene-polymer nanocomposites are not exempted.

The WEEE regulation, 2002/96/EC, mandates that producers of electrical and electronic equipment make arrangements to collect but also recycle or reuse these items at the end of their useful lives throughout the European Union [7]. The WEEE regulation covers home appliances, consumer goods, toys, office and telecommunication equipment's, lighting equipment, electrical/electronic tools, recreational and sporting goods, (bio)medical devices, monitoring and control equipment, as well as automatic dispensers.

European Parliament and Council Directive 2000/53/EC addresses recovery and recycling of end-of-life cars for the automobile industry. Each EU Member State is required by this law to reach a 95% reuse/recovery rate by average weight per vehicle by January 1, 2015, including an 85% materials recycling rate [7]. The vehicle owner cannot be charged for the costs associated with recycling the car, and the process must not pollute the environment. The guideline further stipulates that the use of particular hazardous compounds in new cars must be kept to a minimum.

Across both situations, the question for the future use of nanomaterials such as graphene-based nanofillers in industry is: How can we create techniques for recovering, separating, and recycling composite materials while protecting the environment, according to waste management guidelines and meeting the recently discussed upcoming health and safety regulations? If NPs are deemed "toxic," how will we safely recycle the cars and gadgets they were made from?

8.2 Challenges and the Way Forward for Fire Retarding Nanocomposites Materials

Graphene has demonstrated strong thermal and mechanical characteristics in addition to good flame retardancy, making it a promising and efficient flame-retardant. Nevertheless, creating flame-retardants based on graphene still faces several difficulties. First, due to its strong pi-pi force and weak van der Waals forces, graphene with an untreated surface disperses poorly in a polymer matrix and is susceptible to aggregation. Second, the connection between graphene sheets and the matrix of the polymer is so weak that the stress cannot be efficiently transferred to graphene, leading to a decline in mechanical performance. Third, although 5% graphene filler is required to obtain the best flame retardancy, the graphene sheets have a strong tendency to aggregate at this dosage. Additionally, the flame retardancy performance suffers due to the char of graphene-filled polymers not being solid enough to create a fire defense system. When employed alone, these difficulties render graphene or GO ineffective flame-retardants [11].

Although many efforts have been made to create synthetic processes for graphene, it is still challenging to manufacture SLG in quantities greater than 90% or even 100%. MLG or few-layer graphene makes up the majority of the market's commercially available graphene's "FLG." Due to the ease with which graphene sheets can be re-aggregated and the dearth of technology that can effectively exfoliate the sheets completely, the exfoliation process for the creation of graphene cannot obtain SLG on a wide scale. Because it is impossible to completely convert GO to graphene, the reduction methods from GO to generate graphene are similarly unable to produce SLG on a large scale. Therefore, despite the enormous efforts that have been made to synthesize graphene since it was discovered 10 years ago, it is still difficult to generate pure SLG for commercial use. As opposed to MLG, FLG, and GO, pure SLG is what gives graphene its incredibly good qualities. Because the graphene used to prepare composites is not pure SLG, the outstanding properties cannot be seen in polymer composites. This is one of the reasons that the gains in properties of graphene/polymer composites are not as great as they should be. Nearly all of the reported efforts have used modified graphene (such as GO, FG, and RGO) or a modification procedure to generate graphene/polymer composites in order to obtain strong interaction between graphene and polymer matrix. However, altering graphene might significantly alter its inborn properties, which would then have an impact on how well graphene/polymer composites perform. Additionally, the organic solvent employed in the modification process is not suited for industrial manufacturing and is not environmentally friendly. There are currently few techniques available for the mass manufacturing of pure or chemically unaltered graphene sheets. It is so difficult to create graphene/polymer composites that retain as many of the natural characteristics of pure graphene as feasible while simultaneously achieving a favorable interaction between the two materials through environmental means.

As was already noted, there are three primary techniques for creating composites made of graphene/polymers. Although they may come with possible pollution issues due to the use of particular reagents, in situ polymerization and solvent blending can generate a fine dispersion of graphene in polymer matrix. The dispersion of graphene in the polymer matrix is poor, although melt processing is environmentally friendly and suited for mass production. The creation of graphene/polymer composites using an efficient approach that is both environmentally friendly, practical for commercial production, and produces graphene that is evenly spread in the polymer matrix remains a problem to date. As a result, it restricts the use of graphene in polymers.

The price of graphene is also a problem; at the moment, the cost of commercial graphene manufacturing ranges from 70 to 200 USD/g. Until graphene can be produced in huge quantities at a reasonable price, practical applications won't be possible. The fundamental questions, such as how to prepare pure SLG, how to obtain good contact between graphene and polymer matrix without modifying the technique, how to make graphene/polymer nanocomposites in commercial production, and how much graphene costs, still merit additional study. A new avenue for the creation of high-performance composite materials with several uses has been made possible by the discovery of graphene. Eventually, new techniques and technology will help graphene/polymer nanocomposites evolve [12].

8.3 Summary

In conclusion, the exploration of graphene-bearing polymer composites in the realms of electromagnetic interference shielding and flame-retardant materials has revealed a landscape marked by challenges and opportunities. The identified challenges, encompassing issues of material integration, property optimization, durability, scalability, and environmental impact, serve as crucial waypoints guiding our trajectory toward the future. However, this study does not merely delineate obstacles; it also charts a course forward. By delving into advanced material engineering, optimization strategies, durability enhancements, manufacturing innovations, interdisciplinary collaboration, and environmental responsibility, we uncover a roadmap for progress. As we stand at the intersection of current challenges and the way forward, it is the collaborative and innovative spirit of future research that will unlock the full potential of graphene-bearing polymer composites, propelling them from the laboratory to transformative applications in diverse industries.

References

1. J.T. Orasugh, S.S. Ray, Functional and structural facts of effective electromagnetic interference shielding materials: a review. ACS Omega **8**(9), 8134–8158 (2023)
2. J.T. Orasugh, S.S. Ray, Graphene-based electrospun fibrous materials with enhanced EMI shielding: recent developments and future perspectives. ACS Omega **7**(38), 33699–33718 (2022)

3. S. Ran, F. Fang, Z. Guo, P. Song, Y. Cai, Z. Fang, H. Wang, Synthesis of decorated graphene with P, N-containing compounds and its flame retardancy and smoke suppression effects on polylactic acid. Compos. B Eng. **170**, 41–50 (2019)
4. J. Kinaret, A.C. Ferrari, V. Fal'ko, J. Kivioja, Graphene-driven revolutions in ICT and beyond. Procedia Comput. Sci. **7**, 30–33 (2011)
5. A. Mirabedini, A. Ang, M. Nikzad, B. Fox, K.-T. Lau, N. Hameed, Evolving strategies for producing multiscale graphene-enhanced fiber-reinforced polymer composites for smart structural applications. Adv. Sci. **7**(11), 1903501 (2020)
6. T. Mumberg, S.F. Hansen, A. Baun, R. Arvidsson, Assessment of graphene-based materials against the substances of very high concern criteria. Sci. Rep. E **2023** (2023)
7. G. Armstrong, An introduction to polymer nanocomposites. Eur. J. Phys. **36**(6), 063001 (2015)
8. S. Arepalli, P. Nikolaev, O. Gorelik, V.G. Hadjiev, W. Holmes, B. Files, L. Yowell, Protocol for the characterization of single-wall carbon nanotube material quality. Carbon **42**(8–9), 1783–1791 (2004)
9. S. Arepalli, S.W. Freiman, S.A. Hooker, K.D. Migler, Measurement issues in single-wall carbon nanotubes (2008)
10. Directorate-General for Employment, S. A. a. I. E. C, Working safely with manufactured nanomaterials Non-binding guide for workers. https://op.europa.eu/en/publication-detail/-/publication/4d51f5b2-545d-11ea-aece-01aa75ed71a1/language-en. Accessed 26 May 2023
11. B. Sang, Z.-W. Li, X.-H. Li, L.-G. Yu, Z.-J. Zhang, Graphene-based flame retardants: a review. J. Mater. Sci. **51**, 8271–8295 (2016)
12. X. Fu, C. Yao, G. Yang, Recent advances in graphene/polyamide 6 composites: a review. RSC Adv. **5**(76), 61688–61702 (2015)

Chapter 9
Conclusions and Prospects

9.1 Conclusions

In this book, we have presented an overview of graphene-based filler-bearing-polymer nanocomposite materials with emphasis on their synthesis approaches, EMI shielding mechanisms, flame-retarding mechanism, characterization approaches, applications, challenges, and prospects. Considering the available literature, we have deduced that graphene-bearing polymer nanocomposite materials have presented properties suitable and, in some instances, superior to their counterparts conventionally/currently applied in desired industries. The book contains vital information that is suitable for emerging researchers within the niche of flame-retardant systems, EMI shield, as well as other related emerging technological important niches where graphene-based nanofillers adoption is paramount.

9.2 Prospects

It is realistic to anticipate that the cost of raw materials will decrease as demand and usage of nano-reinforcements grow and larger-scale production becomes more common. There should be an intentional drive by researchers globally toward reducing the price of synthesizing graphene and its derivatives. This book mainly aims to promote the use of graphene-based fillers in the development of advanced materials such as EMI shields and flame-retardant material systems.

Despite the health and safety risks associated with exposure to nanomaterial-based powders, it's crucial to keep in mind that graphene dispersions are safer and easier to work with than their powder counterparts because they can be handled and processed in the same way as regular micron-sized reinforcements. In truth, nano-reinforcing filler firms have spent a lot of time and money conducting research to create new products and formulas that do not require handling with nanomaterials. Alternative

S. S. Ray et al., *Graphene-Bearing Polymer Composites*, Springer Series in Materials Science 340, https://doi.org/10.1007/978-3-031-51924-6_9

methods include masterbatches of high loadings of nanoclay, NT, or NPs already dispersed in a variety of engineering polymers or off-the-shelf polymer nanocomposites ready for processing into finished articles (e.g., "PLASTICYL"™ masterbatches of 15–20%). CNTs dispersed in thermoplastic resins [1] or "nanoMax"® nanoclay-polyolef [1].

In contrast to fully exfoliated GNPs, GPO can be conceptualized as intercalated graphene platelets. It is known that Haydale Composite Solutions has disclosed that they are collaborating with Alex Thomson Racing to look into the possibility of incorporating functionalized GNPs into polyester and epoxy resins as well as carbon fiber-reinforced plastics for use in ocean racing yachts. Although with regard to graphene-polymer EMI shields, potential uses of these systems include reducing weight in aircraft wings as well as structural components protecting against EM radiation, protecting against popcorn failures in shielded chips, but also thermal heat management of delicate electronic package systems. Future prospects for graphene-polymer nanocomposites are promising since they have the capability to incorporate such unique functional property performance into diverse polymer matrices.

9.3 General Summary

In conclusion, the examination of graphene-bearing polymer composites for applications in electromagnetic interference shielding and flame-retardant materials yields promising insights and challenges alike. The findings underscore the immense potential of these advanced materials to revolutionize industries, offering enhanced functionalities in shielding against electromagnetic interference and bolstering flame-retardant capabilities. However, this journey is not without its complexities. As we stand at the threshold of practical implementation, it is imperative to acknowledge the challenges of material integration, property optimization, durability, scalability, and environmental considerations. Nevertheless, these challenges are not roadblocks but rather stepping stones toward progress. The prospects for graphene-bearing polymer composites are bright, with avenues for innovation in material engineering, optimization strategies, durability enhancements, manufacturing processes, interdisciplinary collaboration, and environmentally conscious practices. It is through dedicated research, collaborative efforts, and a commitment to overcoming these challenges that we pave the way for the widespread adoption and transformative impact of graphene-bearing polymer composites in the dynamic landscape of electromagnetic interference shielding and flame-retardant materials. The future holds the promise of realizing the full potential of these composites, offering solutions that extend beyond conventional boundaries and usher in a new era of advanced materials in diverse industrial applications.

Reference

1. G. Armstrong, An introduction to polymer nanocomposites. Eur. J. Phys. **36**(6), 063001 (2015)